THE NINE LIVES OF
CHRISTOPHER COLUMBUS

OTHER BOOKS BY MATTHEW RESTALL

On Elton John

Ghosts: Journeys to Post-pop

Seven Myths of the Spanish Conquest

Blue Moves

*When Montezuma Met Cortés:
The True Story of the Meeting That Changed History*

The Black Middle: Africans, Mayas, and Spaniards in Colonial Yucatan

Maya Conquistador

The Maya World: Yucatec Culture and Society, 1550–1850

SELECTED COAUTHORED BOOKS

*The Friar and the Maya:
Diego de Landa and the Account of the Things of Yucatan*

The Maya Apocalypse and Its Western Roots

The Maya: A Very Short Introduction

Return to Ixil: Maya Society in an Eighteenth-Century Yucatec Town

Latin America in Colonial Times

The Conquistadors: A Very Short Introduction

*Invading Guatemala:
Spanish, Nahua, and Maya Accounts of the Conquest Wars*

*Mesoamerican Voices:
Native Language Writings from Colonial Mexico, Yucatan, and Guatemala*

THE NINE LIVES OF

CHRISTOPHER COLUMBUS

Matthew Restall

W. W. NORTON & COMPANY
Independent Publishers Since 1923

Copyright © 2025 by Matthew Restall

All rights reserved
Printed in the United States of America
First Edition

For information about permission to reproduce selections from this book, write to
Permissions, W. W. Norton & Company, Inc., 500 Fifth Avenue, New York, NY 10110

For information about special discounts for bulk purchases, please contact
W. W. Norton Special Sales at specialsales@wwnorton.com or 800-233-4830

Manufacturing by Lake Book Manufacturing
Book design by Beth Steidle
Production manager: Gwen Cullen

ISBN 978-1-324-08693-2

W. W. Norton & Company, Inc.
500 Fifth Avenue, New York, NY 10110
www.wwnorton.com

W. W. Norton & Company Ltd.
15 Carlisle Street, London W1D 3BS

1 2 3 4 5 6 7 8 9 0

to zz

We all fall prey to preconceptions that make us take certain things for granted. This is a dangerous thing. . . . I believe that, in an investigation, even the most elementary assumptions must be broken down and examined afresh.

L'homme est comme l'Océan, il a une tendance au mouvement et un poids naturel à l'immobilité.

Contents

	Introduction	xi
LIFE ONE	The Genoese	1
LIFE TWO	The Admiral	29
LIFE THREE	The Remains	63
LIFE FOUR	The Saint	89
LIFE FIVE	The Lover	119
LIFE SIX	The Local	143
LIFE SEVEN	The Iberian	163
LIFE EIGHT	The Adam	187
LIFE NINE	The Italian	217
	Epilogue	245
	Acknowledgments	253
	Notes	255
	Bibliography of Sources	295
	Illustration Credits	315
	Index	317

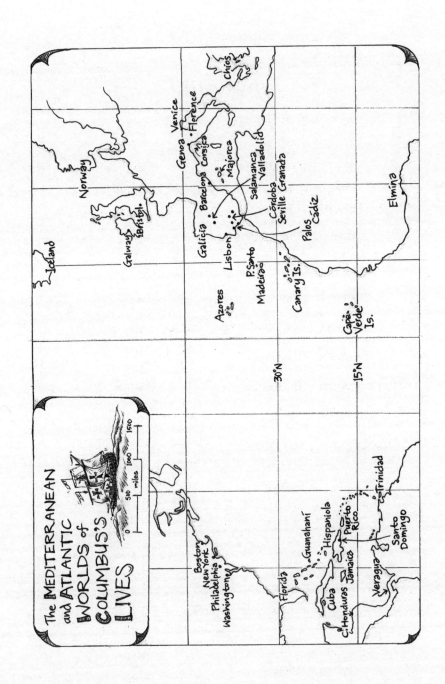

Introduction

CHRISTOPHER COLUMBUS IS STILL ALIVE.

Strictly speaking, he was a mere mortal, a fact as sure as his names at birth and at death, as the where and when of his earthly arrival and departure. For the evidence is incontrovertible that he arrived in the world as Cristoforo Colombo, in the autumn of 1451 in the Mediterranean port city of Genoa, and he left the world in the Spanish city of Valladolid on May 20, 1506, as don Cristóbal Colón. And yet, more than five centuries of battles over his legacy have maintained him in a liminal half-life—or half-lives, as there are many Columbuses, a veritable transatlantic congress of the undead.

As a result, millions of people—billions, even, on multiple continents—have learned and remembered his name. Most have some sense of what he did, that it was momentous, a great achievement, an act of primacy. But was his achievement "great" because he "discovered America" and thus made possible the hemisphere's "great" nations? Or was it an apocalyptic catastrophe for tens of millions of Indigenous and African peoples because he initiated centuries of enslavement, population collapse, and colonialist exploitation? Or should Columbus neither be blamed nor credited with much at all of what happened in the world after 1492—let alone all of it?

The history of Columbus is not only wrapped up in such vast and thorny questions. It is also shrouded in numerous mysteries that have been woven, displayed, disputed, and "solved," many of these "solutions" remarkably imaginative and astoundingly far removed from historical reality. In their totality, they constitute a far-reaching cultural phenomenon, one that I call Columbiana.[1]

These manufactured mysteries center on Columbus's name, nationality, place of birth, ancestry, education, religion, intellectual vision, moral fiber, sexual proclivities, and current resting place—all challenged and reclaimed, all made battlegrounds, all apparent mysteries with multiple solutions over which to fight with astonishing fierceness. Over two dozen nations, islands, provinces, or cities claim to be his birthplace. His bones spent time in four or five cities; three of them claim the remains are still there. His prowess as a lover, or sexual predator, or covert devotee of bestiality, has been proclaimed. Some have argued—and continue to do so—that he was secretly Jewish. Others have insisted that he was one of Catholicism's greatest heroes. He was for long a controversial candidate for official sainthood, the campaign approved for a time by two popes, and he continues to be both lauded as a divine messenger and derided as an agent of Satan. Monuments to him in city squares are fought over, today more than ever, with passion and sometimes violence. He is hailed as the heroic founder of democratic nations and the evil architect of genocide.

Some facts are too well-evidenced to be questioned—that the man "sailed the ocean blue" (as the ditty goes) from Spain to the Caribbean in 1492, for example. But how we describe and talk about that simple fact has been itself made into a field for combat. Should we change "famous" to "infamous" and add "European" to "discovery"? Or should we dispense with "discovery" completely, or reverse the agency of encounter to have Indigenous Americans discovering Columbus? All of which is a precursor to defending Columbus Day or renaming it as Indigenous Peoples' Day, to preserving or pulling down statues, to denouncing or celebrating the entire enterprise of European imperialism in the Americas and the Atlantic world.

While nobody has claimed Columbus cheated death in the literal sense—there have been no sightings to compare to Elvis or Tupac being spotted at the local 7-Eleven—he has in numerous ways become more alive, more present, through the past five centuries, and increasingly so in the last two. Amid all the mysteries, real and imagined, the greatest one is surely this: Columbus survived. He lives. Why?

I STOOD AT THE edge of a large chicken coop and an impressive vegetable garden, admiring the view across the verdant valley. It was my first day as a guest of a wonderfully hospitable local family. We were in the Spanish province of Galicia, in an ancient village just a few miles from the coast of the Atlantic Ocean. A neighbor finished mowing her lawn, then wandered over to the bushes that loosely marked her property line. I was introduced as a historian. Of what? Spain, Mexico, conquistadors, empire, that sort of thing. Ah! The faces of the neighbor and her husband lit up with understanding. No wonder I was visiting Galicia. I surely knew that this was where it all began. Yes, of course, I said, hesitant but playing along as politely as I could. Not in this valley, they explained, but not far from here. My host took her cue: ah yes, she said, his castle! And she offered to drive me there, to the ancient *castelo* where he had grown up and lived, the local man—the Gallego hero— who later changed his name, who later discovered America.

I smiled nervously. Were they teasing me? Was this a bit of Gallego humor? Well, that's an interesting twist on an old tale, I said weakly. Oh no, they all said, repeating a phrase I was to hear often whenever Spanish speakers shared with me their truth about Columbus: *hay muchos datos*; there is much evidence.

This was not a gentle hazing, but an earnest imparting of a treasured belief, one that served as a small but significant thread in the fabric of local community identity. They weren't ridiculing me; they were risking exposing themselves to *my* ridicule. Fully aware that the outside world does not believe that Columbus was really a Gallego nobleman named Pedro, they stuck their necks out, hoping that my mind was open. Or at least that I was courteous enough to pretend that it was.

After all, *hay muchos datos*.

I like to think that I approach any historical topic with an open mind, for that is crucial to writing anything interesting about it. But the world of Columbiana is a topsy-turvy one, where belief precedes evidence, and where openness of mind means shutting off the laws of logic and dismissing the rules of veracity. It is a confounding place to be.

I was therefore torn, my mind both open and skeptical, as I ventured to that local castle, the Castelo de Soutomaior. I found it restored, welcoming, its gardens manicured, its rooms adorned with simple museum presentations on local history. Prominent was the Galician Columbus (el Colón Gallego), displayed unambiguously and as clear as oversized print. He was, visitors are told, in reality the fifteenth-century local lord Pedro Álvarez de Soutomaior, nicknamed Pedro Madruga ("Peter Early Morning," because his legend includes a tale of him rising before dawn to best a rival). A children's guide to "Secrets of Soutomaior Castle" stopped short of the common Galician claim that Pedro Madruga mysteriously disappeared in 1492 by changing his name to Cristóbal Colón. But it nonetheless stated that "some people think Christopher Columbus was related to the Sotomayor family," followed by a puzzle and map showing Columbus leaving Spain to "discover a new continent"; Genoa was neither mentioned nor appeared on the map.

The castle prompted more questions than it answered, opening up a whole new world of belief. Here were facts underpinned not by evidence, but by faith; we might call it "faithistory." I sensed that the answers to this extraordinary phenomenon lay not only in Galicia itself, but in the larger picture that bridged the Atlantic Ocean and stretched over half a millennium. Leaving Soutomaior, I could not help wondering whether the ghost of Columbus stalked its battlements; or was I going as mad as the Genoese (or Gallego?) navigator had himself apparently gone.

Thus hooked—how many Columbuses were there, why did people keep inventing them, what were all their stories and how did they constitute pieces of a puzzle that might form a clear picture?—I ventured further, to libraries and archives, monuments and cityscapes, in a dozen countries on both sides of the Atlantic. Meanwhile, digging into my own office files, into the boxes of xeroxed documents and microfilm rolls and crammed notebooks assembled over the past thirty years, I found Columbus everywhere.

While still a university student, I had annotated the brilliant biography of the navigator by my undergraduate mentor, Felipe Fernández-Armesto. The query "why?" was prominent in my marginalia. Three decades later, as a research fellow at the U.S. Capitol in Washington, D.C., I tried to make sense of the myriad Columbus references in print and image throughout the building and the city (work that became a small part of my book, *When Montezuma Met Cortés*, whose protagonists proved easier than Columbus to pin down and keep in the past). Across the decades in between, I had regularly read, written, and lectured about him. But for every word published or uttered in a classroom, I had scribbled a hundred that sat waiting for a framework, a home, a place where they might tell anew a story already told too often. This book had been germinating, it turned out, all along. It had been waiting for that moment in a Gallego garden, for me to discover another Columbus, to discover how many lives he had.

In Arabic, a cat has six lives; in Spanish and other Mediterranean languages, seven. Staring up at that castle where a Gallego Columbus is still, in local minds, a lord of superlative accomplishments and many lives, I experienced a eureka moment. It suddenly struck me that this Columbus demanded to be taken seriously—the Gallego people deserved as much—but within the context of all his other manifestations, all *siete vidas*. Except he has had more than seven lives, more like the nine that a cat enjoys in English. And none of the thousands of books written about him have sought to understand *all* those Columbuses—from Genoa to Ohio, from mausoleums to monuments, from parades to statues beheaded or thrown in rivers, from the campaign to canonize him to TikTok indignation over his supposed crimes of bestiality (because the claim—the *belief*—that he raped manatees demands to be explained within the larger Columbiana story).

No writers have threaded these lives together and sought to answer the whole cluster of questions prompted by Columbus's lives. Why, over five centuries after his death, are graves being opened in search of "the real" Columbus? Why are statues being toppled and Columbus Day protested? Why are so many people on both sides of the ocean he famously crossed invested in claiming him as their own, while so many others are outraged by his alleged crimes against humans and animals and the world? This book presents and contextualizes all these phenomena, seeking to explain why Columbus has yet to die.

YOU HAVE CERTAINLY SEEN Columbus's face. You may even recognize him as the somewhat surly, broad figure of the portrait by Sebastiano del Piombo, hanging in the Met (New York's Metropolitan Museum of Art) since J. P. Morgan donated it in 1900. Except there's a catch, one very representative of Columbiana. In the late twentieth century, Piombo scholars determined that the Venetian artist did indeed paint the portrait—but of a clergyman in Bologna, not of Columbus, whom Piombo never met. Likely created around 1534, the painting's date of 1519 was added later, as was the mendacious inscription identifying its subject as the famous Ligurian. But facts play a curious role in the world of Columbiana. The Piombo rendering has been, in the Met's words, "the authoritative likeness of the explorer" since the 1590s, and the museum is reluctant to surrender that connection, labeling it "Portrait of a Man, said to be Christopher Columbus." The anonymous clergyman will have to continue carrying the burden, as he has for centuries, of being said to be Columbus.[2]

But that Bologna cleric's face is not the only one borrowed to represent the Genoese explorer. Columbus visages imagined by artists adorn some one hundred eighty monuments to him in the United States alone (less than a quarter of them removed since 2018), as well as thousands of paintings and drawings. Just one example is the so-called Medici portrait. Florence's Uffizi Gallery holds the original, which Thomas Jefferson had copied in 1784. He later hung it in the White House and then in his drawing room at Monticello, inspiring many of the rival portraits that were placed equally prominently during the nineteenth century in state senate chambers, governors' mansions, and historical societies. At the time of the 1892 Quadricentennial, Columbus images were on display everywhere—including seventy-one at the World's Columbian Exposition in Chicago and another seventy at a Smithsonian-supported exhibit in Madrid. Claims to authenticity were—and remain—common, with owners insisting their portraits were made from life, or copied faithfully from a lost such artwork; the Met's hedging is an old Columbiana tradition. Because we need to *see* Columbus, we are easily tempted to take that leap into faithistory, to believe that those are his eyes staring back at us.[3]

Said to Be: A portrait by Sebastiano del Piombo (1485–1547), dated 1519, inscribed "This is the likeness of the Ligurian mariner Columbus, the first in the world to enter the antipodes" (*Haec est effigies liguris miranda Columbi antipodum primus rate qui penetravit in orbem*). In reality, the portrait is of a Bologna clergyman, painted after 1519, with the date and inscription added later still.

In fact, there survives not a single portrait of Columbus made from life. Nor is there evidence that one was created. Our sense of what Columbus looked like is vague; he was a fair-skinned northern Italian of medium height and build, his hair either red or blond but going white in early middle age, his eyes either blue or some other light color. All those portraits in galleries and books are no more accurate than the hundreds of sculptures in Europe and the Americas. The point is not that Columbus's appearance is yet another enigma to be solved, but that we have always seen Columbus the way we want to see him. He has been painted, drawn, and sculpted in the image of the creator—and that is true of his depiction in writing too.

Between the late eighteenth and early twentieth centuries, biographies of Columbus were as numerous as sculptures and paintings; and another flurry was written around the end of the twentieth century. By this century's third decade, at least eighteen hundred books on Columbus in all languages had been published; with printed essays and articles and online pieces in the thousands. One writer recently asserted that there are more works devoted to Columbus "than to Alexander the Great, Leonardo da Vinci, or Adolf Hitler." He may be right. While nobody claims that Columbus has inspired more books than anyone (Jesus, Lincoln, Shakespeare, and Napoleon are the usual suspects for that claim), he certainly appears high on such lists, especially if we restrict them to the history of the Americas. Likewise, he comes third only to Washington and Lincoln as the name memorialized in U.S. toponyms, buildings, roads, rivers, and mountains—some six thousand instances across every continental state, and that's not including businesses and other nonpublic examples.[4]

The vast majority of Columbus books have drawn upon previous ones to advance particular positions, most to defend, some to denounce. Some turn Columbus's life into a ripping good yarn. Some are billed as historical novels, many more as histories, but it is often hard to tell the two apart—a tradition that began with the first Columbus biography, attributed to his son Fernando Colón. Many are blatantly "fictographic," to borrow a term from novelist William Boyd. Quite a few are for children, beginning as early as the eighteenth century. Some cling furiously to one moment in the story, most obviously the First Landing on the First Voyage. Columbus was "a crank," asserted one biographer, and thus he "has attracted cranks"; to write on him requires cautious steering "to elude the cranky theories and undisciplined speculations" that like "siren voices rise on every side." Creating yet another Columbus book would seem a venture both superfluous and foolhardy.[5]

But almost all of the books are instances of Columbiana. Most claim to say something new. Very few do. Most are driven by partisanship; objectivity tends to be crushed by conviction. Still, a few stand out as being relatively free of premeditated agenda, written by authors with "mythistory" radar and no horses in the race. The series of books by

Fernández-Armesto is one example. Another, written a century earlier, is the series of Columbus studies by Henry Harrisse.[6]

Yet even Harrisse could not escape the mythology of Columbiana. By Harrisse's time—his *Christophe Colomb* was first published in Paris in 1884—the key elements of that mythology had become the very frame through which Columbus was viewed. Harrisse outlined them not as distortions or mythistories to be pulled apart, but as core themes of the subject:

> The obscure origins of Christopher Columbus, his challenging beginnings, his stubborn fight against prejudice, crowned by the most dazzling of discoveries, his ephemeral popularity in the country whose empire and wealth he had multiplied tenfold, his death in the midst of universal indifference, added to the marvel of an enterprise begun simply in a calculation of distance.[7]

If Harrisse—with his knack for impaling scholarship that lacked rigor or objectivity—could not see a century and a half ago how mythistorical Columbus had become, it is no wonder that writers are still bamboozled. A sharp mind like that of Tony Horwitz saw the contradictory nature of Columbus literature but drew the wrong conclusion when he noted that "despite all the attention—and, often, because of it, the real Columbus remains elusive." In fact, the "real Columbus"—the historical figure whose life can be reconstructed through careful reading and cross-examination of written evidence—"is better known and more accessible than any comparable figure of his day" (Fernández-Armesto's words).[8]

Contrary to the common claim, therefore, that Columbus remains shrouded in obscurity because we have "so little solid information" about him, his life is in fact abundantly evidenced. Some of that evidence even comes from Columbus himself. Although, despite popular assumptions, his writings do not include unfiltered, personal works such as shipboard logs or diaries, they do comprise dozens of documents written between 1493 and 1506, as well as numerous marginalia (or postils) in books he owned. His writing was far from elegant; most surviving examples are in Castilian Spanish, his third language after Portuguese and his native Ligurian dialect, and the rest are in a poor Latin. But they tell us much, despite, often because of, his autodidactic literary awkwardness.[9]

Indeed, all of the best known of his seafaring contemporaries—Cabot, da Gama, Magellan, Vespucci, and so on—are better candidates for elusiveness and mystery. The greatest canard of Columbiana is that "what is clear is that virtually nothing is known of Columbus." That exact phrase prefaces a historical mystery novel, 2012's *The Columbus Affair*. That hybrid genre ideally lends itself to muddling fiction with fact. Columbiana is particularly fertile ground for such a muddling. The promotional copy for *The Columbus Affair* sets up the book's plot with the claim that "many questions" about Columbus exist: "Where was he born, raised, and educated? Where did he die? How did he discover the New World? None have ever properly been answered. And then there is the greatest secret of all." That secret is an obvious invention—that Columbus was a Jew who buried treasure in Jamaica. But the reader is invited to believe it has some historical basis, simply because of the assertion that Columbus's history is full of questions never "properly answered."[10]

In a world where postmodernism and the internet have turned once-well-rooted concepts such as evidence, fact, and truth into moving targets, to be fired upon at will, turned inside out, or simply played with, who knows what is clear and what is really known? We shall find out. For the purpose of this book is not just to "properly" answer questions about Columbus; I aim neither to fall into the same old trap of identifying Columbus as a mystery, nor to hoodwink you into being persuaded that I have consequently solved the mystery. Instead, I show that those questions *have* been properly answered, and I explore why—considering that the "real" Columbus is *not* "elusive"—hundreds of historians and other writers forged the myth of the mysterious Columbus. Why do so many still seek that license to speculate? Why have they been keeping Columbus alive?

I have already used two terms to be occasionally used again: "myth-istory" is history intertwined with a particular myth or legend or misconception that has its own history but lacks historical evidence; and "faithistory," a term I have invented, refers to history based not on evidence but on firmly held belief that a particular event occurred. In the chapters to follow (identified as Lives), I shall explain and explode four threads of mythistory that run through Columbiana.

The first such thread is the tired old notion that Columbus's life is shrouded in mystery. At the heart of this myth is the misconception that Columbus's birth, ancestry, and origins are not known for certain, that they are thus a mystery that calls for suggested solutions, dramatic discoveries, and ingenious explanations.

Like any other human being, Columbus had a distinct personality and life experiences marked by unique moments. But the idea that he was in every respect atypical in his day, able to achieve what nobody else could because he was so unlike them all, is our second myth—that of uniqueness. Columbus is better understood as being more typical of the Iberian and northern Italian men of his day than atypical; as another merchant mariner who became an explorer, slave trader, and conquistador-settler. Furthermore, thousands of men similar to him were also protagonists in the enterprise of sailing west to trade, enslave, conquer, settle, and in one way or another achieve wealth and status.

Closely related is the notion that Columbus was a heroic visionary. This mythical Columbus struggled valiantly against those with less wisdom and more power, until he was able to realize his extraordinary above-mentioned vision; he confronted, was disparaged by, but then proved wrong, "the medieval ignorance of learned clerics at the Spanish court." The best-known element of this (a submyth, if you like) is that of the Flat Earth, but all its elements evoke well Columbus's contrasting Lives.[11]

Fourth is the idea—deeply rooted, widely held to this day, and responsible for much of the warfare waged on the battleground of Columbiana—that Columbus can be credited or blamed for all that happened after 1492, that his first Atlantic crossing determined subsequent world history. Indeed, for centuries, for better or worse, he has been held responsible for everything from European colonization to globalization, from slavery to genocide, from the spread of Christianity to the creation of the United States. He was for so long "a symbol of empire in the Western imagination" that in time he was seen as its architect too. In this century, the myth of determinism has made Columbus responsible for "the greatest genocide in human history" (as one U.S. politician recently put it), a common association that hinders not helps us to understand how Indigenous populations declined after Europeans reached the Americas.[12]

The tradition of Columbian responsibility stems most obviously from the fact of his primacy as "the Discoverer." But it has also been nurtured by debate over opinions he wrote and decisions he made, especially during his years in the Americas. As Columbus was an "inherently imperial" historical figure, the tradition grew over the centuries, shifting from credit in the age of empire (sixteenth to mid-twentieth centuries) to blame in the postcolonial age (late twentieth century to the present). Whether your inclination is to defend or denounce Columbus, to champion or bemoan European colonization in the Americas, I hope to convince you that the cause-and-effect connection between the two—Columbus and world history since 1492—is a convenient but misleading myth.

You will find in this book, therefore, a biography of Columbus, if an unconventional one. If my depiction of the original Columbus of the fifteenth century seems unsympathetic, that is because he emerged as such not by intention but through the sources and in contrast to some of the other Columbuses depicted here (for example, he was, despite nineteenth-century claims, no saint). That contrast I explain in various ways: through, for example, Christianity's cult of saints; and through the histories of imperialism, nationalism, localism or micropatriotism, and migration.

Regardless of how many historical explanations we can find for the plethora of beliefs and convictions surrounding Columbus and his Lives, there will always remain a corner of Columbiana that seems to defy explanation, flourishing in the equivalent of the "rough corner" or "wild corner" that some gardeners like to allow. Recognizing Columbiana's rough corner helps us to understand the opinions within it, however unacceptable or wildly unsupported by evidence they may be. It means that there is place in Columbiana for the Colón Gallego faithful, for example, however much we remain unconverted by their faith. Perhaps a recent advocate for "those who believe in weird ideas" is right that "we should all be cultivating a healthy Rough Corner in our minds . . . ragged and free."[13]

THE NINE LIVES OF
CHRISTOPHER COLUMBUS

Homeland: The monument in Genoa, completed in 1862, dedicated "To Christopher Columbus: the Homeland."

LIFE ONE

THE GENOESE

AS SUSANNA CRADLED HER NEW BABY BOY, AND HER HUSBAND, Domenico, noted with satisfaction how much little Cristoforo looked like him, they could not possibly have imagined that five centuries in the future his name would be known to hundreds of millions. Told such a tale, they would not have believed it. Shown the statue of Cristoforo that would be raised in Genoa four centuries later—let alone its reproduction on a postage stamp in a hemisphere he would discover for Europeans—they would have been utterly bewildered. And if they had been told that in that future, other places—more than ten in Italy and twice as many elsewhere—would claim that Cristoforo was actually born in *their* city or country, to different parents, with different names, they would have been dumbfounded. And rightly so. Such a tale would have made no sense. Indeed, it challenges us today to make sense of it.[1]

For at the time of Cristoforo Colombo's birth in Genoa in 1451, and throughout his first four decades of life, his biography lacked mystery, controversy, or anything very remarkable. His life before 1492 was not exactly *ordinary*, in that his travels at sea in his twenties and thirties allowed him to see things denied the field-toiling European majority; and the same might be said for the thoughts inspired by the books he

Commemorative: The monument in Genoa was widely reproduced on paper, especially at the time of the Quadricentennial commemorations—as on this postage stamp issued in the Central American republic of El Salvador.

taught himself to read as an adult. But his life hardly rose to the level of *extraordinary*. Not yet.

We shall explore later how and why such misinformation and invention came to clutter the history of these early decades in the life of the man eventually known in English as Christopher Columbus. For now, we focus on the written evidence of his origins and deeds through to his fortieth year. And to best understand what was unexceptional about Cristoforo Colombo—to demythologize him; to normalize him, if you like—we need to have some grasp of his Genoese context. Four aspects of late-medieval Genoa matter in particular.[2]

The first of these is the port city's location on a natural harbor, the most significant resource available to the surrounding region of Liguria. A common Latin maxim, *Genouensis ergo mercator* (Genoese, therefore a merchant), reflects the city's deep history of mercantile expansion and prosperity. By the thirteenth century, Genoa was one of the Mediterranean Sea's leading powers. It drew in immigrants from the poorer rural areas of northern Italy. The Colombos were typical in this way: Columbus's great-grandfather Antonio Colombo migrated from the countryside to a Ligurian mountain village near Genoa; his grandfather Giovanni moved down to the coast, to a nearby satellite of Genoa called

Quinto; his father, Domenico, moved from there into Genoa itself; and one would expect that if Domenico had sons, some or all would take to the sea and end up in places where Genoese merchants settled—such as Lisbon or Seville—which of course is exactly what Cristoforo and his brothers did.³

That the young Cristoforo would choose to go to sea is not surprising. Genoa faces the sea, its back clinging to the steep Mediterranean mountain coast, its buildings climbing the hillside like a jumbled staircase. In the words of Genoese senator Paolo Taviani, "Genoa cannot give its children anything, but throws them back onto the sea, which is everything to the city and from which it receives everything."⁴

But as a boy, Cristoforo could only *see* the sea, apparently lacking chances to take to it; or perhaps, denied literacy and much education, his yearning for other places developed slowly. Claims by later biographers

Sea Views: Woodcut of Genoa in 1492, as imagined by a German artist, first published in Hartmann Schedel's *Weltchronik* (World chronicle; 1493).

Another Sea View: The Columbus monument sits today between a sea of motorcycles and the hillside that Genoa straddles. Rain regularly threatens, in this century as in the fifteenth.

that Cristoforo sailed as a child are pure fantasy. His father, Domenico, apprenticed as a cloth weaver, taking up his trade in Genoa in 1440. From 1447 to 1451 he also served as warder of one of the city's gates. He married Susanna Fontanarossa, from a nearby village. Cristoforo, the first of four, was born in Domenico's final months as a gatekeeper. Cristoforo and his two brothers inevitably apprenticed in the wool trade. In 1470, when the boys were teenagers, the family moved thirty miles along the Ligurian bay to the small town of Savona, a Genoese dependency where Domenico made ends meet trading in wool while running a tavern. Cristoforo began cropping up in local records as his business partner—both "a wool worker in Genoa" and a buying agent in wine for the tavern. In 1472, twenty-one-year-old Cristoforo still appeared in the archives of Savona as "Christoforo de Columbo, lanerio de Ianua" (wool merchant from Genoa).[5]

He was mentioned again in 1473 in records relating to his father's wool business. In that year, or the one before, he seems to have finally taken to sea as an agent selling wool for his family; he would not be mentioned in the Ligurian archives for six years.[6] The eldest of Cristoforo's brothers, Bartolomeo, would also be drawn to the sea, but likewise would not take to it until his twenties, later joining Cristoforo in Portugal and Spain. There is some evidence that Bartolomeo became particularly fascinated by cartography and navigation. The youngest brother, Giacomo, would not leave Genoa for Spain until after Cristoforo returned from his first Atlantic voyage and became don Cristóbal Colón (Giacomo then became don Diego Colón).[7]

The second relevant aspect of late-medieval Genoa was the shift in its mercantile empire in the decades surrounding Cristoforo's childhood. A significant transformation took place when he was an infant. The Ottoman Empire had been steadily expanding into the eastern Mediterranean, and in 1453, Constantinople was captured and turned into the Muslim city of Istanbul. In the previous century, Genoa had competed aggressively with Venice for control over Mediterranean trade, especially in the sea's eastern half, maintaining an extensive network of colonies and trading posts in the Mediterranean and Black Seas; access to luxury goods from Asia was central to Genoa's imperial economy. But in the fifteenth century, with access to luxury Asian goods gradually reduced, that trade shifted to bulk goods, such as woolen cloth and items like alum related to the industry in which the Colombo family worked. Political instability also undermined state power, creating a vacuum into which mercantile families stepped to forge corporate empires in shipping and banking. It was those families for whom Cristoforo worked, as he pursued success at sea in his twenties. By the sixteenth century, Genoa was one of Europe's most innovative and influential banking centers; its Casa di San Giorgio (Bank of San Giorgio, or St. George) would play a crucial role in the Columbus story.[8]

Third, during its heyday in the eastern Mediterranean, the Genoese had not ignored the trading possibilities offered by the Atlantic, playing a central role in the late-medieval creation of what historians call the Mediterranean Atlantic. That seemingly contradictory label conveys the fact that, from the late thirteenth through fifteenth centuries, merchant mariners tied together by sea the economies of northern

Europe and the Mediterranean. They used the Atlantic Ocean to do it, testing the boundaries of the seas as far as their ships would permit. At the forefront, closely intertwined, were the Portuguese, the Castilians, and the Genoese.[9]

As early as 1291 the Vivaldi brothers of Genoa, Ugolino and Vadino, embarked on an expedition, funded by the city's Doria banking family, to reach Asia via the Atlantic Ocean. It is not clear whether they tried to round Africa's Cape of Good Hope (as Vasco da Gama would do) or to sail west (as Columbus would), but they never returned. In the two centuries between the Vivaldis and the Colombos, thousands of Genoese, Portuguese, and Castilian sailors and merchants built the colonies and trading networks of the Mediterranean Atlantic, precursors to the transatlantic trade routes of subsequent centuries.[10]

The final aspect of late-medieval Genoa that matters here was its factional politics. Beginning in the 1390s, the Duke of Milan and the king of France both coveted Genoa's wealth and vied to control it. Throughout the fifteenth century, two dominant factions in Genoa allied with either the Milanese or French, resulting in periods of Genoese independence interspersed with periods of "protective" rule by France or Milan. The situation worsened in the early sixteenth century as Spain joined in, having emerged as a major imperial power in Europe—although that brought stability to Genoa after 1528, when it became an aristocratic republic indirectly controlled by Spain. Navigating the fifteenth-century factional swings and roundabouts was not easy for locals of any rank, and Domenico Colombo did not always find himself on the winning side. That may have helped nudge Cristoforo to seek his fortune at sea. Ironically, the country where he would end up settling and founding a noble dynasty was the one that would come to control the city of his birth.[11]

The last time Columbus set foot in Genoa was in 1479; he was about to turn twenty-eight and had spent most of his twenties at sea, based in Lisbon rather than Genoa. He returned in 1479 as an agent for wool and sugar merchants. The nature of early modern archival records is such that often only when things go wrong do we gain insight into what has been happening all along. That is the case here: two of Genoa's most powerful merchants, Lodovico Centurione and Paolo Dinegro, had given Columbus a large sum of money to buy sugar in Madeira; failing to procure and sell it, he was thereby in debt to them. Columbus appeared before a notary to untangle the mess, and the next morning he left his hometown for the final time.[12]

Columbus may have made brief return trips to Genoa between 1473 and 1479, for which no records have survived, but few trips if any, and none after that. He corresponded with family members and bankers in Genoa, but after 1479 he is mentioned in the Genoese archives only as a relative living in Spain—still *Christoforum de Columbo*, but now also (in a 1496 example) *Armiratum regis Ispanie* (Admiral of the king of Spain). As for that deal-gone-wrong with the Centurione and Dinegro families, it lingered unresolved: on his deathbed, Columbus left 125 ducats "to the heirs of Luigi Centurione Escoto, Genoese merchant" and 50 ducats "to the heirs of Paolo Dinegro, Genoese."[13]

On April 2, 1502, Columbus wrote from Seville to the San Giorgio bankers in Genoa, beginning with the declaration that "bien que el coerpo ande acá, el coraçon está alí de continuo" (although my body walks here, my heart is still there). That may well have been true. But he would die four years later having not set foot in his hometown for the last half of his life.[14]

AS A CONSUMMATE READER, Columbus's younger son, Fernando Colón, was keenly aware that legend formation needed larger-than-life moments. The mundane details, messy storylines, and pedestrian motivations of ordinary life needed to be replaced by dramatic incidents that anticipated a higher destiny. As he strove to write about his father in the language of legend, don Fernando faced a challenge: how to present his father's transformation from a weaver's son working for Genoese mercantile families, with mixed success to boot, into a Lisbon-based explorer of impressive connections pursuing a providential vision of glorious ambition? Something symbolic was required to turn the grubby reality of Columbus's garrulous quest for social mobility into a triumphal turning point. Columbus was a talker, but he seems to have talked only of his ambitions and achievements, not of his early life—including to Fernando, who spent little time with his father before he died when Fernando was seventeen. Fortunately, Fernando left us a clear trail showing where he found that necessary symbolic story.[15]

Marcantonio Cocci was a Venetian historian who took the Latin name Sabellicus. He died the same year Columbus did, but he left published histories, written in Latin, of Venice and of the Mediterranean world. A copy of his *Enneads* ("ab orbe condito ad inclinatonem Romani Imperii" [from the world's founding to the decline of the Roman Empire]), printed in 1498, was acquired by Fernando Colón, who began reading it—he tells us—on August 3, 1534, forty-two years to the day after his father had left Spain on his First Voyage across the Atlantic. There he found Sabellicus's account of a pirate attack on four Venetian galleys off the coast of Portugal. The coincidental name of the corsair leader, *Columbus iunior archipirata illustris* (the famous archpirate, Columbus the Younger), inspired Fernando to borrow the tale and insert it into his father's life. In the margin of his copy of the *Enneads*, Fernando drew a large manicule (a pointing index finger), repeated the phrase *Columbus iunior archi / pirata illustris*, and a few pages later drew another manicule with the margin note *Christophorus colo[m]bus pater meus* (Christopher Columbus, my father).[16]

Fernando began his paragraphs on his father's years in Portugal by crediting his arrival there to Columbus the pirate, citing Sabellicus. Fernando made almost no effort to hide the trick, such is his delight at having come up with it. He imagined his father not as "Columbus Junior" himself, but as a member of his namesake's pirate company, engaging in hand-to-hand combat all day long, until the Venetian galley on which he was fighting caught fire, forcing him to leap overboard. Clutching a fate-given oar, the young *Christophorus* swam two leagues (equivalent to seven land miles) to shore, where he collapsed half-dead—still breathing only because "God was saving him for great things." His recovery over a symbolic period of three days, followed by his walk to Lisbon, where he was welcomed by the Genoese community "with such courtesy and warmth that he took up residency and was married there," was an unabashedly imagined resurrection. As a recent scholar of Fernando noted, referencing this created incident, "resurrection is among the most powerful narrative devices ever invented."¹⁷

Indeed, the story can still be found in almost every book on Columbus, having lived a long life of repetition and embellishment, mutated from fiction into fact by the alchemy of Columbiana. In the sixteenth century, the famous Dominican friar Bartolomé de Las Casas (whose writings, as we shall see later, are a crucial source on Columbus) added the possibility that the Genoese mariner was wounded during his day of fighting; in recent centuries, that imaginary battle wound, along with the whole heroic story, has been repeated as unquestioned fact in numerous popular accounts. Like so many other myths and creations of Columbiana, careful scholars made little impact with their detailed and convincing deconstructions of the story's invention. Writers and readers preferred their Columbus to be adventurous and swashbuckling, tested by fire and chosen by God.¹⁸

The resurrection arrival story also served as necessary filler for biographers unable to find real evidence of exactly how Columbus landed in Lisbon. His Portuguese decade was part of his humble past, before the sudden ascendancy in social rank that accompanied his return from the Americas in 1493. He therefore kept quiet about it. Arriving illiterate in Portugal, he wrote nothing in those years that has survived. Yet restricting ourselves to well-evidenced facts, we still know half a dozen solid things.

We know that Columbus settled in Lisbon between 1471 and 1476 and then left for Castile in 1484 or 1485; a reasonable estimate for his Lisbon years is thus 1474–84. We also know that the Portuguese called him Christovam or Christoval, keeping his surname as Colombo or shortening it to Colom or Colón. And we know that for part of that decade, his brother Bartolomeo was with him. Most writers, instinctively giving priority to Columbus, claim Bartolomeo joined him, but Bartolomeo just as likely arrived first.¹⁹

It is also certain that Columbus married. Probably in 1479, the date uncertain because the earthquake and fire of 1755 destroyed Lisbon's archives. For the same reason, we don't know exactly when in 1580 they had a son named Diego. But the entire Colón family tree, up to the present, branches out from Diego. An obvious step would have been for Columbus to marry into one of the older, well-established Genoese families in Lisbon. That surely occurred to him during his first few years there. In fact, the marriage he was able to contract was not far off that trajectory: his first and only wife, dona Filipa Moniz, was probably the granddaughter of a merchant from Piacenza (eighty-five miles northeast of Genoa).[20]

The marriage was not the leap into aristocracy that biographers later imagined and relished. Nonetheless, on her mother's side, dona Filipa was a noblewoman, of modest wealth and connections yet living in a high-class nunnery before her marriage. Her late father had been in service to the Portuguese Crown, with lordship over an island in the Atlantic; but Porto Santo was "one of the smallest, poorest, and remotest fiefs in the Portuguese monarchy," a stunted sibling to Madeira. Yet considering how humble were Columbus's origins, the marriage was a great boost to his social ambitions. It further stimulated them and his growing obsession with the romantic notion that seafaring derring-do could bring a man a fiefdom on his own island.[21]

We shall return in Life Five to Columbus's romance—or lack thereof—with dona Filipa. Suffice to note here that when their son Diego was a toddler, Filipa died; we know not exactly when or how, but her grave still exists in Lisbon (in a chapel that survived both the 1755 earthquake and the 1834 conversion of the adjacent ruined convent into the National Museum of Archaeology). Biographers have long imagined that during their short marriage, Christovam and Filipa lived in Lisbon, on Madeira, on Porto Santo, or all three; there is no evidence either way. Fernando Colón clearly invented the story that Filipa's father, Bartolomeu Perestrelo, had helped discover Madeira and Porto Santo, and that he was thus the founding governor of the latter. Portuguese sources reveal that Perestrelo was no mariner after all, not even present at those voyages of discovery. Instead, he was given the right to settle Porto Santo because both of his sisters were "mistresses" to the powerful archbishop of Lisbon (by whom they had two children each, all four later legitimized). Perestrelo proved to be a poor governor, ridiculed in the chronicles of the time for introducing rabbits to Porto Santo, causing the new colony to fail within a couple of years. A second attempt also quickly failed.²²

There are grounds, then, for speculation regarding the marriage of Christovam Colom and dona Filipa Moniz (or Perestrelo e Moniz). But, as usual in the world of Columbiana, most of it is excessive: Samuel Eliot Morison's depiction of Filipa as the daughter of a penniless widow, for example, grateful for the tie to a brilliant and "up-and-coming" young adventurer; and Las Casas's picturing of the future admiral devouring his late father-in-law's invaluable, private collection of "maps, *portolanos* (navigation manuals), commentaries, and manuscripts" (in Taviani's words). More likely, dona Filipa's rank and connections were more impressive than her family income, but it was her family history that resonated with Columbus's self-image as a seafaring adventurer worthy of medieval chivalric romance. And while his track record as a mariner working for Genoese merchants promised much, it would soon be undermined by his relentless ambition to be the central and sole figure of such a chivalric romance.²³

As for the maps and manuals, even Las Casas sensed that Fernando Colón was using his imagination (the account attributed to Fernando even garbles Filipa's father's name), softening the claim to this: Perestrelo "must have had instruments." Columbus surely heard tales, but there was no cache of inspiring papers. That has not stopped biographers from breathlessly describing such documents as real and "more valuable to Columbus than any ordinary dowry."²⁴

Another solid thing we know is this: surviving records in Genoa prove that in 1478 Columbus made the above-mentioned trip to Madeira, then to Genoa and back to Lisbon. There are no comparable records of other voyages, but Columbus later claimed vaguely that during these years he sailed in the Atlantic Ocean to multiple places, in addition to Madeira: south down the African coast as far as Guinea; north to the English port of Bristol, to the Irish port of Galway, and up to Iceland. The shifting of his base from Genoa to Lisbon may have reflected his ambitions, but it was a predictable and logical step, not a bold one. For many generations, Genoese merchants had profited from growing Mediterranean Atlantic trade networks; those for whom Columbus worked had agents up and down the Atlantic, including in the places he mentioned. His multiple claims to have seen Elmina, the Portuguese trading fort on the Guinea coast, likely reflect one or two voyages there between 1482 and 1484. He infers that he brought enslaved people from there to Portugal, which seems highly plausible.[25]

Thus, while those other voyages are not undisputed facts, it is reasonable to accept all of them except Iceland as examples of places where a mariner working out of Lisbon for Genoese and Florentine merchant families would sail. The Iceland claim is less likely; its source is Fernando Colón, who provides nonsensical details (such as the wrong latitude, imaginary vast tides, and a February voyage). The story that bodies of an Asian couple were washed ashore in Iceland, and seen by Columbus, was likely a tale heard in Bristol or Galway—and the bodies were likely Inuit caught in a storm off Labrador. Of course, biographers have turned the Iceland story into fact, unable to resist its daring leap into "difficult and dangerous" waters, a bold voyage "fundamental to the conception of the undertaking that would make him famous."[26]

Biographers have not only overapplied their imaginations in making Columbus's pre-1492 years foreshadow his later ones, but they have also been overly credulous with apparent evidence of his intellectual development. The best-known example of this is the Toscanelli letters. Paolo dal Pozzo Toscanelli (1397–1482) was a prominent physician, mathematician, and cartographer in Florence, where he also helped manage his family's trade in spices and hides. Most biographers of Columbus make much of three letters that Toscanelli allegedly wrote—one in Latin to a Lisbon cathedral canon named Fernaō Martins, two in Spanish to Columbus himself—as crucial to Columbus's intellectual development. Taviani is a good example: he accepts the letters as authentic, stating that the 1474 letter to Martins was in Columbus's hands by 1480, convincing Columbus "that Asia lay beyond the Atlantic" and inspiring him to write immediately to Toscanelli. Moreover, the learned Florentine insisted that the distance to Japan (Marco Polo's "Cipangu") was only 2,400 miles (it is really six times as far), and thus sailing to it was feasible.[27]

Not so fast: a close look at the three letters reveals a peppering of red flags. Most historians dismiss the two Spanish copies as apocryphal while accepting the Latin copy. Yet the latter is in Columbus's handwriting (or that of his brother Bartolomeo), and in poor Latin (of "rude orthography and doubtful construction," as one early scholar rightly judged); Toscanelli was fluent in Latin, the Columbus brothers were not, knowing it well enough to make a "doubtful" composition or translation. There is no substantiating evidence of Canon Martins's existence. All three letters have the same date (June 25, 1474), a nonsensical detail if they genuinely constitute a correspondence series. The letters refer to a voyage west being "shorter than the one you take by Guinea," a reference that only makes sense after 1488, when Bartolomeu Dias rounded the Cape of Good Hope (but when Toscanelli was dead). References to "the land of the spices" are similarly anachronistic; as of 1474, voyages down the African coast sought gold and slaves. Other details in the letters suggest that their common source was written in the sixteenth century. In 1903, Henry Vignaud conducted a book-length study of the letters, concluding that there was no evidence that Columbus and Toscanelli ever corresponded or knew each other, that the letters were all sixteenth-century forgeries, and that Columbus "gave expression to the ideas found in the letter to Martins only after his last voyages." I think Vignaud was right.[28]

There is one more thing we know with certainty regarding Columbus's Portuguese decade: he pitched to King João II, who had succeeded his father on the Portuguese throne in 1481, a plan to sail west into the Atlantic. No further details are known. Las Casas imagined that Columbus claimed he would find it all—"he would discover great lands, islands and mainland, all very prosperous and very rich in gold and silver and pearls and precious stones, and in an infinite number of people"—a hard sell that sounds plausible in light of the personality that would later emerge. Las Casas also claimed that Columbus made demands similar to those of his later contract with Spanish monarchs, including the claim to a new title, that of Grand Admiral of the Ocean Sea (as Iberians called the Atlantic). Las Casas may have just copied those later demands, as some have alleged, but the overreach sounds like Columbus, and it likely influenced the outcome of the pitch. For we also know that King João said no.[29]

If Columbus's demands were indeed "outrageous" and "exorbitant" (as historians have called them), enough to prompt the king to deny permission and support, there were likely two further reasons for his decision. One is that Portuguese expeditions had already attempted to sail further west into the Atlantic, and they would continue to do so. And none of their captains were Genoese mariners demanding to be elevated to the Portuguese nobility; they were royal agents like Diogo Cão and Bartolomeu Dias. The other reason is that Columbus's geographic knowledge was faulty; we have no details of his pitch to João II, but in every subsequent report of what he said, Columbus insisted the world was far smaller than it really is. Whether he was claiming to reach Cipangu or a legendary island near it (variously called by late-medieval Europeans Antilia or Antillia, Brasil, or the island of Seven Cities), his calculation as to its distance was wildly wrong—and the king's advisors knew it.[30]

Historians have assumed that the pitch to King João failed in 1484, in the months between his return from a possible trading voyage to Guinea and his frustrated departure from Portugal to Spain. The guess is reasonable. More certain is the fact that on March 20, 1488, the king wrote to "*Christovam Collon*, our special friend in Seville," responding positively to a request (now lost) by Columbus to return to Lisbon. Columbus had asked for guarantees, probably because he had left debts, and he offered his services. The king reassured him that "we will have great need of your ability and fine talent," and that during his stay he "will not be arrested, detailed, accused, summoned, or prosecuted for any reason whatsoever."[31]

But since that failed 1484 pitch, things had not gone well. At thirty-seven, he was dependent on Genoese and Florentine friends in Seville. His Portuguese wife had died, and he had left their eight-year-old son with the boy's aunt in Huelva, while a teenager in Cordoba was carrying his second child. He needed royal safe passage to Portugal because of his creditors there. And when he finally made it back to Lisbon in December, he arrived just in time to see Bartolomeu Dias return from the first voyage to the Cape of Good Hope—thereby discovering a sea route to

Asia, rendering Columbus's ill-conceived proposal a pointless fantasy. Just one Genoese mariner among many in Lisbon and Seville, the grasp of his ambitions had exceeded his reach.

IF COLUMBUS'S PORTUGUESE YEARS have been mythologized by his biographers, that is even more the case with the next seven or eight years (roughly his thirties) spent in Spain. The mythistory of Columbiana has dramatized what was a frustrating period for the Genoese mariner. His pitch to the Spanish monarchs—Isabel, queen of Castile, married to Fernando, king of Aragon—was never rejected to the extent that Columbus was expelled from either kingdom. But nor was his proposal accepted and placed under contract before 1492. As a result, Columbus spent these years hoping to change royal ambivalence into acquiescence, as he followed the court around from one kingdom and city to another (the court was itinerant, as not until the next century would the kingdoms evolve into a loosely centralized state with a capital city).

Considering Columbus was Genoese, why did he not forget about trying to win royal support—in the Iberian kingdoms or elsewhere—and raise private capital to fund his voyages? After all, Genoa was a greater banking center than any city in the Iberian Peninsula. In fact, most of the funding for the First Voyage did come from private investors, contrary to the popular legend of Queen Isabel pawning her jewels. The issue was less funding than it was royal patronage, which gave the individual who made the discovery—be it a new trade route, new trading partner, or new lands to be claimed and exploited—protection from rivals and the right to lucrative and prestigious offices of state. Upward social mobility was Columbus's driving motivation, one he pursued obsessively and then fought to defend with paranoid preoccupation. Neither for Columbus, nor for any Spaniard in the era of conquest and colonization, could such a rise in rank be achieved without a written and notarized contract of royal support.[32]

Columbus embraced the system of royal patronage and license with the enthusiasm of the neophyte and convert. The excessive privileges he demanded from a royal contract reflected that attitude. He seemed to see his proposed voyage west as his one chance to achieve respect and rank, not just for himself, but also for his sons—especially the eldest (one suspects Fernando was the "spare" in case Diego died prematurely). After all, Cristoforo Colombo was always going to be the parvenu, an upstart Genoese sailor, denigrated behind his back. The changing of his name to Cristóbal Colón—"for more comfortable pronunciation," as royal chronicler Antonio de Herrera snidely put it in 1601—could not hide his humble origins from the Castilian and Aragonese who watched him scurry after the itinerant court, doggedly pursuing royal patronage. All for what? So that his sons might grow up not as wool-weaving, cheese-making Colombos, but as representatives of a newly minted Castilian noble house of Colón.[33]

The reasons why Isabel, Fernando, and their councilors were for long intrigued yet not convinced by Columbus's proposition to sail west to Asia are fairly straightforward. But before we get to them, let us explode some mythology regarding these years campaigning for support in Spain. Why did it take the Genoese mariner so long to win royal approval for his voyage west? According to conventional mythistory, Columbus was a heroic, forward-thinking visionary who was mocked and dismissed by the court's supposedly wise clerics—whom he later proved wrong. That interpretation has rested on a trio of fictions. The first—to translate from a 1946 Spanish history—was that "the great and extraordinary achievement of Columbus" was "to formulate a methodical and scientific hypothesis" regarding sailing west, and then to pursue it "with a truly superhuman perseverance and fortitude." Such phrases expressed a tradition firmly established in the previous century, whereby Columbus was intellectually ahead of his time, a founder of the modern world; his was "essentially a Renaissance enterprise," one "responding to man's new ideals." None of that was true, matching only Columbus's conceit in his final decade.[34]

The other fictions were that the medieval centuries were the Dark Ages, with Spain still mired in ignorance, and that those not-so-learned clerics gathered in a dramatic Inquisition session at the University of Salamanca to hear and unwisely dismiss Columbus's brilliant ideas. That session is often called the Council of Salamanca or the Talavera

Ridiculed: Nicolò Barabino's *Cristoforo Colombo davanti al Consiglio di Salamanca* (1887), in Genoa's Palazzo Orsini, often reproduced with titles such as "Columbus Ridiculed at the Council of Salamanca," "The Despair of Columbus," or "Columbus Refuting the Dominican Friars in the Conferences at Salamanca" (as in this October 8, 1892, *Illustrated London News* version).

Commission, after the friar who chaired it, Hernando de Talavera. In fact, it had no such grand title. Columbus did present his plan to a committee appointed by the Spanish Crown, but that has been misleadingly embellished, dramatized, and mythologized.

The following snippet of dialogue from a 1979 play titled *Christopher Columbus*, despite being historical fiction, captures how the commission's meeting has been commonly depicted in generations of history books and other media:

> COLUMBUS: The Earth is not flat, Father, it's round!
> THE PRIOR: Don't say that!
> COLUMBUS: It's the truth; it's not a mill pond strewn with islands, it's a sphere.
> THE PRIOR: Don't, don't say that; it's blasphemy.[35]

American middle-school children who paid attention to the widely used 1980s textbook *We the People* learned that a thousand years ago, European sailors believed "that a ship could sail out to sea just so far before it fell off the edge of the sea." The many people who read *The Discoverers*, a mid-1980s best seller by former librarian of Congress Daniel Boorstin, were likewise told that a "legion of Christian geographers... afflicted the continent from AD 300 to at least 1300," suppressing European knowledge of the earth's true shape. One extraordinarily wrong-headed chapter of Boorstin's was titled "A Flat Earth Returns."[36]

In the opening scene to Ridley Scott's 1992 movie, *1492: Conquest of Paradise*, Columbus (French-accented, as played by Gérard Depardieu) sits on a rocky shore gazing out to sea. The setting is reminiscent of nineteenth-century images of Columbus as a visionary (José María Obregón's 1856 *The Inspiration of Christopher Columbus* is a good example). Scott's film adds to that picture one of Columbus's young sons, told to watch a ship disappear over the horizon by his father, who meanwhile peels an orange. With the ship gone and the fruit peeled, Columbus exclaims, "What did I tell you? It's round. Like this. Round!"[37]

Movie viewers around the world thereby had their grasp of medieval ignorance and Columbus's genius confirmed, as the Flat Earth myth was deep and widespread. By the time of the Quincentennial, the Flat Earth myth was almost two centuries old, having been invented and spread mostly by writers of the early United States. As historians recently noted, "the image of Columbus as the lone adventurer challenging clerical ignorance fitted the contemporary self-image of Americans." The myth flourished in the fertile ground of nineteenth-century writers systematically denigrating earlier periods, especially the medieval, so as to show how progressive their own era was.[38]

The twentieth century was no less welcoming, being an era in which people "loved to hear of professors and experts being confounded by simple common sense" (in Morison's words), even if it was clear that "the whole story was misleading and mischievous nonsense." At a 1982 United Nations meeting, Ecuador's representative Miguel Albornoz introduced a resolution to commemorate the Quincentennial of "an event which is perhaps the most significant in world history," noting that one of Columbus's great achievements was to prove that the world was round. Albornoz—like Boorstin and Scott—was merely expressing what he and millions of others had learned in school, read as adults, or seen in movies.³⁹

This "Flat Error," then, was "not the alleged medieval belief that the earth was flat, but rather the modern error that such a belief ever prevailed"—as defined by Jeffrey Burton Russell, whose book-length exploding of the myth in 1991 was the best, but far from the first or last. Back in 1942, Morison had pointed out that the Salamanca Flat Earth debate was "pure moonshine." But although Morison's book found a wide audience, so did Chu Berry and Andy Razaf's jazz song "Christopher Columbus," a hit for Fats Waller in 1936, with its memorable rhyme:

> *Since the world is round,*
> *we'll be safe and sound.*
> *'Til our goal is found,*
> *we'll just keep a-rhythm bound.*⁴⁰

In reality, Columbus knew the earth was round because that had been common knowledge for many centuries. Flat earthers were no more plentiful in medieval Europe than in the world today (there will always be someone to believe anything). The irony of Columbus's orange in Scott's *1492* is that medieval texts are full of food metaphors denoting the world's shape: in the 1246 *Image du monde*, the sky was the earth's eggshell; in the 1266 *Li Livres dou Trésor (Book of Treasure)* by Dante's teacher Brunetto Latini, the world was "round like an apple" and also egg-like, its core like a yolk; William Caxton wrote in 1480's *The Mirrour of the World* of the globe as "a round apple." Even the "simple people" that John of Mandeville laughed at in his *Travels* of 1370 knew of the world's sphericity; they just had trouble understanding why people in the antipodes didn't fall off.⁴¹

Columbus's thinking was that of an uneducated but self-taught late-medieval man. The handful of books we know he read repeatedly and annotated were not surprising choices—such as Ptolemy's *Geography*, a second-century compendium rediscovered and published in Europe in the fifteenth century; and *The Book of Marco Polo*, loosely familiar to Columbus by 1492 but not acquired and read closely until 1496. Polo's book helped make mainstream the idea that travel would uncover marvels, as did Pius II's *Historia Rerum ubique Gestarum*, a compendium of actual and imagined geographic features of the world (marked with 861 postils by Columbus and his brother Bartolomeo). The most heavily read book Columbus owned (898 postils), Pierre d'Ailly's *Imago Mundi*, was mined perhaps in the 1480s but certainly in the 1490s for support for his bizarre theories—on his supposed 1492 discovery of a route to Asia, for example; his 1498 discovery of the Garden of Eden; and his 1500 claim that his discoveries were God's sign that Christ's return was imminent. Columbus's worldview was informed by the larger phenomenon, going back four centuries to the First Crusade of 1094, of European expansion and the steady accruing of theories and knowledge on how and where to extend trade networks. But, closer to Mandeville's simple people than the visionary genius of modern invention, he struggled with how to reconcile the shape of the world to his understanding of what would happen if you tried to reach the other side.[42]

The so-called commission headed by Hernando de Talavera was simply a committee, an informal, ad hoc body, made up of both ecclesiastical and lay officials. It met irregularly, just three times (in Cordoba, summer 1486; in Salamanca, that December; and in Seville in 1490). It was not a commission of the church or of Dominican friars or of the Inquisition, nor was its Salamanca meeting an assembly of university scholars. It simply used university space for one of its meetings because committee member Rodrigo Maldonado de Talavera had taught law there before assuming an administrative post at court. The question of the earth's roundness was not mentioned once, at any of the three meetings.[43]

Instead, the discussion focused on the *size* of the earth. Columbus insisted that the earth was small enough for him to reach Asia by sailing west into the Atlantic. The committee was concerned that the distance was too far, and after its final gathering in 1490 it made a recommendation of rejection to the crown on the basis of that concern. As Maldonado de Talavera later recalled, "we all agreed that what the Admiral was saying could not possibly be true, and against [our] opinion the Admiral determined to go on the said voyage." Or, as Russell put it, "the committee's doubts were understandable, for Columbus had cooked his own arguments."[44]

Columbus and the committee concurred that there was no unknown landmass to the west—just some islands on the way to Asia. But they disagreed on the distance. The committee accepted the old Ptolemaic reckoning of the continents comprising 180 degrees of longitude, leaving a vast ocean between Spain and Japan of another 180 degrees. Such a voyage was, in the 1480s, impossible. Knowing this too, Columbus realized that he needed to reduce that 180. He did it by claiming that d'Ailly was right in estimating the ocean at 135 degrees of longitude in size; that Marco Polo's travel accounts showed Asia was bigger than Ptolemy knew, thus arbitrarily dropping the number from 135 to 107; and that Japan would be reached sooner by sea than China would, thereby cutting the number further to 77. And as he planned to sail from the Canary Islands, not Spain itself, the distance was really 68 degrees of longitude, he argued. In a final sleight of hand, he asserted that d'Ailly's degrees were oversized anyway, and therefore the final number was 60.

The true distance from the Canaries to Japan is about 200 degrees of longitude, even farther than the Ptolemaic estimate. But Columbus had more cooking to do. D'Ailly had followed the ninth-century Arabic astronomer Al-Farghani in sizing a degree of longitude at 56 and 2/3 miles. Columbus decided that Al-Farghani meant the shorter Roman miles, at 45 to a degree. But as he was sailing above the equator, he could assume 40 miles. In the end, he cut the distance to less than 2,770 miles; in reality, it is five times farther, some 13,700 miles. Columbus's calculations were outlandish, completely wrong, and insufficient to fool Talavera's committee.[45]

The real story therefore turns the mythistorical one on its head: the Castilian officials of church, law, and court correctly understood the size of the earth; the ignoramus was Columbus. All of them, on both sides of the discussion, knew the earth was round, and none of them knew or imagined that a vast continent unknown to them lay to the west. Does this adequately explain why it took Columbus so long to win royal support for his First Voyage? And if so, if European knowledge of the world did not change between 1482 and 1492, why did Columbus eventually gain such support?

In fact, there is a crucial larger context that helps answer both those questions: economic competition, diplomatic tension, and periodic warfare between the kingdoms of Portugal, Castile, and Granada—two Christian, one Muslim. For much of the fourteenth and fifteenth centuries, Portugal led Castile in exploring Atlantic islands and the West African coast. Lagging in terms of the quantity and success of expeditions, Castile played the papal angle, arguing as early as 1345 at the court of the pontiff for rights to the Canaries—claiming they could be justly *reconquered* as Muslim kingdoms could be. Castile gained little beyond the conquest of some of those islands, obliged to elbow into the edge of Portugal's African trade, especially starting in the 1440s, prompting complaints and periodic conflict. Then, in 1474, Portugal challenged Isabel for the Castilian throne. The ensuing five-year war was primarily fought in the north of the peninsula and in the seas of the Canaries and the adjacent African coast. The Genoese in Castile were quick to acquire privateer licenses to break by force into the Portuguese-African trade, while the Genoese governor of Portugal's Cape Verde Islands defected to Castile, and Genoese merchant families in Seville, such as the Pinelli and Rivarolo (the latter would bankroll Columbus's Fourth Voyage), financed the campaigns that made the Canaries Castilian not Portuguese.[46]

The peace treaty of 1479 left Portugal with the Azores, the Cape Verde Islands, and a virtual monopoly on developing trade with coastal Africa. But it left Castile free to complete the conquest of the Canaries, enacted with single-minded brutality, slaughtering and enslaving the Indigenous population, achieving full control by 1496. As a result, not only did the island of Gomera fall fully under Castile's thumb by 1483, but its deep-water harbor of San Sebastián was now available as a new launching point into the Atlantic. As late as 1492 nobody knew—Columbus included—that it would soon prove to be virtually the only way to cross the ocean all the way to the Caribbean islands.[47]

Castile's final conquest of the Canaries took two decades because Isabel and Fernando meanwhile focused on the simmering conflict with Granada. Ruled by the Nasrid dynasty, Granada was the last Muslim kingdom in Iberia, representing an Islamic presence there since 711. Sitting along the sierras and Mediterranean coast east of Seville and Gibraltar, the 12,000-square-mile kingdom survived into and through the fifteenth century through a combination of diplomacy (periodically paying tribute to neighboring Castile and Aragon) and warfare (fending off major attacks, while maintaining almost constant frontier raids and skirmishes).[48]

Relations with Castile were relatively calm between 1465 and 1481; Granada was distracted by internal political instability, and Castile was distracted by conflict with Portugal. But in 1479, Fernando became king of Aragon, effectively uniting it with Castile, to whose queen he had been married for ten years; at the same time, civil war in Castile and its conflict with Portugal were resolved. Frontier raids gradually intensified until, in 1481, the conflict slid into outright war, as Muslim and Christian forces crossed into each other's kingdoms to attack, seize, and plunder cities and towns. Levels of violence were shocking, with slaughter and enslavement routine practices. In 1482, Abu Abdallah (hispanized as Boabdil) became Sultan Muhammad XI by deposing his father (as his father had done to his grandfather); and Pope Julius II reissued a 1479 crusading bull calling for a holy war against Granada.[49]

Over the next decade, Boabdil would himself be deposed as sultan by his father, then by his uncle, before regaining the throne for five years—all while civil strife tore apart his capital city and persistent warfare steadily shrank and destroyed his kingdom. In 1483, Boabdil was captured by Aragonese forces; as a condition of his release, his infant son spent his boyhood as a hostage at King Fernando's court (where Columbus would have seen him). In the few years between Columbus's meetings with Talavera's committee, the war intensified in the Granadan kingdom's west, culminating in the bloody siege of the Muslim port city of Malaga in 1487 (when Columbus was living in Cordoba, just a hundred miles to the north; he requested, and was granted, royal permission and funds to visit the siege camp). When

Malaga fell, its Muslim inhabitants were enslaved and sold. Young women and girls, publicly auctioned by the thousands, received the highest prices. Seville- and Cordoba-based Genoese merchants, who continued to trade in Granada throughout the war, played central roles. In 1491, Fernando and Isabel built a massive siege town, named Santa Fe ("Holy Faith"), outside the city of Granada; to spare the starving inhabitants mass enslavement, Boabdil capitulated. Castile and Aragon publicly promoted their war as holy, as a crusade. But its true rationale—to acquire territory, wealth, slaves, and regional power—was blatantly obvious.[50]

The war against Muslim Granada has tended to be ignored or marginalized in Columbiana literature, coming in only when Columbus, as a hanger-on at court and in Santa Fe, witnesses Boabdil's formal surrender in January of 1492. Yet it totally overshadowed Columbus's years in Spain, determining the outcomes of his ambitions. It also offered crucial lessons in how to use the language of religious motives and justifications to disguise political ambition, economic goals, and the inhumane treatment of non-Christians.

Nor did those lessons end in 1492, although Columbiana literature treats the year as a hard stop, after which Granada is apparently irrelevant. Not so. The war was over, but Castile's conquest and colonization of the Muslim kingdom had just begun. As late as the 1570s, Granada's ruling council—reeling from a recent Muslim rebellion and fearing attack from North Africa—was still reminding the king that theirs was a *ciudad frontera*. More important, therefore, than Columbus's witnessing of Boabdil's surrender was his awareness, for the rest of his life, of how Spain used complex and violent interaction between Christians and non-Christians to wield imperial power. And the violent subjugation of Canary Islanders, dragging on during the years of his First and Second Voyages, offered Columbus a parallel example. "Just war," crusading, and *Reconquista* ideologies could underpin colonies forged with warfare, slave trading, and the imposition of feudal systems of labor exploitation.[51]

So, why did Columbus excite "only intermittent flickers of interest outside Spain," winning the support of the Spanish monarchs only after "long and unremitting effort"? Because of his muddled misunderstanding of the world's size; his excessive demands for compensation and preferment; the fact that Portugal was already developing a successful sea route to Asia; and the delay caused by Castile being mired in competition with Portugal and in a long, grinding war with the Muslim kingdom of Granada.[52]

That final factor—geopolitical instability—was also, paradoxically, the reason for Queen Isabel's sudden change of mind in 1492. The resolution of that instability, the alignment of developments in Castile's favor, was dramatic. The occupation of Granada left no remaining Muslim kingdoms in the peninsula and strengthened Castile in relation to Portugal, allowing Castile to complete its conquest of the Canary Islands, giving any voyage west a base from which to launch. That development was a stroke of good luck for Columbus. Had the Canaries become Portuguese, the Americas would have been found in the 1490s by them—not by Columbus and the Castilians.

IN AUGUST OF 1492, Cristóbal Colón, not yet with the honorific "don," but no longer Cristoforo Colombo or Christovam Colom, sailed west from Spain to end his life—to end it completely, as others before him had done in the storms of the Atlantic, or to end his relatively ordinary Life One. The transformation would come in the spring of 1493, upon his return. And while his Life Two would be far shorter than Life One, it would bring the social mobility that he craved, even if his achievements never seemed quite enough for him. That contrast between his ordinary first and extraordinary second lives, combined with his incessant drive for status and rank, left Columbus increasingly embarrassed by his origins.

Consequently, in everything he later wrote (and everything he said that was recorded by others), there is very little about those pre-1492 years, and virtually nothing about the Genoese years of his childhood and youth. Amidst so little, there are even some outright lies (the best known being the tossed-off fib, "I am not the first admiral of my line").

In a world rife with prejudice of all kinds—not just racism and ethnocentrism, but deep classism and regionalism—Columbus faced ridicule for being a wool merchant's son and a sailor from an Italian port. To be Genoese in Castile was not shameful per se. For centuries, Genoese merchants had been marrying into noble families in Castile, hispanizing their names, acquiring political office, and becoming aristocratic themselves; three-quarters of Seville's noble families had Genoese or hispanized-Genoese surnames by the end of Columbus's life. Genoese men in Seville and other parts of Castile, especially Andalusia, thus tended to belong to deep-rooted merchant families that were often wealthy, educated, even aristocratic. But Columbus was none of those things.[53]

It is thus not surprising that he developed something of an inferiority complex, which he sought to offset with fantastic ambitions and, later, an increasingly grandiose conception of his own purpose in life—given and guided by God, no less. One of the priests he befriended, Andrés Bernáldez, who later chronicled the age, noted that Columbus had "great intellect but little education." Modest origins of which he was ashamed, a lack of education, a keen intelligence, an autodidactic drive, and social ambition all combined to create in Columbus narcissistic and confabulist tendencies, tempered by insecurity and dissatisfaction.[54]

It is also hardly surprising, then, that he sought to disguise his origins, to fog them up with the early haze of Columbiana. As Spanish historian Salvador de Madariaga put it, "like the squid, he oozes out a cloud of ink round every hard square fact of his life." Not even his own sons could escape such a cloud. He denied them the joy of knowing their Genoese relatives (aside from their two Genoese-turned-Castilian uncles) or even of grasping their Genoese roots; they grew up as Castilians, pages in the royal court, upstart nobles of the new Colón dynasty, told no more by their father about his origins than anybody else was. And *that* mattered because a biography of his father attributed to the youngest, Fernando, filled in those gaps with imaginative claims and fictional details—inventions that seeped into all the early published accounts of Columbus's life, and from there into Columbiana's great book pile.[55]

Therein lies an ironic twist to the tale that will unfold throughout the next eight Lives: In trying to cloud over his Genoese origins, Columbus sought to throw more light on his status as a Castilian nobleman and high official of what we call the Spanish Empire. Instead, he generated a tradition of mystery that would centuries later nurture speculations and allegations that would have appalled him.

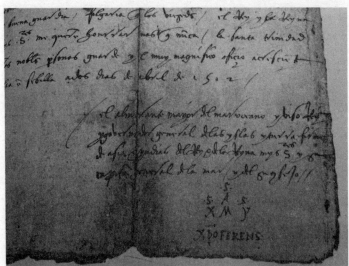

All at Sea: The First Voyage, whose completion would win Columbus his title of admiral, as depicted in the Quincentennial U.S. stamp series. Closing an April 2, 1502, letter sent from Seville to Genoa's Bank of San Giorgio, Columbus signed off with a full rendition of his extravagant titles and acronymic signature, as "The High Admiral of the Ocean Sea and Viceroy and Governor General of the Islands and Mainland of Asia and the Indies of the King and Queen, My Lords, and their Captain General of the Sea and [member] of their Council. S. / S. A. S. / XMY / Xpo FERENS."

LIFE TWO

THE ADMIRAL

THE OCEAN SEA—THE ATLANTIC—IS CRUCIAL TO THE COLUMBUS story. Without it, there is no obstacle to be overcome, and thus no story at all. The crossing gives the story its momentum, as well as its traditional hero. It also gave that putative hero his goal, which was never such a thing as "the discovery of America." Rather it was the right to be addressed, as he was in March of 1493, in a letter sent by King Fernando and Queen Isabel as he approached their court in Barcelona, as "don Cristóbal Colón, nuestro Almirante del mar Océano" (our Admiral of the Ocean Sea). That, indeed, had been his self-declared motive, as he confessed to the monarchs in Barcelona that spring: he had sailed to the *Oriente* and to *Yndia* "so that from that moment on I might call myself *don* and be the High Admiral of the Ocean Sea ... and my eldest son should succeed me as such."[1]

There was more to the title. It continued with "Viceroy and Governor of the islands that have been discovered in the Indies"—meaning, according to Columbus's claim, Asian islands—and the office of admiral gave authority not over a permanent fleet (there was none) but over the monarchs' subjects across the seas, with the right to judge and arbitrate whatever trade and commerce those subjects might conduct in those islands. Yet there was a resonant geographic ring to "Admiral of the Ocean Sea." Columbus made much of it. For example, in a private missive sent from Seville in 1502 to the Bank of San Giorgio in Genoa, he unnecessarily used the full version, including his viceregal rule over "Asia and the Indies," a blatant boast from a local boy made good to his hometown's old élite.[2]

Persisting to this day as an aristocratic title, passed down through twenty generations to the current don Cristóbal Colón, the title evokes the triumphal narrative that has come to dominate Columbiana. In that story, the admiral did not merely cross the Ocean Sea, he conquered it. The three small ships of his First Voyage bobbed on a sea that was frighteningly empty, a vast churning void whose threatening, undulating surface barely hid a bottomless violet-black graveyard. The ninety men who sailed on the First Voyage believed that the ocean promised islands—eventually, somewhere. Yet it is not hard to imagine the terror of such a trip, and the foolhardy audacity of those who made it.

It is tempting, therefore, to see the ocean as a fearsome antagonist in this story, and those who sailed across it as brave heroes, as conquerors of seas that Columbus would by his Fourth Voyage come to describe as violent, threatening, "boiling," and "ugly." But that is to slip into seeing Columbus through the lens of later Lives, as the bold adventurer who "dreads no tempests on the untravell'd deep" (as a poet in the North American colonies put it in 1770). Indeed, Columbus's Four Voyages are the siren call that has lured the vast majority of Columbiana writers to their subject. The voyages provide the structure for the traditional Christlike narrative of Columbus's life, his rise and fall and redemption, his tortured but triumphant passage from obscurity to misunderstood genius to mistreated hero to apotheosized founder of the New World. This Life Two comprises only the last fourteen years of Columbus's actual mortal life, but his biographies focus overwhelmingly on those years; without them, Columbiana would not exist. By way of resisting the narrative narcotic of the Four Voyages, I thus offer here a mere outline.[3]

FROM THE DAY HE set sail from Palos in August of 1492, Columbus spent two-thirds of his remaining life either at sea (in the Atlantic or Caribbean) or in the Americas. His First Voyage was the shortest (thirty-three days from the Canaries to the Caribbean, seven months between leaving and returning to Spain). The other three were much longer, of similar duration to each other (thirty-three, thirty, and thirty-one months, respectively). Most of those months were spent on land, trying to administer a new colony, for example (as in 1495–96 and 1498–1500), or being stranded on Jamaica (for a year in 1503–4). His time spent back in Spain between voyages (six, twenty-three, and seventeen months, with eighteen months between his final return to Spain and his death) was devoted to campaigning to secure the rewards that motivated the voyages: titles, income, political offices, and the legal founding of a new dynasty.

The schedule was grueling, exhausting to merely contemplate, reflecting Columbus's obsessive personality. He was as relentless during these fourteen fortunate years as he had been during the previous fourteen frustrating years. He seldom seemed satisfied, always wanting more. The toll on his mental health revealed itself in his surviving writings, while his lifestyle was not easily reconciled with the untreated gout that eventually killed him.

The First Landing was in the Bahamas, most likely the small island known variously as Guanahaní, Watlings, and San Salvador—although the debate will remain tediously endless because of the vagueness of the *Diario*, the log of Columbus's that survives only through Las Casas's redacted and edited version. On the First Voyage, the ninety travelers on three ships (only two ships and half the men returned) found the Bahamas and the east end of Cuba and Hispaniola. The 1,200 on the seventeen ships of the Second Voyage revisited those islands, exploring the whole south coast of Cuba and Jamaica, adding the run of islands from Puerto Rico to Dominica. The Third Voyage was smaller (three hundred men and thirty women on six ships), as Columbus's marginalization in the Spanish colonization process had begun by its 1498 departure. It added a very small part of the South American coastline—where the islands of Trinidad and Margarita sit—because Columbus used the Cape Verde Islands as the launchpad to catch the winds further south.

The Fourth Voyage comprised just four old and small caravels. It added another mainland coastline to Columbus's "discoveries" list, that of Central America from Panama to Cape Honduras, although by this time thousands of Spaniards and other Europeans had reached numerous islands and mainland coasts, including Brazil.[4]

Despite the misleading impression of maps ubiquitous in Columbiana tracing the Four Voyages across an empty ocean, Columbus's monopoly on the transatlantic trip lasted one voyage. Beginning with the middle of the Second Voyage (1493–96), oceanic voyages without Columbus were being made, including between Spain and Hispaniola. Typically marginalized in Columbus-centric narratives of the Four Voyages are those thousands of other people involved—both the Europeans who also crossed the Atlantic and the Indigenous people at the heart of every chapter of the story.[5]

The highpoint for Columbus was the three years between April 1493—that triumphal moment when the monarchs Fernando and Isabel greeted him in Barcelona in person with the promised titles—and his 1496 return from his Second Voyage to controversy and criticism. During that peak period, he was lionized, acclaimed as a St. Thomas, whose spread of the faith eastward was surpassed by Columbus's westward feat. Had he lived in ancient times, gushed Peter Martyr, he would have been made a god. As Fernández-Armesto more recently put it, "the fugitive from the weaver's shop in Genoa and the tavern in Savona was elevated to the highest ranks of pagan and Christian hero-worship."[6]

Centuries later, all that praise would help forge a new Columbus for a new era; but in the immediate future, his rapid rise presaged a series of falls. Aside from the First Voyage's cinematic moments of primacy—its series of firsts—the Third Voyage brought the greatest dramatic contrasts. Sailing that more southern route, he reached the coast of South America, conceding for the first time that he had found "a very great continent, which until today has been unknown" (he also claimed to have discovered Terrestrial Paradise). Later that summer of 1498, he landed again in Hispaniola, where he asserted his authority and suppressed a pair of rebellions by Spanish colonists.

But in the autumn of 1500, he was arrested by don Francisco de

Bobadilla, a judge sent from Spain. Dispatched in chains to the monarchs, they received him well and permitted him another transatlantic voyage; yet he was banned from the new colonies, never to recover their rule. Mobs of returned settlers, angry over colonial failures and their unpaid wages and shares, protested outside Granada's Alhambra (taken from Boabdil by King Fernando), mocking Columbus's two boys as they came and went: "Look at the Sons of the Admiral of the Mosquitoes, of him who discovered the Land of Vanity and the Land of Deceit!" In the traditional narrative (especially its nineteenth-century rendering), Columbus's disgrace is positioned as a disgraceful act by the Spanish monarchs, with Fernando portrayed as a kind of Pontius Pilate.[7]

Columbus would thus die—as Washington Irving imagined it at the close of his influential 1828 biography—beset by "the chills of age and cares of penury, the neglect of a fickle public, and the injustice of an ungrateful king," unaware of his future resurrection, of "the splendid empires which were to spread over the beautiful world he had discovered" and thus "revere and bless his name to the latest posterity!"[8]

But fall is a relative concept. Calculating the modern-day equivalent of wealth from past societies is an inexact science, and so we cannot be sure how wealthy Columbus became, what value to assign a *maravedí* of the 1490s, and how to adjust for relative buying power. That said, the admiral estimated when he drew up a will in 1497 that his annual income was 4 to 8 million maravedís—roughly equivalent to $5 million today. That wealth was further secured by the time he died in 1506. "The most consistent single purpose to which" Columbus "was dedicated was the desire to found a noble dynasty of his own." That goal was achieved; its foundation in the set of contracts, privileges, or *Capitulaciones* granted by the monarchs was so solid that it underpinned his noble dynasty's survival into the twenty-first century. Posterity, indeed, and so much for penury and neglect.[9]

The Four Voyages are a siren call, then, because they are a tempting triumphalist trap—complete with the narrative rhythms of rise and fall—that draws us away from two crucial contexts. Those contexts, the focus of the rest of this chapter, help us to understand better the significance of the voyages. The first is prompted by the misleading image of those three ships, forlorn in an empty ocean—accurate in the immediate context of the First Voyage, but in the larger sense giving the false impression that Columbus did it all alone, that he initiated the great oceanic turn in world history. In reality, that turn *produced* Columbus. His voyages were but one manifestation of it. The Ocean Sea was not seen as empty and infinite, but as a place where wealth and opportunity might be found. The sea stories that inspired voyages into the ocean were in fact *island* stories, and they drew thousands of people during the dozen years of Columbus's Four Voyages alone—tens of thousands on hundreds of ships, if we include the decades before and after his voyages, and almost all of them part of a larger pattern of conquest and colonization. We might think of this context as the myth of Columbian uniqueness.

Naked and Afraid? The title-page illustration to the published 1493 letter announcing the discovery of Hispaniola and other "islands in the sea" by *Christoforus Colom Oceanice classis Praefectus* (Admiral of the Ocean fleet). Despite its fanciful inaccuracies, the woodblock carving centers the most significant aspect of Columbus's "discovery," one missing from the new admiral's titles: the islands' Indigenous peoples.

If the ocean was not empty, nor were its islands. What made them places of opportunity was the fact of their prior inhabitation. It was people who were to be conquered and colonized, not empty territory. If there is no story here without the ocean, then the story has no lasting meaning, no bite, no light and dark, without the Indigenous peoples who were encountered right from the very first moment of "discovery." Columbus's grand title, despite all the words that follow from the word *almirante*, makes no mention of the millions of people already living across the ocean. His intention had never been to find "new" people, only Asian rulers and their subjects with whom to trade and, in time, subdue. Yet their existence was an obvious conundrum, as reflected in textual and visual accounts that from the very start presented an image of Indigenous islanders as entirely lacking the traits of human civilization. That misunderstanding (or that racist lie) underpinned the tragedy that would unfold in the lands visited by Columbus during his Four Voyages.[10]

RUMORS STALKED COLUMBUS IN the months after his return from his First Voyage, his presentation to the monarchs in Barcelona of Indigenous people and other evidence of islands discovered, and the rapid reprintings of the 1493 letter announcing his success. The most prominent of those rumors was this: Columbus knew exactly where to go because someone had made the return trip before him, a navigator whose name was lost but who had shared everything with Columbus before dying—in his house, some said in his arms. Moreover, before expiring, the Unknown Pilot, as he was eventually dubbed, shared logs and charts, explaining how he had navigated across the Ocean Sea and back. Columbus never revealed the source of such knowledge.

Las Casas first heard the Unknown Pilot story as early as 1502, on Hispaniola, although he seems not to have written about it until after Royal Chronicler Gonzalo Fernández de Oviedo detailed it decades later in his *Historia general y natural de las Indias*. Oviedo's version has the pilot of a caravel taking merchandise from Spain to England blown by "such mighty and violent tempests" that the ship was forced westward to "these regions and Indies," where the crew "went ashore and saw naked people." The crew starved to death on the long sail home, with the Unknown Pilot the sole survivor, sharing "in great secrecy" the details of the journey with his "very intimate friend" Columbus. Who knows whether this tale, believed by "the common people," was true, ruminated Oviedo. But "as for me, I hold it to be false."[11]

Oviedo was right, of course. But he was ahead of his time—at least on this point. The story was repeated dozens of times across the centuries that followed, with believers far outnumbering skeptics, until modern historians felt obliged to bust the blatant myth of "this Tale of the Ancient Mariner," the "fairy tale" of a "fantasized pilot." Even then, numerous modern writers swallowed it "hook, line, and sinker." Many followed Las Casas, who recounted the story in great detail—adding that Indigenous people on Hispaniola remembered the pre-Columbian visit of a bearded white man (surely the Unknown Pilot)—while refusing to pass judgment on the tale's veracity. In his hit 1940s books, Samuel Eliot Morison asked why the story was so widely believed for so many centuries, imagining an old sailor dying in Columbus's house, prompting those keen "to pluck at the laurels of the great" to whisper that "the old fellow" told the Genoese navigator how to get to Hispaniola.[12]

Morison, I think, was close to hitting the nail on the head. The First Voyage had been Spanish in every respect save for the identity of its chief captain. Three Spanish brothers, the Pinzóns, played essential—some said equal, or even greater—roles. But the eldest, Martín Alonso Pinzón, had died upon his return, allowing his Genoese partner to claim all the glory. Giving credit to the Unknown Pilot, whom some claimed was Portuguese or Galician or Basque but most said was from the Pinzóns' home province of Andalusia, confirmed that the achievement was really all Spanish. The motive, then, was more than mean-spirited laurel-plucking; it was patriotism. Then, once the story had been repeated enough times, confirmation bias took over—as it has always done with the myths and made-up tales of Columbiana.

But Morison danced around an even more important explanation. Folk histories tend to reduce complex historical patterns to simple, personified tales. The migration of thousands over hundreds of years becomes the journey of one "tribe" over one lifetime. Waves of incursions become a single invasion led by one charismatic leader, often a fictional amalgamation of multiple historical figures. The Unknown Pilot was surely such an amalgamation, an imaginary personification of the hundreds of sailors who explored the Ocean Sea in the fifteenth century, many telling tales of seeing evidence of islands to the west—even catching sight of the islands themselves. And Columbus is such a figure too (the Known Pilot, if you will): a vast, complex pattern of movement, all inevitably resulting in European invasions of the Americas, reduced to a manageable, simple tale of one man defying the odds to singlehandedly turn history's tide.

If we zoom out to view that vast pattern through the broadest of lenses, Columbus's voyages are a small chapter in the sweeping history of European expansion from the late-fourteenth century through the early modern centuries. Zooming in a little, we find Muslim expansion into the eastern Mediterranean pushing European trade west, with banking and merchant houses from Italian city-states—especially Genoa—playing a leading role in financing such a shift. The turn to the Atlantic is traditionally dated to 1415, when the Portuguese capture of Ceuta became the first permanent European possession on the African mainland. But the western shift accelerated significantly after the Ottoman seizure of Constantinople in 1453—when Columbus was a baby.[13]

For a poor, uneducated Genoese boy, the omnipresent sea presented a horizon of dreams—most immediately the Gulf of Genoa, then the Ligurian Sea, then the Mediterranean, and beyond it the Ocean Sea. As Columbus educated himself in early adulthood, his maritime dreams were fed by a centuries-old cultural tradition that saw the Ocean Sea as a place of islands whose alchemy could transform ordinary men into lords and princes. Jean de Béthencourt was proclaimed king of the Canary Islands on the streets of Seville. One of those islands, Lanzarote, was named after its fourteenth-century discoverer, Genoa's own Sir Lancelot—Lancelotto Malocello. Conquest and service to the Portuguese Crown in the Madeira archipelago brought upward mobility to men such as Columbus's father-in-law. One of the most popular chivalric romances of his day (a genre that appealed greatly to Columbus) was Joanot Martorell's *Tirant lo Blanch*, first published in Valencian in 1490. Its main protagonist, Tirant the White, saves the Byzantine Empire from the Turks—just the kind of extravagant medieval superhero fantasy that Columbus would, as his delusions deepened, come to take seriously—and includes a witty subplot in which a king of the Canaries launches from his islands an invasion of Europe.

Some islands of the Ocean Sea were real, some fictional; some famous voyages really happened, others were imaginary. The blurred line between fact and fantasy matched how Columbus saw the world. The dangers of the open sea were at first part of what made it exciting. After all, such risks had helped the ocean keep its secrets. And in his early writings on the ocean, to attract further investment and consolidate royal support, Columbus stressed his triumph over those dangers, promoting the ocean as a place where the air improved as one sailed west and reached the paradisical promise of the new islands he had discovered. Only later, as he lost control of the new colonies, did the balance between promise and danger shift.[14]

By the time Columbus settled in Portugal, the Portuguese had spent two generations exploring the Atlantic—learning to navigate by

A Fishy Tale: The depiction of St. Brendan's mythical ninth-century search for Earthly Paradise in the Atlantic, in Honorius Philiponus, *Nova typis transacta navigation* (1621). St. Malo is rowing separately, the Canaries to his west. St. Brendan is on an island that turns out to be a whale. The island named after him is spectral; it disappears upon being seen, as seemed to happen all too often to European sailors.

the stars, showing that the tropical seas could be sailed, mapping the West African coast, and discovering the ocean's islands. The legend of Ireland's St. Brendan, like that of the Unknown Pilot, personified earlier phases of exploration into the Atlantic. It was widely believed that islands named in the story, or islands like them, existed to the west—whether or not the legend as a whole was taken in the fifteenth century to be factual (and details such as the Irish sojourn on an island that turned out to be a whale or sea monster were surely taken literally by few who heard or read them). A 1439 expedition ostensibly to find Brendan's isles resulted in the Portuguese discovery of seven of the nine Azores islands. After that, voyages of exploration left Portugal every few years, and virtually annually by the 1470s.[15]

In 1452 (when Columbus was just months old), Flores and Corvo, the westernmost Azores islands, were found by Diogo de Teive, a captain of Prince Henrique (Henry the Navigator); Flores is roughly halfway between Lisbon and Newfoundland. Searching for the mythical island of Brasil, Teive sailed through the Sargasso Sea and as north as the latitude of Cork. There were more islands to the west, he was sure. One of his pilots, a Spaniard named Pedro de Velasco, was still alive in 1492 and living in Palos—where he helped recruit sailors for Columbus's First Voyage. A Portuguese pilot named Vicente Dias was convinced that when a trip from Africa to the Azores blew him too far west, he saw coastline. Another, Gonçalo Fernandes, had a similar experience, and so convinced the king that the monarch granted the yet-to-be-discovered island to his own brother in 1462.

From that year through 1488—when Diogo Cão and Bartolomeu Dias reached the southernmost tip of Africa—a steady string of contracts and charters were issued by the Portuguese monarch to find western islands. Some had yet to be named, others were given names from stories such as the St. Brendan legend. Some of those toponyms would remain legendary (like the Seven Cities), but others would end up attached to real places (like the Antilles and Brazil). Columbus named one Caribbean archipelago after the "Virgins" of St. Ursula (who in one legend had sailed west). Not one of these documents asserts or suggests that the islands and possible mainland to the west were Asia or adjacent to it; that muddleheaded notion was Columbus's, helping to explain why he never became directly part of the persistent Portuguese process. Indeed, the plethora of Portuguese voyages in the fifteenth-century Atlantic has inspired claims, periodically bubbling up for centuries, that Portugal's navigators actually did find the Americas before Columbus.[16]

Bristol's merchant sailors had meanwhile begun crossing the northern Atlantic. In 1498, the Spanish ambassador at the English court wrote to the Spanish monarchs that English voyages west into the Atlantic had been mounted repeatedly since the previous decade, making specific reference to a compatriot of Columbus's, Giovanni Caboto (known to English-speakers as John Cabot), who started sailing from Bristol in 1491.

> For seven years those of Bristol have equipped each year two, three, four caravels to explore the island of Brasil and the Seven Cities, with the imagination of this Genoese.... I am certain that Your Majesties have heard that the King of England has formed a fleet in order to discover certain islands and mainland which he has been assured were found by men who sailed from Bristol for the same purpose last year; I have seen the letter made by the discoverer, who is another Genoese like Columbus.[17]

From 1480, if not earlier, regular expeditions sailed from Bristol to profit from expanding trade with Iceland, to access massive northwest Atlantic cod shoals, and to reach and gain further knowledge of islands in the ocean. Tales of land sightings abounded. Seemingly vast islands were given names, versions of the same names on the lips of Portuguese navigators or inked on Venetian charts—like Hy-Brasil, the Seven Cities, and Antillia. The Genoese cartographer Andrea Bianco, for example, included actual islands with mythical ones and a statement that an "authentic island" lay some fifteen hundred miles west into the Atlantic; Newfoundland is almost two thousand nautical miles from Bristol, some twelve hundred miles from the Azores.[18]

Did Bristol mariners, sailing west with cargoes of salt to preserve fresh cod, land on the shores of such islands—legendary places that we know today as Labrador and Newfoundland—and thus "discover America" before Columbus? Very possibly, although there is no evidence for it. More to the point, westward voyages were being made from multiple European ports by multiple merchants and national interests in the fifteenth century. Inevitably, some reached the Americas and returned; equally inevitably, colonization resulted by century's end.[19]

All these voyages are typically seen as precursors to Columbus's, significant only for explaining the sources of his knowledge and inspiration for his vision. But if we look at this history in a less Columbus-centric and more balanced way, the voyages emerge as mutually inspiring and informative, with Columbus's 1493 return from islands previously unknown to Europeans as a—not *the*—milestone moment. Far from an original idea in the fifteenth century, the notion of sailing west or south into the Atlantic to find new islands was arguably the *least* original thought that an ambitious merchant mariner might entertain. As Fernández-Armesto observed, "almost every element in the thinking that underlay [Columbus's] enterprise was part of the common currency of geographical debate in his day." Throughout the century, speculative assumptions regarding the islands and peoples and riches of a seemingly infinite Atlantic filled maps, works of geography, the meeting rooms of merchants and monarchs, and many a mariner's dreams.[20]

A set of interrelated developments fueled those dreams and turned ambition into incipient empire. One was Portuguese–Castilian competition. As mentioned earlier, rivalry between the two kingdoms had turned into war in the Iberian Peninsula and the Atlantic Ocean, motivating monarchs and merchants to invest in new ways to exploit African coasts, Asian trade routes, and new lands in and across the Ocean Sea. Another was the slave-based sugar plantation economy. Initially developed in the eastern Mediterranean, that economy likewise moved west, first to regions such as the Algarve coast of southern Portugal, then out into the Atlantic islands (the Azores, Madeira, Cape Verde, and the Canaries), and finally into the Americas. Slave labor was acquired wherever it could be, by trade or by force—on the West African coasts, in the Canaries, and on the Caribbean islands, beginning in 1492 (a topic to which we shall return momentarily).[21]

Ships and the winds upon which they depended were two more developments. During the fifteenth century, the Portuguese and other Europeans learned how to build ships capable of sailing far into the Atlantic Ocean: the quest for wealth to the south and west, and the development of bigger, stronger, faster ships, were mutually reinforcing. At the same time, navigators developed a better understanding of the oceanic gyres, the great circular, clockwise wind and current system of the Atlantic Ocean. Columbiana is full of claims that the Genoese navigator's decision to sail west from the Canaries was "original and brilliant," that "Columbus not only discovered America, but he discovered the two great maritime routes of the North Atlantic, which the next four centuries of navigation by sail would verify are the fastest and best." In fact, many navigators sought "to crack the Atlantic wind code" in the fifteenth century, Columbus playing a role (albeit an important one) in the collective cracking of that code in the 1490s. He simply followed "in the steps of others who had plied the same waters but for shorter distances."[22]

He also had no choice in choosing where in the ocean to sail west. Although Portugal and Castile were at peace in 1492, and Columbus could theoretically have sailed from any of the Atlantic islands, it made the most political sense to use as his launching point the ocean's sole Castilian possession, the Canaries. His outbound trip of 1492 confirmed what had hitherto been suspected, that the westerly winds swept across the southern seas (at roughly latitude 30 to 35 degrees north; the Canaries are ideally positioned at 28 to 29 degrees). On the 1493 outbound trip of his Second Voyage, Columbus caught the gyre with a tad more efficiency, sailing southwest instead of due west. He was thus out of sight of land for only thirty days—a speed that would remain typical for centuries. The clockwise wind system obviously permitted a return to Europe as well, and the Portuguese had learned during the second half of the fifteenth century that at the latitude of the Azores, the winds blew east (frustrating their efforts to use those islands as a base to sail west). On his first return, in 1493, Columbus sought those exact winds. It took some trial and error to find them, but he knew, as other navigators did, roughly where they were. It would be twenty years before Spaniards discovered that by counterintuitively sailing north up the Florida coast, Europe-bound ships could catch the sixty-mile-wide, two-thousand-mile-long eastward curve of the Gulf Stream.[23]

One could argue that Columbus was lucky rather than "original and brilliant." One could also argue that a voyage from the Canaries to the Caribbean was inevitable in the 1490s, with return voyages at the latitude of the Azores likewise inevitable. The peculiarity of the Atlantic is that its gyres, its trio of currents and trade winds, "reach out to the rest of the world"—to Africa, and thence to Asia, as well as to the Americas—an opportunity that Europeans worked to seize throughout the century, finally grasping how to do so in its final decade. The route that Columbus helped pioneer would become Spain's oceanic highway—the *Carrera de Indias*—for centuries governed by a combination of promise and danger. The winds were a vital friend, but they could also be a deadly enemy, obliging Atlantic crossing to be a collective enterprise. Growing competition among Europe's empires, at war more than at peace during the three post-Columbus centuries, made sailing in fleet formation all the more important. The great Spanish fleet—*La Flota*—would comprise dozens, then scores, of ships crossing together, growing to hundreds in the eighteenth century. The Ocean Sea of *La Flota* was the same one on which Columbus had faced the forces of nature and the treachery of men.[24]

Such treachery (as Columbus saw it) came both from those who sailed and those who competed with him. For during his lifetime, thousands sailed west into the Ocean Sea, most of them reaching and returning from the Americas. Among those thousands were members of his own family, which is exactly what we would expect on the basis of Italian and Iberian patterns of exploration, conquest, and colonization. An emphasis on family—exclusively male relatives—underpinned Columbus's motives and goals, as well as the undertaking of his ambitions and plans. After all, only family could truly be trusted. Bartolomeo was with him in Lisbon, from the start or close to it. As don Bartolomé Colón he would sail to the Americas with his brother and help him attempt to govern the first colonies there. A third Colombo, Giacomo, joined his brothers in time to cross on the Second Voyage (as don Diego Colón).

When word reached Genoa in 1493 that Cristoforo had become the family's arriviste in Spain, other kinsmen jostled to join him. Two cousins, Giovanni Antonio and Andrea, left Genoa to participate in Columbus's Third and Fourth Voyages. His younger son, Fernando, joined him as a teenager on the latter trip. Relatives of his sons' mothers also sailed with him—a brother-in-law via his Portuguese wife, for example, and a brother and cousin of the Arana family of Cordoba—although, as we shall see, their lives were worth less to the admiral than those of his blood kin.[25]

The expanding inclusion of Colombo/Colón men in Columbus's activities contributed to growing Spanish resentment of the whole clan. But that stemmed from Columbus's initial success in finding the route across the ocean and then his shortcomings as an administrator; nepotism and the corporate use of kinsmen was neither unusual nor offensive. The Genoese mercantile empire was structured entirely along family lines, being the sum of its dynastic companies. That was also how Castilians and other Iberians engaged in economic and political enterprise. In bringing in brothers and male cousins, Columbus was behaving exactly as Spanish conquistadors did in their slaving and colonizing enterprises in the Atlantic and the Americas.

Indeed, the family who served as principal partner to the Colombo/Colón clan in 1492–93 also centered on a fraternal trio. The Pinzón brothers were from Palos, the Spanish port that made the First Voyage possible by providing ships and sailors, and the port from which the three ships of that expedition departed. Martín Alonso Pinzón was a wealthy shipowner who contributed ships and funds; he captained the *Pinta*, and his youngest brother Vicente Yáñez Pinzón captained the *Niña*. Martín Alonso chose their middle brother, Francisco Martín, to be the master on the *Pinta*, an important position (with responsibility for the crew and their provisions) that would typically fall to the ship's owner (in this case, Cristóbal Quintero, who was also aboard). Columbus may have had the contract with the Spanish Crown, but the Pinzón family were the key investors—of ships, expertise, and a reputation that allowed them to recruit men to sail.

Tension was inevitable, especially considering Columbus's ego. The

Diario suggests that Columbus mocked the elder Pinzón for prematurely claiming the prize for sighting land; but then Columbus took credit from the young sailor who *did* spy land. The depiction of Martín Alonso's actions in the *Diario* is increasingly negative, supporting the argument that Las Casas enhanced such criticism to support his vision. In that vision, Columbus was God's agent, aware of how amenable "the Indians" were to conversion. But Spanish explorers, conquistadors, and slavers such as the Pinzón brothers and all their greedy ilk ruined everything (*cudiçia* and *cudiçioso*, "greed" and "greedy," is a motive assigned almost exclusively to the elder Pinzón in the *Diario*, never to Columbus). Exaggerated or not, Columbus and Martín Alonso Pinzón clearly fell out during the First Voyage and were separated at sea for a while; it may well have been Pinzón and his crew who first saw Hispaniola.

Because the *Santa María* ran aground and was consequently broken up, Pinzón took the *Pinta* back to Spain, beating Columbus and the *Niña*. Later testimony by survivors of the voyage strongly suggests how different the history of "the Discovery" might have been if Pinzón— supported by his kin and all the men of Palos and Moguer—had told his version to the monarchs first, or even at all. But the monarchs wanted to see Columbus first, denying a prior audience to Pinzón, who died a few months later. He achieved some revenge from the grave, as the crown later used a reconstruction of the Pinzón perspective to undermine Colón family claims. The 1519 granting by Emperor Carlos V of a Pinzón family coat of arms, featuring three caravels and a pair of feather-clad Indigenous islanders, was also some consolation.[26]

Was the crown right, that Pinzón was equal to Colón as Discoverer? Perhaps, but Columbus defenders have maintained that the Pinzón family and the crown unfairly sought to deprive the admiral of his due, just as Amerigo Vespucci had deprived the continent of its rightful name. Undoubtedly, the First Voyage depended on the financial support, navigational knowledge, and participation of the Pinzón brothers, as Las Casas essentially argued—although he concluded, in the words of a later Spanish historian, that such a role did "not justify crediting Pinzón with the initiative of the discovery." Meanwhile, although Vicente Yáñez Pinzón remained loyal to Columbus on the First Voyage, he refused to sail with him again, going on to make his own voyages in 1499 and 1500—being the first European to see the mouth of the Amazon (where he grabbed and enslaved several dozen local people) and to land on the South American coast below the equator. In the end, his Atlantic crossings outnumbered those of Columbus, and in 1508 he and Juan Díaz de Solís came very close to finding the strait that would later be found by, and named after, Fernão de Magalhães (Magellan). Far less is known of the subsequent career of his brother Francisco Martín, although some accounts have him sailing with Vicente in 1499, then joining Columbus's Third and Fourth Voyages—but drowning during the latter. His nephews—Martín Alonso's sons Arias and Juan—both made transatlantic crossings, as did other relatives and those (such as the Niño family of Moguer) connected to the Pinzóns by blood, marriage, and enterprise.[27]

Participants on the Four Voyages saw them as collective enterprises, theirs as much as Columbus's. For example, Michele de Cuneo of Savona (the same Ligurian town where Columbus had lived in his early twenties) wrote home on October 15, 1495, that "sailing toward the island of Española, I was the first to discover land," and upon dropping anchor at a cape, "the Admiral . . . gave it the name of the Cape of San Michele Savonese in my honor, and he noted thus in his book." Similarly, tension between sets of brothers was not restricted to the First Voyage; on the Fourth, Columbus and Bartolomé were frequently at odds with the Porras brothers, Francisco and Diego.[28]

Columbus, then, was far from alone. Hundreds sailed with him, many on multiple voyages, and thousands traveled on scores of other expeditions that crossed the ocean during these same years. The fact that he was the first to cross from Europe to the Americas and back—putting aside the Norsemen in Newfoundland five centuries earlier, as well as other claims—has obscured the fact that he was part of a process involving many men, thereby elevating his primacy status and generating a myth of uniqueness. But he is better understood not as unique, or even atypical, but as a more or less typical southern European merchant seaman seeking to use ocean-bound mobility to achieve socioeconomic mobility. In doing so, he became an explorer, slave trader, and conquistador among tens of thousands of his era.

IS IT PROVOCATIVELY RHETORICAL to call Columbus a slave trader and conquistador? Was he really even an explorer? One historian has argued persuasively that our sense of "explorer" is imbued with modern assumptions of scientific curiosity, motivations that were developed in the eighteenth and nineteenth centuries and then back projected onto the likes of Columbus. To be sure, fifteenth-century navigators were highly curious and keen to record observations in ways that might seem to reflect the intellectual developments of the age (to which historians like to give grand names, such as Renaissance and Scientific Revolution). But the *primary* motivations of such men—Columbus included—were overwhelmingly "a desire for economic advantage" (in the form of trade with local peoples and, if possible, their conquest), a search for known places (not unknown ones, even if such knowledge was vague and possibly legendary), and "a collateral desire to spread their spiritual values" (among local populations).[29]

In other words, despite all the talk of old and new lands, these "explorers" were looking for people—people with whom to trade, people to subdue and convert, people to guide them to sources of wealth and to work for them, people who could be, if necessary and in compliance with highly elastic European laws on the topic, enslaved. Seldom employed or directly controlled by the crown, such men created independent companies, blessed with crown approval, that sought profit that could be leveraged into rewards from the crown. Such rewards included political power and social mobility, of which Columbus's positions, titles, and dynastic legacy are a classic example. They were armed, willing to torture, kill, and enslave, and usually expected to do so. Explorers were conquistadors, slave traders, and colonizers—in the sense of their motivations, methods, and goals, but also in that they were often the very same men.

The Spanish Crown–Columbus agreement, the Capitulaciones de Santa Fe, was not overtly a contract for colonization, but neither was it a mere charter to explore and trade; in effect, it was a license for imperial expansion. His jurisdictional rights as admiral, the title and office confirmed after the First Voyage, empowered him to conquer and administer lands and people—the term *almirante* having entered Castilian from Arabic in the thirteenth century, its root, *amir*, meaning a commander in war. Against whom, then, would such war be waged?[30]

The "pointless bickering over what, if anything, Columbus can properly be said to have discovered" has often drawn attention away from *whom* Columbus discovered. For beginning with Columbus's return to Spain in 1493, Europeans became aware of the peoples that we call Native or Indigenous Americans. And as Europeans became aware of the extent and variety of those peoples new to them (who, in 1492, numbered perhaps seventy million), so did Indigenous Americans become more aware of each other and of Europeans—a great and complex mutual discovery. Yet within one hundred fifty years of 1492, that Indigenous population had fallen by 70 to 90 percent, a loss of some fifty to sixty million people, representing almost 10 percent of the human population at the time. How did that happen? Can it accurately be termed a genocide, and if so, was Columbus responsible?[31]

From the moment of the First Landing, the very beginning of the European presence in the Americas, Indigenous peoples were present—as was European ignorance and denigration of their humanity.

Presenting People: It is unclear whether the "Natives" that Columbus brought back to Spain, and "presented" to the monarchs, were voluntary travelers—the first Indigenous Americans to discover Europe—or the first Indigenous people to be enslaved by Europeans and forcibly removed from their homeland. This scene was frequently reproduced on public monuments, paintings, in book illustrations, and on postage stamps.

The inhabitants of Guanahaní, and of all subsequently reached islands, were called *indios* (Indians) because Columbus insisted that they were off coastal Asia, or *la India*. Lured or coerced into coming aboard Spanish ships, some Taínos (as we call them, a shorthand for a complex web of overlapping ethnic groups) were put into service as guides, and up to a dozen were brought back to Spain to be presented to the monarchs.[32]

Queen Isabel ordered Columbus to return to Hispaniola, whose seemingly innocuous Indigenous inhabitants were to be converted to Christianity and treated as her new subjects. Columbus had imagined he would be ruling over trading posts, like the Portuguese model on the African coast but on the Asian coast instead; the Spanish model was based more on its very recent conquests of Granada and the Canary Islands. No matter. The way forward had already effectively been decided by the thirty-nine Spaniards left on Hispaniola, at a camp built from salvaged wood and iron from the grounded *Santa María*. The tiny settlement was made possible by help from local Taíno ruler Guacanagarí (exhibiting "the humanity of the Indians," Las Casas later lamented, "toward the tyrants who have exterminated them").

By the time Columbus returned, those Spaniards had fought among themselves and with the Taíno, and to a man they had died. Most of the twelve hundred men on the Second Voyage were conquistador-settlers. Armed with arquebuses and steel swords, and using war dogs and horses, they set about vengefully conquering the island. The situation was complex: Spaniards were dependent for food upon local communities, in which some invaders therefore settled; while local rivalries, such as that between rulers Guacanagarí and Caonabó, were exploited by Spaniards. But the bottom line was this: as long as the Taíno remained docile, willing to submit to panning for gold and working for the Spaniards, they were considered new colonial subjects; resistance of any kind permitted Columbus and the Spaniards, under Spanish law, to slaughter or enslave.[33]

Even so, the typical telling of the rapid decline of the early colony into violence tends to be influenced by the histories that, in the late eighteenth and early nineteenth centuries, built the foundations of the modern Columbus legend. Most important, these histories created a moral separation of Columbus, his motives and goals, from the Spanish settlers. Yet a close reading of sources such as the *Diario*—a profoundly pro-Columbus document, with its complex Columbus and Las Casas authorship—reveals an obsession with finding, describing, and convincing the monarchs of the abundance of two things: gold, and pliable Indigenous peoples. Of the roughly ninety dated entries in the *Diario*, from the day of the First Landing, October 12, 1492, until the start of the return voyage three months later, almost half mention gold, and more than three-quarters mention "the Indians." Likewise, the twin foci of the letter Columbus wrote at sea in 1493 to royal minister Luís de Santángel were gold (its abundance) and "the Indians" (the ease with which they could be manipulated and used).³⁴

It is easy to be distracted by how this voyage differed from those before and after it—by, for example, the fact that it was the first European voyage to the Americas and back, after which mass colonization began; and by the frequent mentions in the *Diario* of India, the Grand Khan, and Cipangu (Japan), with Columbus's obsessive insistence that he had reached Asia. Yet the focus on gold and local peoples made the voyage no different from fifteenth-century voyages along the African coast and sixteenth-century expeditions to numerous parts of the Americas. Gold (and its sibling, silver) were fungible sources of wealth that could immediately be used to pay debts, fund future expeditions, and buy favor with the crown; the sole other source of immediate profit was the local population.³⁵

Indigenous families were also the source of long-term profit for a successful colony—through their provision of labor and the payment of tribute in foodstuffs and other products. Columbus and Las Casas emphasized their docility and suitability for quick conversion as proof of colonization's legitimacy, but for conquistadors it made colonial settlement viable. In the *Diario* "the Indians" are almost like local fauna, repeatedly described as naked, lacking any religion or other attributes of human culture, easily startled, easily seized but prone to running off, easily tricked—like animals, even running around "like chickens." In the 1493 Santángel letter, their gullibility is so readily exploited they are "like beasts" (como besti). When Pope Alexander VI, in a 1493 Bull of

Donation, granted the new islands to his compatriots, the Spanish monarchs (Alexander was Aragonese and beholden to them), he was technically granting the people as heathens to be converted, not giving away lands. And when he noted that its people "walk about naked and eat no meat," he is, from the fifteenth-century European perspective, barely describing people. They were still human, a requisite to being converted subjects of the crown, and indeed their apparent unworldliness was seen by evangelizers as suitability for Christianization. But that made them seem subhuman or quasi-human to some colonists, and the superiority and surrender fantasies of conquistador culture are evident from the start. Short-term profits tend to trump long-term considerations, and the descriptions of Indigenous encounters in sources such as the *Diario* and the 1493 letter are heavily laced with the language of capture—of *tomar*, "to take," and its derivatives ("tome p[or] forza algunos dellos" [I took some of them by force] is an example).[36]

Enslaving non-Christians was an integral part of Columbian, Spanish, and other European voyages of exploration and colonization in and across the Atlantic—despite the counterproductive impact that it had on long-term efforts to establish colonies built upon Indigenous communities. The slaving impetus in fifteenth-century Europe, from the wars against Muslim kingdoms and the quest for enslaved Africans, was powerful and overwhelming; it colored the First Voyage and rapidly came to dominate Spanish activity in the Americas before Columbus's last crossing. Columbus himself oversaw the shipping to Spain of several thousand enslaved Caribbean men, women, and children; the total number of such Indigenous victims in his lifetime was several times that amount. By 1540, between one-third and one-half million Indigenous Americans has been enslaved; their bondage, combined with that of enslaved Africans, underpinned the development of imperial Spain and Spanish America.[37]

To blame or try to excuse Columbus and his brothers for all this is to miss the point: they did not start it; they certainly participated in it; they were powerless to stop it. Is it accurate, therefore, to characterize Columbus in the Caribbean (to quote one popular present-day historian) as "a tyrant and a bully, as a slaver, an unrepentant imperialist and a man of immense avarice and self-promotion"? To fully grasp the Columbiana phenomenon, all opinions must be respected, and I can see obvious elements of truth in such a judgment. But it needs context. Having left his brothers Bartolomé and Diego in charge of the colony when he sailed to Spain in 1496, Columbus returned in 1498 to find anarchy and violence; Francisco de Bobadilla, sent by the Spanish Crown in 1500 to investigate, found the same, soon arresting all three brothers. Some of those modern accusations would have resonated with Bobadilla—his fellow colonists found Columbus to be a tyrannical governor, and officials assumed the lack of profits from the islands was due in part to the avarice of the Columbus brothers. But Columbus's obligation to the crown—indeed, one of his earned privileges—was to promote the enterprise of gold-seeking and (if necessary) people-seizing empire, sanctioned by royal law and papal edict, requiring no repentance by its participants—including Bobadilla, who replaced Columbus as governor. The crime for which the three Colón brothers were arrested was failure to peacefully and profitably exploit the crown's new Indigenous subjects.[38]

The crown, and Queen Isabel in particular, repeatedly stated that "the Indians," as "her vassals," were not to be enslaved, a ban often cited by historians. When Columbus wrote to the monarchs from Hispaniola in 1494, proposing a scheme to exchange enslaved islanders for cattle, the crown rejected the idea. But that rejection was more of a shelving, on economic rather than moral grounds. While Columbus awaited a response to his letter, the enslaving of Indigenous families and the public execution of leaders continued apace.

For the crown's ban was full of holes. "Indians" deemed to be already enslaved (by their own people) could be "redeemed." But the term *rescatar*, "to redeem" or "rescue," also carried a sense of "recapture" (as its Latin etymon means). Redemption thus tended to mean a transfer from the Indigenous slave system to the transatlantic Spanish one, the two being very different. In Taíno communities, rulers (*caciques*) and nobles (*nitaínos*, hence the name given by Europeans to the whole nexus of ethnic groups) received tribute in labor and goods from dependents called *naborías* (these terms are all hispanized). Spaniards classified *naborías* as chattel slaves, a willful failure to understand the Taíno household system. As a result, most *rescate* captives were free people being enslaved for the first time. Columbus and the Spaniards had seen, if not participated in, the trading of Granada's enslaved Muslims, Canary Islanders, and captive Africans, and they were thus quick to add Indigenous Americans to that "legal" source of profit.[39]

Another loophole comprised "those who by our command are declared slaves," being those "who are called cannibals"—in the words of a 1504 license given by Queen Isabel to Juan de la Cosa to raid islands in the Antilles. The transatlantic trade in enslaved islanders, dubbed Caribs for the very purpose of their captivity, was already well established. More than five hundred had been shipped to Spain with the 1496 return of Columbus's Second Voyage; the slave trade soon scattered them across southern Europe, and although they rapidly died from disease (or, like several on a ship off Genoa, at sea), they were in demand as "the best slaves found so far" (as one Venetian nobleman enthused).[40]

Why were the cannibal tendencies of these enslaved families not a problem in Europe? Because Indigenous cannibals were a product of the Spanish slaver-colonist imagination. Indeed, an entire "race" (the Caribs) was invented and the whole region thus named (the Caribbean). The forming of the legend, beginning in 1492, is transparent in the *Diario*: most of "the Indians" were placed in the docile, noble-savage category described above; but the men of the First Voyage came expecting, and believed they saw (or heard credible tales of others seeing), Indigenous peoples in various monstrous or abnormal categories, from dog-headed to one-eyed people, from islands of women only to—above all—cannibals. Columbus was dismissive of some such tales in his 1493 letter, noting that "I have not found the monstrous men" (hombres monstrudos) that some expected, aside from an island "where the people are born with tails," but he excitedly reported on whole islands of human eaters of human flesh. Why were Columbus and the Spaniards so quick to believe stories of *caníbales* and to classify entire communities as such (even though not a shred of evidence was found, or has since been found, to support the notion of "cannibal tribes"; all "evidence" being hearsay or acts of resistance to being captured)?[41]

The answer is slavery. But Spaniards did not enslave indiscriminately. Similar to how societies today create both legal and moral systems to protect workers *and* loopholes in those systems (such as classifying workers as "illegal" or "alien" or "undocumented"), so did the legal and moral systems of Spaniards and other Europeans require loopholes to permit enslaving "Indians." Those who resisted Spanish incursion—including captive-taking—were thereby classified as *Caribe* or *Caniba*. The category was used to justify attacks on the inhabitants of whole islands and coastlines, their resistance confirming the prejudicial view of them as savages suitable for "redemptive" enslavement. From the very start, in that 1493 Santángel letter, Columbus enthused that the cannibal islands would produce "as many slaves as they may order to be shipped" (esclauos quãtos mãdaraen cargar). "You have arrived at a good moment," the colonists on Hispaniola told those who reached the city of Santo Domingo in April of 1502, in the largest fleet yet to sail from Spain to the Americas. "There is to be a war against the Indians, and we will be able to take many slaves." As recalled later by Las Casas, "this news produced a great joy in the ship." (Columbus, completely side-lined by this point, was in Spain preparing to launch his Fourth Voyage.) Whole coastlines of the American mainland were designated as Carib country so that their inhabitants could be legally enslaved, while a 1508 edict by King Fernando added the category of "useless lands"—those that apparently lacked gold and docile people to mine it, whose "Indians" could therefore be seized, branded, and sold.[42]

Yet another loophole was that of interpreters. In the 1503 license Isabel granted to Cristóbal Guerra—to return to the South American coast he had scoured for pearls, gold, and slaves in 1499—she conceded that he "could take, wherever he may find them, Indian men and women as interpreters for those lands, as long as it is not to be slaves nor to do them evil or harm, and that they be taken as much according to their will as possible." A whole ship of captive families could be sailed through such a hole. Indeed, Cosa and Guerra enslaved more than six hundred "cannibals" on those two expeditions alone, from which Isabel would have received a 25 percent tax on the profits (a fifth was more typical) had storms not shipwrecked and stranded the expeditions—prompting the starving Spaniards to butcher and eat one of their Indigenous captives (according to Oviedo, who takes the irony with a lightness bordering on approval).[43]

The debate over the morality of Columbus's actions in the Caribbean has tended to focus on what some authors have dubbed a "list of charges"—often reduced to three. The first is Columbus's taking of Taínos from Hispaniola to Spain in 1493, in order to present them to the monarchs as evidence of his discoveries. This has been rather weakly defended as a voluntary visit, and the return of one Taíno man to his home later that year is likewise taken as a sign of the admiral's "good will." Nor is there evidence that any of them were later sold as slaves. Yet true or not, Columbus ruthlessly prevented them from escaping, and most seem to have soon died (as did the Aztecs brought by Hernando Cortés upon his first return to Spain—for similar display purposes). Furthermore, there is no escaping the larger colonialist context, in which the perceived overwhelming superiority of Spanish over Indigenous culture is seen as justifying the taking and moving of Indigenous people—be it for show-and-tell, to better exploit their labor, ensure their religious conversion, force them into sexual slavery, or to sell them in the growing Atlantic slave market.[44]

Second, upon returning to Hispaniola in 1493 to find the settlement of La Navidad destroyed, Columbus sent Spaniards to make sorties into the island. The ensuing violence—the invaders abused, raped, tortured, and enslaved hundreds—is blamed on the man who held the titles of admiral, governor, and viceroy. Even Las Casas, always keen to show that Columbus was God's agent, bitterly noted that 1494 was a turning point, marking "the beginning of the spilling of blood, later to become so copious first on this island and then in every corner of these Indies." On the other hand, Columbus apologists have stressed that he was ambivalent about slavery, opposed to the mistreatment of Indigenous peoples, and equally disciplinarian with his own people (hanging both his own steward and his majordomo, for example, for allegedly stealing and selling supplies, and having his brother-in-law fatally tortured).[45]

If that defense seems feeble, it is better than the one common in the nineteenth century. According to that defense—as articulated by

The Climb: The 1897 Columbus statue in front of the cathedral in Santo Domingo imagines Indigenous islanders as barbarians forced to climb—probably in vain, yet apparently willing to try—toward the civilizational pinnacle symbolized by their European discoverer.

a prominent historian in 1903 using Columbus's own arguments—removing from their homelands those classified by Spaniards as Caribs obliged them to stop eating human flesh; resulted in the saving of their souls; and encouraged others to submit to colonial rule. Such a defense was effectively an omelet-making-requires-egg-breaking apologia for the entire enterprise of European colonization in the Americas:

> Those who criticise [sic] Columbus and lay at his door the horrible cruelties which followed for more than three centuries and a half in the wake of the slavery system should remember the age in which he lived, and particularly the work he was called upon to perform.

Called upon by whom? By God and the imperative of Christian civilization, an argument predicated upon the belief—as prevalent in Columbus's day as in 1903 and beyond—that Indigenous people suffered under heathenism and barbarism. In the age of empire and the White Man's Burden, doubling down on the Columbian invention of the Caribs helped drive this defense home, further dehumanizing them as the worst kind of "savages," who "fought as beasts fight, for the love of killing," addicted to cannibalism, treating children "like capons, in order the better to fatten and to improve the flavor of their flesh."[46]

The third charge relates to the system of labor put in place to acquire gold tribute from the local population. To enforce compliance, Indigenous workers were terrorized. There were beatings, hangings, hands were severed, other tortures enacted in public, families were broken up, and ships returning to Spain were loaded with the enslaved. Some allege that this was Columbus's evil nature manifested, others that his incompetence as an administrator allowed this all to happen—and was thus his fault. The contrary argument observes that neither the demand for gold nor the system of local labor provided as "tribute" to new lords was invented by Columbus; and it is surely the case that had the admiral died on the outward Second Voyage, the same system would have been imposed, with similar or identical results. Las Casas and Oviedo seldom agreed, but they did in their judgment of Columbus here, and Fernández-Armesto followed their lead: his crime was "misjudgment rather than wickedness."[47]

To what degree this excuses Columbus is open to debate, but it surely places him in between monster and saint. More to the point, it reinforces my earlier emphasis that he is neither the architect nor instigator of the colonial system that arrives with him on his earlier voyages. Confusing the issue is the contradictory nature of that system—its dependence upon the Indigenous communities that it paradoxically set about destroying. To denounce the colonial system as a genocide is to sidestep that contradiction. As is the depiction of it as "settler colonialism"—the focus of which is the displacement of Indigenous peoples, if not their genocidal removal, from lands taken by Europeans for settlement. The phrase is often misapplied to the regions of the Americas where Spaniards sought to build colonies with Indigenous populations, exploiting their local knowledge, resources, and labor. The treatment of Indigenous peoples in the Caribbean and circum-Caribbean was undeniably brutal—shockingly so, by the standards of any era—beginning in 1492 and extending through the century that followed. The fact that their systematic raping, maiming, slaughtering, displacing, and enslaving largely took place outside Columbus's control or after his death does not excuse his involvement. But nor can such practices and patterns of exploitation be credited to or blamed simply on him; at the root was not a Genoese mariner, but the logic of empire and the violent, exploitative bigotry of colonialism.[48]

IN 1892, THE UNITED States of America was consumed by a largely uncritical and often giddy celebration of Columbus. As the Quadricentennial was really a patriotic celebration of "America" and its place in the world in the 1890s, criticizing Columbus was seen as unpatriotic. It was thus with delicate phrasing that one member of the Massachusetts Historical Society addressed his peers in October of that year, offering a mildly contrary position: he argued that had the Genoese navigator "and his whole company gone to the bottom of the sea," the "discovery could not long have been deferred." Larger developments made it inevitable in or before 1500, he insisted. After all, "it is of little consequence whom the light strikes first."[49]

The Massachusetts man was right. Columbiana's obsessive disputation over every tiny detail of Columbus's life and deeds has inflated the importance of those details. A false determinism has resulted, whereby the phenomenon that altered world history was not the rise of European empires in the Americas and then the world, but the minuscule moment when Columbus first set foot on New World soil—"the most important event after the sublime magnitude of the sacrifice of Christ" (as imperial apologists put it). That event is thereby transformed from being of great symbolic significance to being deterministically unparalleled, the root and cause of all that has happened since.[50]

That elevation of the First Voyage began within a decade of the navigator's own lifetime. A Genoese merchant named Barnaba Caserio, for example, writing to the doge of his hometown in 1616, and petitioning for support for his enterprises in Spain, argued that "if Henry VII, King of England, had listened to the Genoese, Christopher Columbus, half the world would now belong not to the Spaniards but to the English." The theme has persisted ever since. For some, the world-changing moment preceded even the First Landing. For example, the *Diario* noted that on October 7, 1492, Columbus saw "a great multitude of birds flying from north to southwest, so one could believe that they were going to sleep on land," and therefore he changed course from west to west-southwest. A long sequence of historians, beginning two centuries ago with Washington Irving, has proclaimed that had the three ships not changed course that day, the first arrival point for the discovery of America would have been Florida—and therefore, *to this day*, Spanish, not English, would be the first language of the United States, whose population would be "Spanish and Catholic instead of English and Protestant."[31]

Thank God for Columbus, suggested such assertions in the nineteenth century. But even as the tone shifted in the twentieth century toward condemnation, the determinist logic remained. In his lectures on the Quincentennial, award-winning environmental writer Barry Lopez was keen to connect 1492—and thus 1992—to centuries of environmental and cultural destruction in North America. Yet by falling back on Black Legend clichés, quoting Las Casas's most lurid depictions of conquistador violence, Lopez pointed his finger specifically at Spaniards and at the man who opened the door to them: "What followed for decades upon this discovery were the acts of criminals"; "Columbus began" all their slaughter, rape, and theft. By personifying invasion and colonization as Columbian, Lopez unwittingly redirected outrage over its practices and ramifications. Even the finest scholars have done the same; the careful Anthony Pagden, for example, could not help judging Columbus "responsible for ushering in a wholly modern, hitherto unknown form of criminality, and with it a wholly modern form of political domination."[32]

Regardless of where our sympathies lie, to follow the assumptions of Columbian determinism is to take a short cut to a dead end. For such assumptions rest upon a weak grasp of colonialism in the Americas and of historical process in general. The vast amount of time and ink spent debating the precise location of Columbus's First Landing is justified in a narrow sense—being significant to those invested in the debate, and to those islands who have tried to leverage their claims into tourist revenue. But the larger picture—the great transatlantic surge of Spaniards and then other Europeans, along with the Africans they forced to cross the ocean—would have occurred in almost the same way (and perhaps, as far as we know, the *very* same way), regardless of whether Columbus's First Landing was at Florida or Guanahaní or another Bahamian island. Or whether the date was 1492 or 1482 or 1502. Or whether the expedition was led by Columbus or another Genoese, a Portuguese, or a Castilian. Had Columbus turned south and hit Cuba, or sailed up to Virginia, and returned from there or perished at sea with all three ships, the impetus of colonialist expansion might have been slightly different in the 1490s and perhaps the first decades of the sixteenth century. But only perhaps; the momentum that then took Spaniards into the Aztec and Inka Empires would have been unaltered. The political, religious, and linguistic map of the Americas today would still be the same.

The same, too, would be the destiny of the Ocean Sea to become a graveyard—regardless of who became its admiral. The Palos sailors who signed up in 1492, then backed out, feared as much. Some who sailed on that First Voyage, heard halfway across making mutinous mutterings, narrowly escaped being hurled into an oceanic grave. But the North Atlantic gyre would soon become an oceanic highway, and deep below its churning waves the sediment of the ocean bed would incorporate many thousands of bodies. Increasingly large vessels, even whole fleets, would regularly drag their human cargo down into the ice-cold depths. Even more often, the diseased, the disobedient, and the dead would be thrown overboard.

And for every European given an oceanic grave, there were a growing number of Indigenous and African passengers, forced into making the voyage as victims of the massive centuries-long transatlantic slave trade. As early as 1495, almost half of the five hundred Taínos enslaved and shipped to Spain died at sea, including Caonabó, a cacique or local ruler who had gamely fought off Spanish incursion before capture; "we cast their bodies into the sea," Michele da Cuneo dispassionately noted. A 1499 "exploring" expedition by Alonso de Ojeda and Amerigo Vespucci "seized 222 people for slaves" in the Bahamas; the mortality rate aboard, more than 10 percent consigned to the deep, would prove to be tragically typical. It wasn't only the dead, but the troublesome and sick who were thereby disposed of. Of an estimated 650,000 enslaved Indigenous Americans taken to Europe between 1493 and 1600, and the 300,000 enslaved Africans taken in the opposite direction (a number that would top 12 million by the early nineteenth century), some 10 percent would end their days tossed into the ocean, dead or dying or defiantly living, to become part of the mass grave beneath the *Carrera de Indias*. As British-Guyanese novelist Fred D'Aguiar said of the Atlantic, "the sea is slavery."[33]

We can argue ad nauseam over Columbus's role in all this, over the fact of his primacy—over who was "the first" and the applicability of the term "discovery." But in the end, that is all irrelevant to the simple fact that Columbus did not determine the world that emerged during and after his lifetime. This is intended neither as an attack on Columbus (although you may object to him thereby being marginalized), nor is it meant as an apologia (although you may decry this apparent failure to hold him accountable). It is, rather, an objective appraisal of the entire, vast historical record of the early modern age, one that began around the time of Columbus's birth and outlived him by centuries until it evolved into our own era—for better or worse.

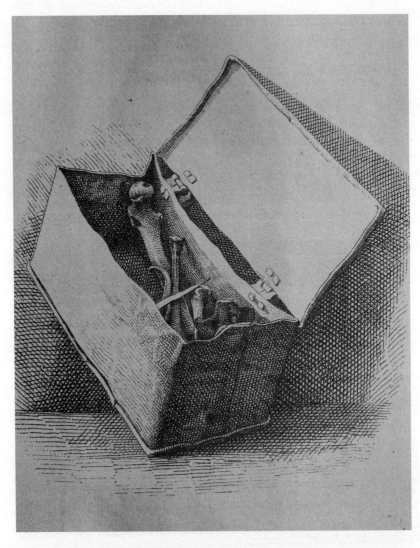

Remainders: The bones of Columbus (or are they merely those of a relative?) found in 1877 in the lead urn once buried beside the cathedral altar in Santo Domingo. This sketch was made that year, before the bones were again interred. The urn has never been opened again.

LIFE THREE

THE REMAINS

ROCCO COCCHIA, OF THE ORDER OF FRIARS MINOR CAPUCHIN, spent most of his years and career in Italy. But a brief stint in the Caribbean city of Santo Domingo caused his life to intersect with the third Life of Columbus—and helped him rise into an archbishopric back in Italy.

Cocchia (1830–1900) arrived as papal vicar in Santo Domingo at the end of the third civil war waged to create an independent Dominican Republic—the Six Years' War of 1868–74. His four years there coincided with the final presidency of Buenaventura Báez, who had earlier tried to sell his country to the United States, and who would be deposed (again, and for good) in 1878. Within the Dominican colony and republic's turbulent nineteenth century, these mid-1870s were relatively calm years, the context for a campaign of repairs conducted in Santo Domingo's cathedral. As those repairs included work on the chancel, where an elaborate Columbus family tomb had sat for more than three centuries, Cocchia persuaded one of the cathedral's local priests, Canon Francisco Javier Billini, to excavate the tomb.

Cocchia was not Genoese—he was born in Cesinali and would end his career as Archbishop of Otranto, both places in southern Italy. But his identity as a native of the new (since 1861) Kingdom of Italy was surely part of his motivation to access the Columbian coffins. For the admiral himself, according to the official historical record, was no longer in the Santo Domingo tomb. It is not clear exactly what Cocchia and Billini expected to find in 1877, but as far as anyone knew, only the remains of Columbus's son Diego and his grandson Luís were still in the cathedral. What they claimed they found, however, was something different, something far more significant.

THE IMAGE OF AN impoverished Columbus dying a lonely death is compelling, but it is a myth. He was wealthy, well connected, and attended by his sons and many servants and other dependents. The image of him expiring "in neglect, almost forgotten" fits later legend but is fiction. He had been in poor health on his fourth Atlantic voyage, most likely a worsening of the gout that had come on in his late forties; untreated and exacerbated by stress and a rich diet, the illness became severe in the autumn of 1504, soon after he turned fifty-three, with his kidneys suffering painful and eventually fatal damage. From that point, he lasted eighteen months. Consumed by his conviction in his role as God's chosen agent—his obsession in these final years was that the gold found on Hispaniola be used to fund a crusade to take Jerusalem and rebuild the Temple there—he attempted to follow the peripatetic court of King Fernando. For that reason, he was not home in Seville but in Valladolid when his ailments proved fatal. The house where he died is, inevitably, branded the Casa de Colón, and the street was also eventually renamed after him.[1]

Upon the admiral's death on May 20, 1506, his body was sealed in a coffin for a temporary interment, as was customary at the time, with the reburial of bones to take place after a few years had passed. Beneath the Franciscan monastery in Valladolid, Columbus rotted away into bones, scraps of clothing, the jewelry with which he'd been buried, and—allegedly—the chains by which he had been shackled on his voyage of recall in 1500. Less than three years later, what remained was transferred to a small casket (there were no chains, of course) and transported by horse and cart to Seville, where it was placed inside the mausoleum of the noble family of Alcalá, in the city's Carthusian monastery of Santa

María de las Cuevas. Columbus was joined in the monastery in 1526 by his son Diego.[2]

However, Diego had ordered in his will that he and his father be buried together on the island of Hispaniola. As the remains of his uncle, Bartolomé Colón, had also been deposited in Santa María de las Cuevas, Diego instructed all three sets of bones—as well as those of his Portuguese mother, Filipa—to be transferred to a new chapel to be built in Concepción on the island of Hispaniola. But Concepción failed as a settlement. Diego's vision was not for a chapel sitting somewhere abandoned and overgrown. Santo Domingo, on the other hand, originally founded by Bartolomé in 1496, had become the colony's church-and-state center. So, in 1537, the king granted space in Santo Domingo's new cathedral for a Colón family tomb. Columbus's surviving son, Fernando, made plans to accompany his father's and brother's bones across the Atlantic, but he too died, in the summer of 1539, before the voyage could happen. A year or two after Fernando's death, the remains of Cristóbal and Diego were disinterred, shipped across the ocean to Santo Domingo, and reburied with great ceremony in an elaborate new tomb built around the main altar in the cathedral chancel. Decades passed. In 1586, an English force under Sir Francis Drake occupied Santo Domingo for twenty-five days. They looted and ransacked a largely deserted city, but no reports of the attack mentioned damage to the cathedral's high altar or the tombs adjacent to it.[3]

At century's end, Cristóbal and Diego were joined by Columbus's grandson Luís, who had overseen the transfer of bones across the Atlantic around 1541. Don Luís had subsequently been convicted of bigamy, sentenced to a decade of military service in the Spanish fort of Oran in North Africa, where he died in 1572. Once his bones made it to Santo Domingo, the three generations of Colón men lay undisturbed as a century passed. The bones of Bartolomé and Filipa, it seems, never made the ocean crossing.[4]

In 1655, Santo Domingo's archbishop grew alarmed by the increasing presence of English privateers in the Caribbean, especially the threat to the colony and its cathedral from William Penn's thirty-eight-ship fleet. He therefore ordered "that the tombs be covered so that the heretics not desecrate and profane them." Had Penn's attempt to take the island not been aborted, London might have become one of the cities claiming today to hold Columbian remains. (Penn's seizure of Jamaica instead would—as we shall see—have a different impact on the Columbian legacy.) But even if Penn had stormed Santo Domingo, his men would likely have missed the Colón tombs; their coverings either obscured their labeling or, more likely, they were already lacking clear markings.[5]

Cathedral records from 1683 show that "according to the tradition of the most elderly of this island," the remains of Columbus were inside the wall to the left of the high altar, "with those of his brother don Luís on the other side." Human memory, ever unreliable, was crumbling in Santo Domingo, along with the great but now abandoned palace built by don Diego Colón: Luís was, of course, the admiral's grandson not brother; Diego went unmentioned.[6]

Another century passed before the bones were rediscovered, during repairs made to the high altar in 1783. Only on one side of the altar was the stone coffin there disturbed, revealing "a lead urn, slightly damaged, which contained some human bones" (une urne de plomb, un peu endommagée, qui contenait des ossemens humains). The extant record is in French because it was copied by French visitor Moreau de Saint-Méry, who had searched in vain for Columbus's tomb since arriving on the island in 1780. Only after the 1783 rediscovery could he talk to the two cathedral canons who found the coffin (a Manuel Sanchez and a Pedro de Gálvez) and translate their reports. Again, there is no mention of identifying labels or inscriptions, merely "the tradition conveyed by the elderly of the land" that one urn contained Columbus's remains, the other "those of his brother don Bartolomeo" (according to Canon Gálvez) or "those of his brother don Bartolomeo or don Diego Colón, son of the Admiral" (according to Canon Sanchez). Such vagueness was sure to foster mystery.[7]

Then, at the end of 1795, the violent events of the Atlantic world caught up with the bones. The uprising of enslaved sugar plantation workers in the neighboring French colony of Saint Domingue (soon to become Haiti) had prompted a violent reaction by the French, Spanish, and British imperial authorities. The war had started to spread and would soon consume the entire island of Hispaniola. Fearing the desecration of "the ashes of this hero" by rebel or foreign soldiers, officials in Santo Domingo exhumed Columbus's remains and shipped them to Havana. Supervised by the Duke of Veragua, don Mariano Colón (officially recognized as the dynasty's senior descendant, the Fourteenth Admiral of the Ocean Sea), the excavators in the cathedral found bones, ashes, and foot-long lead plates. Identifying these remains as being what was left of the First Admiral, they placed them in a new, locked casket, and the commander of the Spanish fleet in the Caribbean, don Gabriel de Aristizábal, supervised their transport on the *San Lorenzo* to Cuba's capital. There—again with much ceremony—the casket was sealed into a niche in the side of the altar of Havana's cathedral.[8]

Diego and Luís had thus been left behind in 1795. Sure enough, when Bishop Cocchia and Canon Billini dug into the tomb in 1877, they found a pair of lead caskets, each containing bones. However, they also found inscriptions, which they announced as identifying the remains of Luís—and of his grandfather, not his father. Havana, it seemed, had Diego's bones. The admiral had been safe in Santo Domingo all along. Cocchia issued a pastoral letter announcing his discovery of the Discoverer. The Spanish government vociferously denied that this was possible, insisting that Havana (the capital of what remained of the Spanish Empire in the Americas) had Columbus, and Santo Domingo (no longer an imperial city) had merely his son and grandson.[9]

The dispute was fueled by the ambiguity of evidence found on those casket inscriptions. Carved inside one casket was the phrase *Ill.^{tre} y Es^{do} Varon Dn Cristoval Colon* (*Ilustre y Esclarecido Varon don Cristobal Colon*, "Illustrious and Famous Gentleman, don Cristóbal Colón"), and on the top and sides the initials *D. de la A. P^{er} A.^{te}*, which were variously interpreted, most obviously as *Descubridor de la América, Primer Almirante* ("Discoverer of America, First Admiral"). When the casket was returned to its place in the tomb in 1878, following the completion of the repair work, a small silver plate was found inside it, inscribed *U.^{a} p.^{te} de los r.^{tos} del pmer Al.^{te} D. Cristoval Colon Des.^{r}* (*Ultima parte de los restos del primer Almirante don Cristóbal Colón Descubridor*, "Last portion of the remains of the first admiral, don Cristóbal Colón, Discoverer").

As well as bones and dust, the casket contained one small, lead musket ball. The other casket's inscription indicated that it contained the remains of Luís, the grandson who had inherited the title of admiral, stating that he had also been Marquis of Veragua and Duke of Jamaica.[10]

The problem was that the inscriptions did not appear to date from early sixteenth-century Spain: the workmanship was too shoddy; and the abbreviations were odd for the time. (Indeed, to my eye, these are not sixteenth-century inscriptions.) Some of the phrasing—"last portion of the remains," for example—suggests the markings were composed long after Columbus's initial burials. Or, if one was being very skeptical, written in 1877 to bolster the claim that such a small collection of bones was really the rest of Columbus. As for Luís, he was in fact the Duke of Veragua and Marquis of Jamaica, not the other way round. Nobody had a good explanation for that error, nor for the presence of the musket ball in the casket that was supposedly Columbus's. And how could that silver plate have been overlooked when that small casket was first examined?[11]

To bolster their indignation, the Spanish government asked the Real Academia de la Historia (Royal Academy of History) to investigate. The resulting *Informe* or two-hundred-page report of 1879 published archival documents that confirmed the details of the periodic posthumous travels of Columbus's remains—from Valladolid to Havana. The Santo Domingo caskets with their inscriptions could not be examined, as they had been reinterred in that cathedral. But the report was adamant in its conclusion that the Dominican caskets were not the original sixteenth-century ones, and thus they did not adequately offset the evidence of Mariano Colón and Gabriel de Aristizábal's rescue of Columbus's bones in 1795.[12]

The relatively measured nature of the official insistence that Columbus's remains were in Havana was followed in 1881 by a withering demolition of Santo Domingo's "pious trickery" (superchería piadosa) by a prominent Spanish historian, José María Asensio. He accused the Dominican clergy of trying to boost the stalled Columbus canonization campaign and to drum up donations to the cathedral. His motive, he insisted, was not "love for one's own country," not "exaggerated patriotism," but a "love of justice" and a belief that "one should not falsify history."[13]

And so, the debate rolled on, kept alive by the onset of the Quadricentennial. But it mattered for another reason, one increasingly pressing by the 1890s: the Cuban wars of independence. When, late in 1898, Spanish officials prepared to abandon Cuba to the independence rebels and invading forces of the United States, they retrieved the 1795 casket from Havana's cathedral. Columbus once again crossed the Atlantic, taken on the *Conde de Venadito* to Cádiz, and from there up the Guadalquivir River on the royal yacht *Giralda*—flag at half-mast. On January 19, 1899, the casket reached Seville, where it was immediately transferred to the cathedral. Yet again, Columbus was buried, again with great ceremony, in an elaborate new tomb behind a carved marble monument inside the cathedral. The great tomb is still there.

And yet there is more. Back in 1878, the Italian consul in Santo Domingo had revealed that during the examination of the Columbus remains the previous year, he had scraped up from the carpet of the cathedral some of the dust that had fallen off the bones. The local minister of justice, a don Joaquín Montolio, had, the consul claimed, also gathered some of the same "particles and dust" (las particulas y polvos) and given it to the Italian, who had placed both samples in a glass vial. What was now described as "a relic" and "the ashes" of Columbus (who had, of course, never been cremated) was given to the city of Genoa, where it was put on display in the municipal hall.[14]

The supposed ashes were still there during the Second World War, prompting Italian officials to hide the glass vial. It was U.S. soldiers who proudly recovered it with great ceremony in 1945. Television news audiences in the English-speaking world were reassured that "the ashes of the great man" were returned to Genoa's city hall when "negro troops placed the urn in the crypt." A (white) American officer gave a speech praising the Genoese "hero." The American "liberators" were given honorary citizenship in the birthplace of "the discoverer of America." Nobody thought to ask the "negro troops" what they thought of Columbus.[15]

Meanwhile, the Dominican claim on the remains was being taken increasingly seriously—not only in the Dominican Republic itself, but

Dust to Ashes: The ashes of Columbus in a glass vial in Genoa, presently in a corner of the city's maritime museum (Museo di Mare), whose label ties these "ashes" to, and thus takes the side of, the Dominican claim.

across the Americas, with a publicity campaign begun in the United States in 1914 and a hemispheric agreement to fund it signed in Chile in 1923 by most nations of the Americas. In 1931, an international board of judges (Frank Lloyd Wright among them) convened in Brazil, choosing Scottish architect Joseph Lea Gleave to design a massive monument in the Dominican Republic, one that would house Columbus's remains and include a great beacon of light. Ground was broken in 1948 on the site of what would be called the Faro a Colón (Columbus Lighthouse). Talk was one thing, money another; with one nation after another defaulting on their funding promises, construction soon ground to a halt. Not until the Quincentennial loomed did construction resume—in 1986, with a Dominican architect adjusting the original plans—allowing the monument to open in 1992.

Dark Lighthouse: The Faro a Colón, or Columbus Lighthouse, in Santo Domingo. The stylistic mishmash of the tomb is located at the interior crux of the brutalist cross- or sword-shaped museum, above it the darkened source of the light beam.

Shaped like a Christian cross, 680 feet long and 195 feet wide, with an elaborate tomb at the center, the $40 million lighthouse projects 157 beams into the sky—or it would if the lights could be turned on without crashing Santo Domingo's power grid. Theoretically visible from neighboring Puerto Rico (and from space), blackouts and protests mean the lights have been dark for decades. Intended as a massive museum to the history of the Americas, with the Columbus mausoleum drawing tourists to Santo Domingo from throughout the Atlantic world, the concrete monstrosity is in fact seldom visited. The barbed-wire-topped concrete barriers shielding tourists from neighboring shantytowns are hardly necessary. The Dominican Republic's insistence that the remains of Columbus are in the lighthouse, and nowhere else, seems not to translate well into tourist revenue.[16]

Since 1899, then, the Spanish government has insisted that Columbus's remains are in Seville, and the Dominican government has likewise insisted that they never left the republic, preserved in splendor in a bizarre concrete monument; while the Genoese government can claim (if the Dominicans are right) to have some small but symbolically crucial sample of his ashes. All through the twentieth century and into the twenty-first, historians, writers, and politicians argued one side or the other. They dug trenches along their battlelines, hurled accusations, and tied the conflict to the larger war over Columbus's legacy. They thereby made the mystery of Columbus's remains part of his life's larger, ever-expanding, constantly reimagined mystery. And they ensured that his third Life would endure into this century.[17]

MERE MONTHS BEFORE THE First Admiral's bones began their journey from Valladolid to Seville, his other remains were disturbed. Just as his physical remains would never be left alone, moved and dispersed and fought over for centuries, so would the great legacy that Columbus left his heirs—the status, titles, property, and income—be constantly dislocated and disputed. It began almost immediately: as don Diego prepared to move his father's bones in 1508, he sued the king. This would prove to be the first of a series of lawsuits by the Columbus heirs, first against the

Spanish Crown, then among each other, that would last, astonishingly, for more than four centuries.

The first phase of lawsuits—stretching from 1508 to 1563—was driven by Columbus's son don Diego (the Second Admiral), his wife doña María de Toledo, and their son don Luís (the Third Admiral). The details were complex, involving the testimony of hundreds of witnesses to the Atlantic voyages and their aftermath in Spain. But the essence of the battle was simple: the Colón family sought to use Columbus's 1492 contract with the crown—the Capitulaciones de Santa Fe—to assert massive political and fiscal power in the Americas; the crown sought to limit that power, to curtail the family's rights to administrative and judicial positions in the colonies, and to minimize their access to colonial income.[18]

The fifty-five-year conflict of this first phase was marked by a series of judgments, each named after the Castilian city where sentence was passed by royal justices. Despite the concessions made by the crown, not until 1556 did the Colón litigants cease to appeal every judgment. The story may seem extraordinary: after all, how could the crown be both judge and defendant, and did the crown not exercise hegemonic power in the empire's system of rule? In fact, as top-down as it seemed, the loose collection of kingdoms that composed Spain and its global empire was assembled and maintained from the bottom up—"collectively created by the many." Law and government functioned through a dialogue between the crown and its subjects, manifested in a massive system of petition-and-response and lawsuit-and-decree. Between 1492 and the death of King Phillip II in 1598, the crown issued tens of thousands of decrees, and its justices issued tens of thousands of judgments on matters of rule and law. But far from representing authoritarianism, those hundreds of thousands of notarized pages were in *response* to reports, petitions, complaints, and suits—issued in multiple languages by subjects ranging from enslaved Africans to Indigenous American town councils, from old conquistadors to Aztec and Inka nobles.[19]

In this context, it was predictable, not extraordinary, that the Colón family would view their status as an endless negotiation, with the Capitulaciones as the basis for endless litigation. For if the nature of Spanish rule in its first imperial century provided the larger context to the Colón lawsuits, the Capitulaciones and Columbus's own complaints over its implementation were the proximate cause of the sixteenth-century litigation. The combination of the original agreement with four follow-up royal decrees of 1492–97 broke the cardinal rules of legal contracts—overly favoring one side (Columbus) and "so loosely drawn as to invite controversies and facilitate evasions."[20]

The first court sentence came in 1511. With the help of his wife's uncle, the Duke of Alba, Diego had already been appointed governor-general of the Indies—as Spaniards called their American colonies. The arrival of don Diego and his aristocratic bride in Santo Domingo in the summer of 1509 was triumphal vengeance for the unceremonious removal of his father and two uncles from the island nine years earlier. For this was the post that had been taken from Columbus, which he and his male descendants would insist was rightfully theirs. Diego's 1508 appointment had been temporary, but the Seville Judgment of 1511 made it perpetual, adding the title of viceroy "by law and inheritance forever," and confirming that Diego could take 10 percent of all royal revenues in "the gold and other things" from the Indies. As favorable as this judgment might seem, it was not nearly enough for Diego: he appealed, demanding the right to appoint judges and civil officials in the current colonies and all future ones in the Indies, to appoint the officials of the Casa de Contratación (the Trade Board created in 1503), to be paid salaries as both viceroy and governor, to receive 10 percent of all customs duties in the colonies, and to receive 10 percent of royal revenues on all future colonies.[21]

The crown's response was to question what Columbus had discovered, thereby undermining his legal status as *the* Discoverer. As we've seen, Iberian exploration and expansion across the Atlantic during Columbus's lifetime was a collective enterprise involving thousands of people, most of them Castilian men. Dozens of them testified over three and a half years (1512–15) to the fact that while Columbus did reach two mainland locations—Veragua in Central America, and Paria on the South American coast—many others also discovered islands and mainland coastlines, some allegedly before Columbus. The case for Martín Alonso Pinzón as codiscoverer in 1492 was, as we have noted, particularly strong and favored by the crown, which had effectively purchased Pinzón family claims to the fruits of the Discovery.[22]

The next settlement was the La Coruña Judgment of 1520, which confirmed Diego's existing rights and incomes, adding an annuity of 10,000 ducats (or 3.6 million maravedís, roughly $3 million today). It was surely not coincidental that months earlier Diego had lent the king, Emperor Charles V, precisely 10,000 ducats. Diego spent most of the late 1510s in Spain, frequently following the court, and had charmed the teenage king. But word of the existence of the Aztec Empire had reached Spain, and so Diego persisted, now claiming income from Mexico and Panama and whatever lay between them. Even as the crown censured Diego for abuse of power in the colonies and found ways to underpay him, he fought for ever broader authority and sources of income. The stakes were increasingly and rapidly becoming far greater than anyone had imagined in Columbus's day.[23]

Diego died suddenly, traveling in Spain, in 1526—driven to an early death, according to his friend the royal chronicler Oviedo, by the stress and strain of unceasing litigation. His widow, María de Toledo, continued the appeal, returning to fight on in Spain after two decades in Santo Domingo. The sentences that followed, the Valladolid Judgment of 1527 and the Dueñas Judgment of 1534, failed either to cover all matters or satisfy doña María and her son Luís. Another series of testimonies was elicited in 1535–36, following the assertion that Luís Colón had the right to govern and receive income from all the Americas then found and claimed by Spain—from Florida to Peru. The crown's response was to double down on the Pinzón argument, insisting that the Discovery was a joint achievement, with the Colón family due no more than a half of whatever revenues they may or may not be awarded.[24]

By 1536, the extent of the Inka Empire and adjacent regions, let alone the size of the continent north of Mexico and Florida, made the Colón claims seem increasingly outlandish. It was now thirty years since the admiral's death. Both sides were ready to compromise. That year's judgment, ratified by Charles V himself, took the office of viceroy off the table for good, as well as all rights to appoint officials in the Indies. Diego had wanted to govern; Luís was more interested in titles and income, not responsibilities. He settled for a pension (taken from Hispaniola revenues, but he no longer claimed 10 percent), some lesser rights on Hispaniola, and the titles (Third) Admiral of the Ocean Sea and of the Indies, (First) Duke of Veragua, and Marquis of Jamaica (with the entire island as his fiefdom).[25]

And yet, the legal wranglings dragged on for another twenty years. Don Luís and his mother doña María had been given twenty-five square miles in Veragua (in today's Panama) to own in fief and turn into a colony; they failed dismally. And so, in 1556, they negotiated to give up the land in return for keeping the title of duke—and adding Duque de la Vega to it. Luís also kept the title Admiral of the Ocean Sea, but without authority or income. The annuity was increased from 10,000 to 17,000 ducats. The Columbian right to govern all the Americas was now officially restricted to the island of Jamaica. The family's income from the colonies had successfully been whittled down by the crown. With the closing of the case in 1563, what remained of the social status that Columbus had been so dedicated to achieving was a cluster of titles, prestigious but empty of political power, and an income generous enough to support an aristocratic lifestyle.[26]

THE COURTS WERE NOT free of Colón family lawsuits for long. Litigation among Columbus's descendants (and those claiming to be) soon broke out. It was to last for ten generations—a two-century battle that its mid-twentieth-century scholar would deem "one of the celebrated lawsuits of history."[27]

Columbus had only two children, Diego and Fernando. The latter had no children, but he was excluded anyway from inheriting the estate by his father, who, "in an act of quite unthinkable cruelty," made Fernando the executor of the will that severed him from the Columbian legacy. Diego, Second Admiral of the Ocean Sea, had seven children, and thus the centuries of litigation over the Colón dynasty legacy were all between his many descendants.[28]

When Third Admiral don Luís died, his titles passed to his nephew, don Diego Colón de Pravia, who thus became the Fourth Admiral—but who was also Luís's son-in-law, as Luís's daughter Felipa Colón had married her first cousin. Consanguine marriages, especially between first and second cousins, would feature heavily in the Colón family—a practice embraced by Spain's royal dynasty, the Habsburgs, and other royal and aristocratic families in early modern Europe. Eventually, the practice would cause the Habsburgs to die out, and it may have been a factor in the inability of Diego Colón and Felipa Colón to have children. Their cousin marriage was part of a settlement to end the first internecine inheritance lawsuit that began upon don Luís's death in 1572—a lawsuit that had featured Diego and Felipa (with her sister María) suing each other. The marriage pact allowed them to share the legacy (with him as Fourth Admiral and her as Second Duchess of Veragua). But the solution proved short-lived. When Diego and Felipa both died without heirs four years later, the direct male line of the dynasty started by their great-grandfather came to an end. Another lawsuit was inevitable.[29]

In 1498, Columbus had set up a *mayorazgo* (also called a *majorat* or *fideicomis*), a legal arrangement granting his titles and privileges to his eldest male heir. It was then to pass to that man's male heir, and so on in perpetuity. It could never be divided up. Women were specifically excluded in 1498 unless the male line died out. What Columbus intended to happen after that was "obscure and uncertain," despite the documents being "prolix and repetitious." Consequently, just one day after don Diego's death in 1578, his cousin don Cristóbal Cardona y Colón notified the royal Council of the Indies that he was the heir to the full Columbus legacy; and the very next day, Diego's sister, doña Francisca Colón de Pravia, filed her claim. The suit had been prepared the previous year, when Diego and Felipa became ill, and her lawyers were thus ready to pounce. She would pursue the claim in the courts until her death thirty-eight years later.[30]

At the heart of the legal battle was a three-way conflict between three cousins, all great-grandchildren of Columbus. A judgment by the court in Santo Domingo in favor of don Cristóbal quickly came, in 1579, making him the Fifth Admiral of the Ocean Sea and of the Indies.

But the other two, doña Francisca and don Álvaro de Portugal y Colón, immediately filed appeals. Doña Francisca alleged judicial corruption in Santo Domingo, successfully keeping the case open and having it moved to Spain. She had the strongest case, claiming the legacy through her father, whereas the other two cousins (Álvaro and Cristóbal) claimed through their mothers; she also outlived them by decades. But the two men wielded more power, already holding other titles of nobility; don Álvaro, the wealthiest, was related to the kings of Spain and Portugal. Furthermore, other relatives kept joining the fray. Between 1578 and 1608 (a Columbian Thirty Years' War), a total of sixteen Colón family members, plus two Colombos from Italy, filed claims.[31]

So, who won this war? At first, in 1586, don Cristóbal's 1580 victory in Santo Domingo was upheld; as he had since died, the legacy went to his sister (doña María de Cardona y Colón). But she never benefited, due to the flurry of immediate appeals, and because her husband was then investigated and in 1588 arrested for mutilating and falsifying some of the original *mayorazgo* documents. In the end, he was released, but doña María died soon after, and then other claimants joined the suit. Meanwhile, doña Francisca used the small award she had been given in 1586 to keep fighting until a new judgment was made in 1605. It was not in her favor. Despite the strength of her case, it was the wealthy and well-connected Portugal y Córdova noblemen who won the legacy's titles and incomes. But doña Francisca won something too: because the crown had embargoed the Columbus annuity during the three decades of litigation, it had become a fortune of almost 600,000 ducats (more than $100 million today); the courts awarded more than a fifth of that (130,000 ducats) to doña Francisca.[32]

What exactly did the Colón descendants fight over? What Columbus himself had sought with such determination: social rank of the highest possible order and the income to lend it due luster. Specifically, that meant a set of aristocratic titles (two dukedoms and a marquisate), the unique title Admiral of the Ocean Sea and of the Indies), and various political offices that held significant power and wealth but which—as we saw earlier—were gradually removed from the Columbus legacy by the crown. In terms of income, the two jewels in the Colón diadem were the pension and the island of Jamaica; there was also the Columbus Palace in Santo Domingo, various properties in that city, and various properties in Spain. In the best years, all other sources of income almost equaled that 17,000-ducat pension, making the legacy worth something like $10 million a year today. But most years were not best years, and—like Columbus's bones, regularly shuffled and diminished and endlessly squabbled over—the legacy of titles and wealth was a target both moving and shrinking as the centuries passed.[33]

As bitter and bruising as the Colón battles of the sixteenth century had been—replete with accusations of incest, illegitimacy, and criminal fraud—they resumed in subsequent centuries. Referring to what became the most prestigious of the Colón titles, the Veragua dukedom, a saying became popular by the early seventeenth century: *Si quieres hacerte immortal, hazte pleito sobre el estado de Veragua* (If you wish to become immortal, join the lawsuit over the Veragua estate).[34]

One might imagine that the acquisition of the Columbus legacy by one of Spain's most powerful aristocratic families would end the war. And indeed, it kept the legacy within the Portugal-Gelves family for generations. When Nuño died in 1626, his son Álvaro Jacinto became the Seventh Admiral; *his* son (whose parents were second cousins) became the Eighth in 1636, *his* son became the Ninth in 1673, and *his* son the Tenth in 1710. But the litigation was immortal. The courts had ensured that it would be when, in 1605, they granted all "parties their right to demand justice in the Royal Council of the Indies with reference to the ownership of the property of the said estate." In other words, appeals and challenges were welcome.[35]

The indefatigable doña Francisca Colón de Pravia, the sole survivor from the start of the great litigation war of 1578–1608, took up the challenge in 1609. Before her death seven years later, she set up a pair of *mayorazgos*, trusts that would make payments to descendants only if they took the name Colón. Her eldest daughter—who had married a distant cousin in the rival Portugal camp of the now-sprawling Colón dynasty—kept up the fight, along with her three children, from 1616 to 1664. More than a dozen new claimants joined the fray. Thousands more pages of legal documents were generated.

But no claimant succeeded in wresting the titles and legacy from the noblemen of the Portugal camp. Don Pedro Nuño Colón de Portugal was confirmed as Sixth Duke of Veragua, Eighth Admiral, and holder of all the legacy's titles and income in the 1664 judgment. Eight years later, he became Viceroy of New Spain, a jurisdiction that embraced not only the Caribbean islands governed by Columbus's son Diego—the last Colón to be a viceroy, one hundred fifty years earlier—but also greater Mexico, Central America, even the Philippines. His symbolic ascension was ephemeral. Inaugurated into office in Mexico City in December of 1673, he died just six days later.

Meanwhile, after generations of suing each other rather than the Spanish government, the Colón dynasty again took the crown to court. The cause was the loss of the family's plum possession in the colonies:

Columbian Viceroy: The portrait, hanging in Mexico City's Salón de Virreyes (Gallery of Viceroys), of don Pedro Nuño Colón de Portugal, Eighth Admiral of the Ocean Sea and Sixth Duke of Veragua. The sole heir to the Columbus legacy to attain a viceregal post in the Spanish American colonies, don Pedro died six days after being inaugurated as Viceroy of New Spain in December of 1673.

Jamaica. Except for brief disruptive English raids in 1591 and 1635, the island had been a quiet source of steady income for Colón nobles, thanks to the labor of several thousand enslaved Africans. But in 1655, Admiral Penn and General Venables, having failed to capture Hispaniola, turned instead with rapid success to Jamaica. And suddenly, the plum was gone. For generations, the family hoped for the island's restoration, especially as they struggled into the next century to receive the only remaining source of income: the 17,000-ducat annuity due from the crown via the governor of Panama. Paid sporadically for decades, it was not paid at all for most of the 1690s and 1700s. Years of litigation finally resulted in the crown forcing the Mexican port city of Veracruz to use its sales taxes to fund the Colón family—which it did for a century from the 1710s through to Mexico's wars of independence.[36]

It was the Tenth Admiral, don Pedro Nuño II de Portugal y Ayala, inheritor of the Columbus legacy in 1710, who succeeded in getting the crown to resume those annuity payments. With an illustrious career as a viceroy (not in the Indies, like his grandfather's six-day stint, but of Spanish kingdoms in Iberia and the Mediterranean), and with three children, he seemed to have secured the Columbus legacy for posterity. He might even have managed to claw Jamaica back from the English— his great obsession. But his children died young, and in 1733 he thus died childless, leaving the Columbus titles to his sister Catalina. She had married into English and Scottish royalty (to a grandson of King James II), so the legacy was now held by the Portugal-Stuart branch of the Colón dynasty, with the dukedom of Berwick added to the long list of aristocratic titles. Doña Catalina was Ninth Duchess of Veragua and Eleventh Admiral; hers and the Duke of Berwick's titles were passed down to their son, and then to their grandson, Carlos Fernando (aka Fitz-James) Stuart y Silva. As much as Columbus would have lamented the loss of authority in the Americas by his descendants, he would have relished the Colón dynasty's blood and marriage ties to the royal houses of Spain, Portugal, England, and Scotland.

Don Carlos, as Eleventh Duke of Veragua and Thirteenth Admiral, should have held the titles for life. But the descendants of doña Francisca

Colón de Pravia never gave up. One of her nine daughters had a daughter who married a Larreátegui, and their grandson (doña Francisca's great-great-grandson), Pedro Isidoro de Larreátegui (1695–1770), inherited both trusts set up the previous century by doña Francisca. Using the funds, don Pedro had the old case reopened. It proved to be a Pandora's box, unleashing full-scale litigation by the Larreátegui branch of the Colón dynasty and numerous other claimants.

Don Pedro's son, Mariano Larreátegui y Embrún, took up the fight in 1770, now calling himself don Mariano Colón de Toledo y Larreátegui. Don Mariano had on his side one of eighteenth-century Spain's most eloquent and well-known lawyers: Gaspar Melchor de Jovellanos (1744–1811). A jurist, philosopher, and statesman, immortalized in a 1798 painting by Francisco Goya, he argued passionately on his client's behalf. In his words, the Portugal-Stuart branch of the Colón dynasty had seized the legacy in the previous century through "the cunning of covetousness" and its illegal and immoral pursuit "along the dark paths of influence." Jovellanos's lament over "justice profaned by favoritism" might seem like a weak strategy. But it also seemed well supported by the history of the litigation, as the courts concluded in 1790. By that judgment, the Portugal-Stuart line lost the Columbus legacy, and the Larreátegui descendants of doña Francisca Colón de Pravia regained it.[37]

By the logic of the 1790 court decision, the legacy's line of descent since the death of Columbus's grandson Luís should have been different, making don Mariano the Eleventh Admiral and Ninth Duke. But in fact, he became the Fourteenth Admiral and Twelfth Duke, a tacit recognition that his predecessor, don Carlos, was after all a distant cousin and likewise descended from Columbus. In 1795, it was don Mariano as Fourteenth Admiral who accompanied the remains of the First Admiral (if the Spaniards are right) from Santo Domingo to Havana. The lawsuit was extended into the following century by appeals, but there the litigation over the legacy between members of the Colón dynasty came at last to an end. The legacy has remained in the hands of doña Francisca Colón's descendants to this day.[38]

But the story did not end there. Litigation between the holders of the legacy and the Spanish government—which was where the tale had begun—resumed in the nineteenth century. It was the Spanish Empire's loss of Mexico in 1821 that prompted new legal action by the Colón family. A resulting 1830 settlement provided them with 23,400 pesos a year from Cuba, Puerto Rico, and the Philippines. That saw the continued payment of income to the descendants of Columbus through to the Spanish-American War of 1898. However, just as the U.S. invasion of Cuba had prompted the removal to Seville of whatever bones of Columbus had been in Havana's cathedral, so did the U.S. occupation of Puerto Rico, Cuba, and the Philippines disrupt the Colón income. During treaty negotiations, the United States was asked to assume responsibility for the payments. Ignoring the 16,000 pesos coming from Cuba, the Spanish commissioners proposed that the remaining 7,400 pesos or "hard dollars" keep coming from Puerto Rico and Manila:

> The United States will continue paying to the descendants of the Great Discoverer of America, Christopher Columbus, the portion still payable of the pension they have been collecting since the time of their illustrious predecessor, as proof of the gratitude of modern civilization, which Spain has been paying.

In the 1890s, Columbian fever in the United States was at a high, with the First Admiral revered more there than in any other nation. Indeed, at the time of the 1893 World's Columbian Exposition in Chicago, the Fourteenth Duke and Sixteenth Admiral, don Cristóbal Colón de la Cerda, a Spanish senator and later minister of the navy, was fêted, fawned over, and widely celebrated in the United States as the living embodiment of his namesake ancestor. Nonetheless, as the peace negotiation minutes noted, "the American Commissioners stated that they rejected this article." For the first time in centuries, the Spanish government was consequently obliged to pay the Colón family from homeland revenues. The 24,000-peso (by now, peseta) annuity was still coming to them when Spain slid into civil war in 1936.[39]

Unlike his father, the Seventeenth Admiral was not interested in politics. He refused to leave his Madrid home when the war broke out and the city was taken over by the Republicans (an anti-Fascist Popular Front comprising various leftist parties and groups). The admiral was arrested and in 1937 executed. Childless, he was succeeded by his nephew, who assumed the Colón name and fought for the Nationalists until their victory in 1939. The annuity from the government resumed, continuing during the long dictatorship of General Francisco Franco (1892–1975). Yet its significance had become symbolic, seemingly fading into a token payment, just another modest state pension. Nonetheless, the pension is—as the lawsuits once were—immortal. And, of course, the aristocratic titles persist; in Spain, they matter. The Nineteenth Admiral of the Ocean Sea and of the Indies, don Cristóbal Colón de Carvajal y Maroto, enjoyed a long career in the Spanish Navy during the Franco era, rising to the rank of vice admiral. He was considered such a symbol of the Spanish establishment that he was assassinated by the Basque separatist group ETA in 1986. Would it be a stretch to claim he died because of the Columbian legacy, just as his great-uncle met a similarly violent end? Perhaps. Yet surely the lingering symbolic power of Columbus was to some degree a party to their murders.[40]

The son of the Nineteenth Admiral and vice admiral, also called Cristóbal Colón, is a naval officer too; the Twentieth Admiral of the Ocean Sea and of the Indies, he holds the rank of commander in the Spanish Navy. As the other titles have been held variously by Colón family members across the centuries, the Twentieth Admiral is also the Nineteenth Marqués de Aguilafuente, the Eighteenth Duque de Veragua, and the Sixteenth Marqués de Jamaica; he was also the Seventeenth Duque de la Vega, but passed that title on to his son, who is also a Cristóbal Colón and will eventually be the Twenty-First Admiral. These Colón aristocrats are also businessmen. Like the founder of the dynasty, they sail—although not much, says the Twentieth Admiral, "only in the summer, in La Coruña, where I have a little boat"—and they trade, and they pursue rank and status in Spanish society.[41]

IN THE YEARS FOLLOWING the Quincentennial of the First Voyage, Marcial Castro, a high school history teacher from a suburb of Seville, was convinced that Columbus's remains were in that city's cathedral. And he grew determined to do more than just convince his pupils of that fact year after year. In 2002, he started a campaign to open the Colón tombs in the cathedral to compare DNA extracted from the two sets of bones presumed to be in there. As support grew, a team was assembled, headed by a geneticist at the University of Granada, José Antonio Lorente Acosta. The cathedral mausoleum was opened. As the presence in the tomb of Bartolomé (Columbus's brother) was not disputed, the theory was that a DNA match with his bones and those in the other tomb would prove Columbus was also there.

Sure enough, in May of 2006—the Quincentennial of Columbus's death—the results were announced: an "absolute matchup of mitochondrial DNA" between the two brothers, declared Castro triumphantly. News of the solving of the supposed "century-old mystery" soon circulated online. NBC, for example, declared that "DNA verifies Columbus' remains in Spain" and "Spain's got the right bones." But the NBC story was typical of such reports in inadvertently muddling myth and fact, evidence and imagination, science and politics. The report was unclear regarding which brother was the basis of comparison (Diego or Bartolomé), or even if Diego the brother was being confused with Diego the son. It mentioned that the Spanish team was also working with Fernando Colón's DNA, "because that of his purported father is in bad shape." It failed to address the possibility that the DNA match merely confirmed that Fernando and his brother or his uncle might be buried in Seville, not necessarily Fernando's father. And the report even conceded that the tomb in the Columbus Lighthouse might *also* contain some (if not all) of Columbus's remains. Unsuspecting readers, given murky solutions to imaginary mysteries, were likely to become misinformed rather than enlightened. Which is, of course, the nature of Columbiana.[42]

Nonetheless, the possibility that a few of Columbus's bones are in Spain *and* a few in the Dominican Republic—and thus the dust or "ashes" in Genoa and elsewhere are also a Columbian relic—would seem like a good compromise, a basis for letting the remains remain where they are, undisturbed. There isn't much of Columbus left, but enough to go around. Might that be enough?

Theoretically, yes. But Castro and Lorente developed a broader agenda, driven by faithistory: to prove that Columbus was not Genoese after all, but in fact a Spaniard by birth. Without historical evidence for such a claim, Castro and the numerous other believers in a Spanish Columbus rely increasingly on the language of faith—a true believer remains steadfast and persistent, made stronger by skeptics and nonbelievers—and on the promise of new scientific methods, such as DNA and ground-penetrating radar. They also draw upon the internet-age conceptions of conspiracy (Italians conspiring to steal Spain's glory), mystery (the true story has been hidden and unknown for centuries), and theory (the dismissive term used to reference Columbus's Genoese origins).

Leaning hard on these techniques, Castro, Lorente, and their fellow believers have kept the campaign alive. In 2022, for example, Castro announced that he had discovered Columbus's "first tomb," a find promoted as an "archaeology breakthrough." There has in reality never been any question as to where Columbus's body initially lay after his death in Valladolid; it was placed in the crypt of the city's Franciscan convent, a building long since torn down. But Castro was able to trumpet his "surprise" at how quickly he was able to identify the location of where in the city center the convent once stood, following "a detailed historical investigation, confirmed by ground-penetrating radars." All reports mentioned that Columbus's remains had also been confirmed as now lying in Seville. Revelations of "new evidence" continued to emerge, typically on Columbus Day (Spain's Día de la Hispanidad). On that day in 2024, for example, Lorente declared that new DNA analysis of the bones in Seville—supported by a pseudohistorical documentary titled *Colón ADN* (DNA Columbus)—proved that Columbus was "probably Spanish and Jewish" (as a typical headline put it). The story quickly went viral, spread by the unwary and the believers, before skeptics punctured it. But the story will endure, another manifestation of the immortal war over Columbus's remains.[43]

There are many ironies to be found within that war. But the starkest of them all is the fact that the battles over his death—over his bones and corporeal dust, over the name he adopted and the noble titles he struggled to acquire (and some he never knew)—have kept him alive. Bone fragments and old-fashioned titles may seem like symbols of a dead past, not a living present. But they conveyed real rank, wealth, and privilege. Their symbolism was, and still is, meaningful to many people. They function like holy relics, mementos of death whose ironic purpose is to serve as markers of immortality. The Columbian remains, in all their various manifestations, have always had power.

Saintly: An unknown painter created *La Virgen de Cristóbal Colón* in the 1540s in Santo Domingo. Columbus is piously on his knees, saintlike and chosen by God, posthumously in the company of St. Christopher, Jesus, and the Virgin Mary. The cathedral is in the background, shown in today's side view from the Plaza Colón, over which a pointing Columbus has presided since 1887.

LIFE FOUR

THE SAINT

NIMIRUM COLUMBUS NOSTER EST, DECLARED POPE LEO XIII IN 1892, that peak year of Columbiana adulation: "for Columbus is ours." In an encyclical "to the archbishops and bishops of Spain, Italy, and the two Americas" titled *Quarto Abeunte Saeculo* (Four centuries have passed), the pope declared that the Catholic Church should join in the Quadricentennial celebration of Columbus. Why? Because "by his toil... hundreds of thousands of mortals... were reclaimed from barbarism to gentility and humanity." Furthermore, Columbus was no passive tool of God. "The strongest motive" for his enterprise of discovery was "the Catholic faith," his aim was "to open the Gospel's way into new lands and seas," and he was "deeply absorbed" in his determination "to prepare the way for the Gospel" in the whole hemisphere, "giving all his energies to it, attempting hardly anything without religion for his guide and piety for his companion."[1]

Leo XIII acknowledged that Columbus had other, more worldly, motives. But he dismissed them as excusable in the context of Columbus's overarching providential purpose, his privileging of piety over all other considerations. Perhaps it should not surprise us that the pope took such a view in 1892. Not only were the transatlantic celebrations massive, but also Leo had inherited from his predecessor an international campaign, several decades old at this point, to move Columbus toward official sainthood.

It is clear now, as it was to many then, that such a campaign could never succeed. During the centuries that friars and priests from Spain and other parts of Catholic Europe had endeavored to convert Indigenous Americans to Christianity, the predominant perception of Columbus was *not* that "the Catholic faith" was "his strongest motive." And yet the notion of Columbus's piety, even his extreme piety, goes all the way back to the sixteenth century—back, indeed, to Columbus himself. Campaigners for his canonization in the late nineteenth century did not have to dig very deep to find what could easily be presented as evidence of his holy nature and sacred goals.

A painting of Columbus on his knees, praying, in holy company, created in Santo Domingo in the 1540s, captured the view held in some circles, especially ecclesiastical ones, long before Leo XIII's time. Yet the painting fails to identify the kneeling figure as Columbus; it is simply traditionally taken to be him. Like so many of the tales and details about the Genoese mariner, is this another case of something stated with certainty for centuries, but which, when scrutinized with a careful eye for evidence, turns out to be rather different?[2]

Columbus's piety matters not only for its significance to the canonization story, but because it underpins the reinvention of Columbus as a Catholic hero in the United States at the end of the nineteenth century—a Columbus to whom we shall return in Life Nine. Our immediate focus is on these interrelated questions: What do Columbus's own writings reveal about his faith and piety? Do any of his deeds and actions offer supportive evidence? And can we reasonably evaluate the sincerity of Columbus's religiosity, to determine whether he was motivated primarily by faith and religious goals or whether his piety served strategic, political purposes? The answers lead into the story of the campaign to turn Cristoforo into St. Christopher.

TOWARD THE END OF *The Harp and the Shadow*, the last novel by Cuban writer Alejo Carpentier, the campaign to canonize Columbus collapses. "His beard shaking from rage, Leon Bloy rushed to the exit, snorting, *The Holy Congregation of Rites never had an inkling of the grandeur of this project. Little it cared for a providential mission!*" Léon Bloy is not a fictional character; he was an aristocrat-writer, prominent in the French campaign for the canonization. In Carpentier's imagining of Bloy's rant, the Frenchman asks, "what kind of genius" was Columbus to the cardinals? *"Nothing but a sailor. And what has the Holy Congregation of Rites ever cared for maritime matters?"*[3]

Carpentier also imagines Columbus's ghost witnessing the moment (muttering, "I'm screwed"), and later convening with another ghost, that of Andrea Doria, the great Genoese admiral and statesman (remembered in Genoa today as far more of a heroic figure than Columbus). "They turned me down," he tells Doria, who responds, "Of course. You're a sailor and a Genoan." Columbus rambles self-pityingly, until

> Andrea Doria put an invisible hand on his companion's invisible shoulder, and said consolingly: "And who the hell got the idea that a sailor could ever be canonized? There isn't a single sailor in the whole assembly of saints! And that's because no sailor was born without sin."[4]

Indeed. Transforming a sinning Genoese sailor into a divinely chosen Spanish lord would seem as impossible as sailing west to Asia.

But four centuries before Carpentier wrote his novel, two works that became the cornerstones of our knowledge of Columbus were written for precisely just that purpose. We have met the authors and their books before. The 1571 Venice publication *Historie della Vita e dei Fatti dell' Ammiraglio don Cristoforo Colombo* (the *Vita*, for short) was for long attributed to don Fernando Colón, the admiral's illegitimate younger son, who as a boy sailed with his father on the Fourth Voyage, and who became the greatest book collector and librarian of his day.[5]

The other book, the *Historia de las Indias* (the *Historia*, for short), was a vast manuscript started by fray Bartolomé de Las Casas in 1527, composed on and off until a few years before his death in 1566 (and not published until the nineteenth century). Born in 1484 to a merchant family in Seville, Las Casas was a boy in that city when Columbus was campaigning for support from the Spanish monarchs. There is no evidence they met (aside from the fantasies of biographers), but the young Las Casas would have known of the Genoese mariner; his father sailed on the Second Voyage of 1493, and Las Casas himself crossed the ocean as a colonist in 1502. But he grew repulsed by the colonial world of violent exploitation, entering the Dominican convent in Santo Domingo in 1521, going on to devote the rest of his life to *his* vision of colonialism—a divinely guided enterprise devoted to protecting Indigenous Americans in order to save their souls.

His *Historia* was part of that campaign, a three-volume history of Spanish colonization from 1492 to 1520 (three more volumes, continuing the story, were planned), composed as a polemical, millenarian argument regarding "the Indians" and God's plan for them. The friar borrowed from Fernando Colón his father's manuscript logs of his voyages, copying, paraphrasing, and in various ways embedding them into the *Historia*. It is crucial to remember that those manuscripts were not returned and were lost (when Las Casas died, Fernando had been dead decades and his nephew Luís had plundered the Columbian library); all we have are the excerpts edited and deployed by the agenda-driven friar. The image of Columbus constructed by Las Casas in the *Historia* is thus the one "that posterity has largely retained." Our Columbus is Lascasian. The friar built a flawed but heroic agent of God, whose errors neither detracted from his divinely guided purpose and achievement as "*Christum ferens*, which is to say the bearer or carrier of Christ," nor rendered him responsible for the horrors wrought by Spanish colonists. Las Casas was the sixteenth century's great critic of those horrors, while also being the greatest defender of Columbus's reputation—a contortionist's trick that permanently distorted Columbiana literature.[6]

Let us linger on Fernando's *Vita*. A full-length birth-to-death biography, less than a third covers what I have called Life One, the rest Life Two. "From the moment of its publication," as one of its recent editors noted, "it became the basis for all future biographies of Columbus." Two centuries ago, Washington Irving dubbed it "*the* corner-stone of the history of the American continent" (emphasis mine), incorporating much of it into his influential biography. At the turn of the twentieth century, Henry Vignaud worried that it had become "the most important of our sources of information" on Columbus's life, yet popular biographers such as Samuel Eliot Morison dismissed all debate over the veracity of the *Vita*, insisting it "needs no more discounting than does any biography of a distinguished father by a devoted son." A prominent late-twentieth-century translator of the book into English leaned even further on the notion of filial authenticity, insisting that the *Vita* was not only "a rich and faithful source of information about Columbus," but "a moving personal document" that revealed his world's "moral and intellectual atmosphere."[7]

The biblical status accorded to the *Vita* matters in the broader sense, because the debate that Morison sought to sidestep has simmered away ever since Harrisse pointed out in the 1870s how much the book is "swarming with anachronisms, contradictions, and errors." As another historian noted, the *Vita* "frequently connects a fact with a fancy, confusing the whole." It also matters in the sense specific to our focus here: the saintliness, or lack thereof, of Columbus. For the tone and purpose of the *Vita* is made clear from the very first page, on which it is declared that "the admiral was selected by our Lord for the great thing that he accomplished." For he was "a true apostle [*vero apostolo*] of our Lord."[8]

The *Vita* is heavily front-loaded with assertions of Columbus's providential raison d'être. His parents (Fernando's grandparents, Domenico and Susanna) are completely ignored, with the name Cristoforo Colombo given the momentous meaning of "Christ-bearing Dove." That is one way to "translate" the name. Yet Cristoforo was a common Christian name in late-medieval Genoa, and Colombo was simply the surname passed down from father to son. No, we are told, the name reflects the fact that "Christ must have sent him for the salvation of those people"—*quelle genti*, those of the New World, living in "darkness and confusion." Furthermore, the admiral "wanted his homeland [*patria*] and origins to be uncertain" because "our Lord wanted" him to imitate Christ, who "preferred that his lineage not be well known although his ancestors were of the royal blood of Jerusalem." In Columbus's "modest and graceful seriousness" (*modesta e piacevole gravita*), being "so observant of religious matters" (and strongly "opposed to swearing and blasphemy") he was like Christ, or a saint, or at least "could be taken for a professed member of a religious order." Once the *Vita* turned to the transatlantic voyages, its focus shifted to defending Columbus against Spanish rivals and rebels. Yet Columbus remained a pious figure—"favored" and "guided" by "our Lord," divinely aided "just like Moses," with "Indians" becoming Christians "by the work [*per opra*] of don Cristoforo Colombo."[9]

To make sense of competing visions of the *Vita*—is it a faithful source or is it swarming with errors?—and thus to evaluate how literally we can take its apostolic portrait of Columbus, we must digress briefly to consider the authorship of the *Vita*; the other writings by don Fernando; and the role played by Las Casas.

If the *Vita* had been written entirely by Fernando Colón, as so many writers have assumed, we would expect to find some record of it in his library—if not an actual copy of the manuscript, at least a reference to it in another of don Fernando's writings or in his notes on other books or in one of his catalogs. His library was the first great collection of printed books in European history, and the largest of its kind at the time—some 15,300 volumes—"one of the most exceptional things in all of Europe" (as the 1571 dedication to the *Vita* put it). Housed in a palatial library that Fernando built in Seville on the banks of the Guadalquivir River, the collection was curated and cataloged by using a system of his own devising. After don Fernando's death in 1539, "his wastrel nephew," don Luís, began the centuries-long neglect and breakup of the library; but the third that survives, including the catalogs, suggest that in 1539 the *Vita* did not yet exist.[10]

A close reading of the first third of the book confirms that suggestion. It immediately reveals itself to have been written or heavily rewritten later and by other hands. Some passages must have been written after 1558, and thus long after Fernando's death, because of the dates of voyages and published accounts to which the *Vita* refers. There are also numerous errors that Fernando could not conceivably have made—learned as he was, and familiar with his father in his final years (Fernando was thirteen when he sailed to the Indies with his father and uncle, and seventeen when his father died). Individually, these anachronisms and errors can be explained away as editorial tinkering. But viewed all together, they prove resoundingly that the *Vita* had multiple authors.[11]

Those authors did include don Fernando, however. That much is clear from the book's final fifth, in which he recounts the Fourth Voyage. Even here, there are errors and inconsistencies, but they are far fewer and all explicable in terms of editorial interference, stumbles by the translator (we'll get to him momentarily), and an adult recalling teenage years. These chapters reveal what the *Vita* might have been, had it *all* been written by Fernando.

Don Fernando's other writings reveal more. Beginning when he was twenty-two, about five years after his father's death, Fernando joined his brother Diego in the campaign to promote their father's legacy so as to advance it and profit from it. Proposals submitted to the king in 1511 show Fernando eloquently defending his father's right to rule the Indies (and thus the rights of don Diego as Second Admiral of the Ocean Sea), as well as his father's cosmographic vision. Fernando had crossed the ocean with Diego in 1509, when the older brother took up the viceroyalty in Santo Domingo, but he had sailed back within a month in order to represent the family in legal negotiations with the Spanish Crown. The settlement of 1511 was largely to his credit. For the next fifteen years he toiled tirelessly on behalf of his brother, representing him in Rome over a lawsuit stemming from one of Diego's many sexual affairs, traveling around Europe for several years as part of the new King Carlos's itinerant court, all while collecting books. An unfavorable 1524 judgment in the Spanish Crown–Colón lawsuits—resting on the crown prosecutor's claim that Columbus had not in fact been the first to find the Indies—may have prompted Fernando to start writing an account of his father's voyages.

But whatever Fernando wrote at that time, it was not submitted to crown officials and has not survived. Then, in 1526, Diego died, after which Fernando seems to have largely withdrawn from the legal battle, focusing on his books (the library's construction began in 1526) and his work as a cosmographer (that same year he served as the king's examiner of candidates for pilot positions with Seville's customs house or Casa de Contratación). So, how did such writings end up in some form or part in the *Vita*? The answer lies with Las Casas.

Copies of the famous friar's *Historia de las Indias* had long been used by Spanish chroniclers, but it reached a far wider audience with its first publication in 1875–76. The similarity of its Columbus chapters to the *Historie della Vita* was taken to be a confirmation of the veracity of Fernando's biography. Most assumed that the *Vita* was written first (before Fernando's 1539 death) and the *Historia* second (still being written in 1561). Subsequent debates over the two works tended to fall along nationalist (mostly Italian and Spanish) and partisan (pro- or anti-Columbus) lines. But Las Casas had started writing his *Historia* in 1527, making great use of sources shared with him by Fernando. One of those sources was Columbus's now-lost log, or *Diario*, of the First Voyage. Of the portion Las Casas used, he summarized three-quarters and copied (but "corrected") a quarter—a layering of filters that make modern editions of Columbus's "Ship's Log" or "Diary" misleading. The friar mined the *Diario*, working what he found, to advance his tripartite argument: "the Indians" were meek and convertible; success in the colonies depended on only "good Christians" settling there; and Columbus was deeply pious, an agent of God. Likewise in the service of that contention, Las Casas also collected shipboard logs from the subsequent voyages, letters by Columbus, and other sources—all since lost. As one Columbus scholar has persuasively argued, one of those sources was likely an earlier manuscript by Fernando—a lost third text from which both the *Vita* and Las Casas's work were partly derived.[12]

The writings of Fernando Colón and Las Casas therefore dovetailed into each other, both manifested in the *Vita*: don Fernando's long-lost manuscript was copied, altered, and edited into Las Casas's own *Historia*, which was in turn used by the editors who prepared the *Vita*—which also included further edits and additions. Enter don Luís Colón.

Fernando's reprobate nephew, the Third Admiral of the Ocean Sea, was in Seville in the early 1560s, struggling to pay off his many debts—including 2,400 ducats owed to Baliano de'Fornari, a Genoese merchant nobleman then based in the Spanish city. Don Luís had in his possession the manuscript of the *Historie della Vita*, or some version of it, and likely tried to get it published in Seville, hoping its promotion of his grandfather's legacy of primacy and piety would help him in his ongoing polygamy trial (he had earlier tried to get published a long-lost version of the *Diario*). Unable to publish in Seville, don Luís sold the manuscript to Fornari, who in 1561 or 1562 carried it to Venice—one of the few cities in Italy where books could be printed without the approval or interference of Spanish officials.

The most prominent Spanish translator in Venice was Alfonso Ulloa, who had settled in the city after arriving in 1547 as a young clerk to the Spanish ambassador. Ulloa was tasked by Fornari's publisher with translating the *Vita*. But in 1568, before he could finish, he was imprisoned for involvement in the unlicensed publication of a book in Hebrew. His death sentence was commuted to life in prison. There Ulloa completed the translation, awaiting a pardon request from the Spanish king to the Venetian doge. But while that request was in transit, and before he could polish the translation and correct its errors, Ulloa died. The manuscript found in his cell was published months later.[13]

Why has this detour into the history of the *Vita* been necessary? Because it shows how we cannot take for granted anything found in that book, and therefore anything repeated over and over by writers whose ultimate source is that book. And that includes Columbus's piety and religious motivations. The role of Las Casas in the *Vita* story is crucial because Fernando's *other* writings tend to emphasize cartographic and cosmographic arguments: that his father had a unique vision; that his father was the First Discoverer; his father delivered new lands to the crown; and his father's descendants have rights to govern and profit from those lands on the basis of legal agreements. The religious element is present, but relatively muted. One of don Fernando's 1511 *Memorial* reports, for example, proposed an expedition *para dar la vuelta al mundo*—"to go around the world"—not one to take Jerusalem. But for Las Casas, the providential nature of Columbus's life and deeds was primary; it was ultimately all that mattered. And that agenda infused the *Vita* and subsequent centuries of writings on Columbus.

History isn't exactly written by the victors; they outline the plot and claim the best roles, but it is the storytellers who tell the tales that become history. Colón and Las Casas are those storytellers here. Their eyes were on the victory, respectively, of the Columbus dynasty and of Catholicism, and it is the Columbus they created who has been absorbed by subsequent storytellers—by writers who favored, as tellers and consumers of stories tend to do, a simple narrative through-line to which heroes can cling and villains fall by the wayside. Readers want romance and resurrection; reality need not get in the way. History's messiness lacks the comfort of Columbiana's confirmation bias.[14]

Whether readers and subsequent writers believed Columbus was an agent and instrument of God or merely believed that Columbus himself believed that he was, the Genoese sailor became a figure motivated by profound piety and religious fervor. But Columbus himself wrote a fair amount on the topic; can we clear the fog created by his son and by Las Casas by analyzing the First Admiral's own words?

ON THE VERY DAY that the monarchs of Aragon and Castile marched triumphantly into the Muslim city of Granada, King Fernando wrote a letter that was copied and sent to the leading cities of the neighboring Christian provinces. The letter's purpose was to spread news of the victory, but also to *shape* the news, to maximize its political impact. "That city," declared the Aragonese king, "held and occupied for over 780 years, today, 2 of January of this year 1492, has come under our power and dominion." His "our" was a royal one, as Fernando made no mention of Queen Isabel; the "glorious victory" was his, "the honor and increase of our kingdoms" a sign of God's favor, a divinely sanctioned act of revenge against the Muslim conquest of Constantinople four decades earlier. Fernando's frame of reference was a crusading millenarian mentality used to justify political ambition—as reflected in his reference to Granada as if it had never rightfully belonged to the Muslim families who built and occupied it for eight centuries.[15]

Columbus was paying keen attention. He had spent much of 1491 with the king in the siege town of Santa Fe, and he was among the throng of courtiers and soldiers that saw the January surrender of Boabdil, Granada's last sultan. Columbus surely took to heart Fernando's framing of victory as both his and God's. As he pondered what he experienced in the nineteen months between that day in Granada and the day in Barcelona when he announced his discoveries to Fernando and Isabel, he absorbed that way of looking at things. The triumph of his First Voyage was both his and God's, and thus his as God's agent. The published version of the letter that he wrote in advance of the Barcelona meeting is from the very start full of *yo*, *me*, and *mi*. The "voyage of mine" eclipsed the roles of others: "I discovered," "I took possession," "I brought," "I gave." It was a "great victory that Our Lord has given me." And its purpose, just like King Fernando's victory in Granada, was ostensibly that the newly acquired subjects would "become Christians." Yet it was ultimately so that those people would "give us what they have in abundance and what we greatly need."[16]

As one scholar observed, "Columbus was less a rationalist than a combination of religious enthusiast and commercial entrepreneur." Or, put in Columbus's own words, "Gold is most excellent! Out of gold, treasure is made, and he who possesses it can do with it whatever he wants in this world. And it can even speed the flight of souls to Paradise." That gives the impression that religion as a motivation was something of an afterthought. Yet at other moments it predominates Columbus's comments. So, can we ascertain in his writings the balance between religious enthusiasm and commercial ambitions? And what exactly did Columbus have in mind when he imagined imposing his will on the world?[17]

Above all, Columbus's goal was to achieve social mobility, founding a dynasty for the benefit of his sons and their descendants. In this he was, of course, successful. His religious and commercial motives were therefore methods or means to an end. The balance between those two means ("gold" and "souls") shifted during his later years. Specifically, until 1493, his driving ambition was to sail west to reach Asia and its riches, with his self-image based primarily on his belief (tempered by self-doubt) in his ability to do that. On his return from the First Voyage, the balance began to move toward ambitions and a self-image that were more religious in nature. Beginning with the Third Voyage in 1498, religion predominated. Upon his return from that voyage late in 1500, and for the remaining five and a half years of his life, his religious preoccupations were obsessive and overwhelming.

What prompted that shift? Political problems, ranging from competition with the Pinzón brothers on the First Voyage through to wavering support from the crown, caused Columbus increasing anxiety—especially when such crises seemed to threaten the privileges that were to underpin the Colón dynasty. Closely related was the short-term failure of his commercial and colonial enterprise: it became increasingly apparent that he had not found a route to Asia; the first colonies on the islands were beset by violence, revolt, and high mortality; while the Indigenous population collapsed as a result of slaughter, enslavement, and introduced diseases. Finally, his chronic gout and the threat of death accelerated the tipping of the balance toward a religious mindset.

Failure to perceive that pattern of change, as well as a failure to appreciate the intrusive hand of Las Casas as the filter through which we see most of Columbus's writings, has misled many a writer—while helping others with religious agendas to find apparent evidence of his piety, even saintliness. It was easy for those seeking to promote Columbus as a Catholic hero to see God, not Las Casas, placing in the Genoese mariner's mind an unwavering understanding of his holy mission. In reality (in Fernández-Armesto's words), "his sense of divine purpose grew gradually and fitfully and was born and nourished in adversity." As it grew, Columbus sought to promote himself "as a providential agent," to be accepted as "divinely elected to execute a part of God's plan for mankind, by making the gospel audible in unevangelized parts of the earth." His religious self-regard, like his self-education, neither came upon him fully formed with adulthood, nor were they imparted to him wholesale by some mysterious donor (be it a mythical one-eyed pilot, a dead sage, an anonymous friar, or God Himself). Pumped up by ambition and insecurity, narcissism and self-doubt, triumph and tragedy, they gradually inflated.[18]

The claim of providential agency begins with relative subtlety on the First Voyage, with most of the language styling the expedition as a God-guided mission attributable to Las Casas—as a careful reading of the *Diario* shows. For example, Columbus compares the Atlantic's great swells to "high seas, very necessary to me, which had never appeared save in the time of the Jews when they fled from Egypt to oppose Moses." In his mind (or Las Casas's), Columbus isn't Moses yet, but he's close. Similarly, as it became clear that the Indigenous islanders encountered on the First Voyage were not hard-toiling, easily exploitable denizens of a wealthy Asian empire, their suitability as converts began to grow in Columbus's mind; that was not yet his stated sole or even primary goal, but it would be by the end of the Third Voyage.[19]

On the early voyages, Columbus is quick to see God's hand behind events, and to do so in strategic ways. For example, keen on the First Voyage to establish a garrison of some kind on Hispaniola, but lacking a specific mandate to do so, he insisted that God had made "the men of Palos" (los de Palos) lazy and treacherous, resulting in the *Santa María* running aground—thereby providing Columbus with wood, supplies, and men for such a garrison. And his falling out with Pinzón was not his doing, but the devil working through his former friend; while Columbus's return, not Pinzón's, to the Spanish monarchs was God's work. But by the last two voyages, *everything* was God-guided and every single moment a sign of Columbus's exalted status as a divine agent. For example, God led him on the Third Voyage to an island with three mountains (he named it Trinidad), a signpost to the Earthly Paradise just beyond it (the Orinoco estuary); "the Holy Trinity put into my mind the thought, which later became perfect knowledge, that I could sail to the Indies from Spain by crossing the Ocean Sea to the West." That formula's purpose is still strategic (how can the crown argue against an idea and plan of God's?), but Columbus has begun to internalize it; he is on the path to believing himself to be one of the prophets.[20]

In between the Second and Third Voyages (from the summer of

1496 to the spring of 1498), Columbus read and wrote furiously in his efforts to dispel growing doubts at court regarding his claim to have found a profitable westward route to Asia. Much of this work focused on trade and colonialism—from issues in "the Indies" of administration and trade, mining and farming, to plans to access South Asian spices via the newly discovered islands, to schemes for mounting a military assault on Mecca. But at least some of these months were spent as a guest of the Franciscans in Seville, where he used their library to supplement his own in his search for prophecies and predictions of his voyages in the scriptures and other ancient texts. What seems to have begun as strategy to persuade the monarchs of the deep significance of his voyages soon turned into an obsession. By the time he sailed west again, in 1498, he was convinced—and shared this conviction with Fernando and Isabel—that Isaiah had "in so many places in His scriptures" announced that "these lands" found by Columbus would receive the true faith "from Spain."

Off Hispaniola's shore in 1499, "I took to the sea in a small caravel," Columbus wrote, to escape "the Indians and the wicked Christians" who wished him dead. But he had become convinced God was on *his* side, not theirs, for "then the Lord came to help, saying, 'Oh, man of little faith, be not afraid, I am with you.'" The contrivance places Columbus inside the Bible, echoing the Matthew 8:26 passage in which Christ calms the winds and sea to save the disciples ("You of little faith!"). The Old and New Testaments predict his life, turning him into a chosen one of biblical importance. As Columbus grew more deeply persuaded by his own stratagem, he began to hear voices—sometimes a celestial voice, sometimes God speaking directly to him—and to experience sacred visions. The failure of his voyages was swept aside by their reimagining as sacred journeys, secret missions, to spread the true faith.[21]

The conceit of Columbus's rapidly developing status as God's special agent was limitless: his missions would lead to the seizure of Mecca, the recovery of Jerusalem, the millenarian apocalypse, the return of Christ. The notion that the ultimate goal, in service of both God and crown, was to reconquer Jerusalem was hardly an original one. It had been bandied around during the war against Muslim Granada, and in 1492 Fernando had given himself the title "King of Jerusalem" as a way to promote that ambition; Columbus's later claim that it was he who had suggested the crusade to Fernando in 1492 is typical of his later delusions of global religious grandeur. Adding the conquest of Jerusalem to the promotion of his own goals, he began claiming that his western route to Asia would both fund such a campaign and lead to a global circumnavigation that would allow Islam to be taken from behind.[22]

The "recovery" of Jerusalem for Christendom was a favored dream of the millenarian wings of the Catholic Church, especially the Franciscans—and even more so the Observant branch of the order, independent starting in 1477, allied closely with the Spanish monarchs, and militantly devoted to exterminating heresy and converting "infidels." The influence on Columbus of the messianic and millenarian prophetic tradition of the Observant Franciscans began to show around 1496, when he started wearing a Franciscan habit and rapidly grew into a millenarian fantasist himself. The theory that the return of Christ awaited the discovery and conversion of peoples in some unknown corner of the world, such as the so-called antipodes, was an old one; Columbus found it in various sources, including that much thumbed and most annotated of his books, Pierre d'Ailly's *Imago Mundo*. From around 1500, Columbus insisted that "the Indians" encountered by him were those people, thus elevating him to *the* divine agent making possible—even imminent—the Second Coming.[23]

In that year he sent the monarchs a memorandum with the grandiose title, "The Reason I Have for Believing in the Restoration of the Holy House to the Holy Church Militant." Devoid of strategy, plan, or anything practical, the memo rested on the claim that the Holy Spirit had given *him*, the admiral, the Jerusalem idea, just as he had likewise been given the idea to sail west to Asia; he had been chosen because, as "a lay seafarer, an earthly, practical fellow," he was like the disciples and saints, elevated not by birth but by God. Another open letter written that year (to doña Juana de la Torre) laid bare the self-perception of these later years. Soaking in self-pity (while deftly deploying a marine metaphor), the admiral bewailed how the world "has given me a thousand battles... cruelly throwing me to the bottom." Yet he insists on asserting his identity as the chosen one, given "the spirit of intelligence and great courage" by God, who "made me the messenger and showed me where" to find the New World. Columbus's Christ complex is transparent: monarchs and men have abused him dreadfully, but God has favored and protected him above all of them.[24]

Such writings were part of a furious campaign waged by Columbus between his return from his Third Voyage (November 1500) and his departure on the Fourth Voyage (April 1502). Seeking approval for that last voyage, he aimed to prove he had found a way to Asia, thereby restoring his privileges and political positions. He composed a series of letters to the monarchs, a couple more to Isabel alone, one to the Council of Castile, and one to Pope Alexander VI. The documents of his contract with the monarchs were collated as his *Book of Privileges*, of which four notarized copies were made. He also started compiling d'Ailly quotes that seemed to support his burgeoning belief that his own voyages had been prophesied, writing to ask Gaspar Gorricio—an Italian Carthusian friar in Seville—to find similar passages in biblical texts and other *auctoritates* (authoritative writings). The resulting compilation, an eccentric testament to Columbus's "providential and messianic delusions," was titled by Fernando Colón *The Book of Prophecies*.[25]

That title, created by don Fernando three years after his father's death, is misleading: an unfinished multiauthored draft, it was never intended as a book, merely another letter of petition to the monarchs; and only 10 percent of its eighty-four folios are Columbus's work. The so-called *Book* begins with correspondence between Columbus and Gorricio, who was responsible for the summaries of biblical passages that cover most of the seventy folios. The only substantive section composed by Columbus in his own words is a draft of an unsent letter to the monarchs. That draft reads like an introduction to the "evidence" that Columbus imagined he and Gorricio were compiling, all to prove that "what Jesus Christ Our Redeemer said and had previously said through the mouths of his holy prophets came to be." In Columbus's muddled millenarian logic, his entire life and work have been planned and guided by God, prophesied for centuries, with he, Columbus, the crucial means whereby Spain acquires "the Indies," Jerusalem is recovered, and in 155 years "the world will be destroyed."[26]

Prophesied: This graphic history for children (Mary Dodson Wade's *Christopher Columbus: Famous Explorer*) necessarily simplifies Columbus's effort to compile biblical proof that his voyages were divinely ordained (his so-called *Book of Prophecies* was neither a book nor written by him). Yet it also captures a simple truth: faced with his failure as a colonist, Columbus increasingly drew on religious arguments and fantasies to appeal to the Spanish monarchs for ongoing support.

Columbus was surely sincere, having swallowed his own argument whole. He took the manuscript pages of *The Book of Prophecies* on his Fourth Voyage, adding notations, increasingly convinced of the growing gap between how important he was and how marginalized he had become to Spain's colonial enterprise—a sacrifice that he, like Christ, had been called upon to make by God.

But such convictions stemmed from and supported a decidedly secular ambition. His piety was performative. His claim to be an agent, messenger, and prophet of God was another form of self-aggrandizement, "a curiously egotistical form of self-effacement." Never a man "to do anything for modesty's sake"—as Fernández-Armesto observed—"even the humility he affected later in life... was of a showy, exhibitionistic kind." His obsessive insistence in his final years that he was humbled before God was always tinged by an "outrageous egotism." Even at the end, in a codicil to his will, he asserted that God had given him "the Indies"—more than a mere discovery, the New World was his, "a chattel of my own"—which "by God's will I gave to" the king and queen.[27]

The simple fact of Columbus's primacy in the story of the spread of Christianity to the Americas, combined with his own narcissistic self-imagining as an instrument of God, planted a seed. But that seed fell on relatively stony ground, sprouting irregularly for the next four centuries, before finally flourishing in the nineteenth century. For example, the first sweeping history of the Americas, by Gonzalo Fernández de Oviedo (the aforementioned royal chronicler who had also served as a colonial administrator in Darién and Santo Domingo), took a far more measured position on Columbus than did Las Casas. In the part of his *Historia general y natural de las Indias* published in Seville in 1535, Oviedo did trumpet Columbus's presentation of his discoveries to the monarchs in Barcelona in 1493 as momentous. In the tradition of Saints James and Peter, the admiral introduced "the true faith" to "the Indies." But Oviedo downplayed the providential angle, emphasizing material wealth more than the saving of souls. Weighing up Columbus's legacy, Oviedo noted "all Spaniards owe him much," because so many became so "rich." To be sure, "our holy Catholic faith and the Church of God was planted and practiced" in distant lands "by the means and efforts of the Admiral," but by the same means there came to Spaniards "so much wealth in gold, silver, pearls, and many other riches and goods."[28]

Las Casas was furious. For Oviedo, the story was one of the glorious rise of imperial Spain; Columbus took a back seat to Spanish heroes, and "the Indians" were innately inferior, a marginal concern. For Las Casas, the popularity of Oviedo's book made his "monstrous lies" all the more pernicious; Oviedo was an "utterly vain trifler" (vanissimus hic nugator), blinded by God for his crimes as a colonist, stupidly incapable of grasping "the good disposition of the Indians." Las Casas wrote his *Historia* partially as a corrective to Oviedo. But the book remained unpublished, while others borrowed Las Casas's tactic of rewriting Columbus's life as a providential journey and applied it to Spanish conquest heroes—Hernando Cortés being the most obvious. Much to Las Casas's annoyance, it was the deeply unsavory Cortés who became portrayed as God's great agent by Franciscans in Mexico and chroniclers in Spain.²⁹

Meanwhile, an apostolic Columbus evolved in Italy. As the sixteenth century wore on, and Spain's empire continued to expand, followed by those of other imperial states to Italy's west, Columbus's compatriots Vespucci, Caboto, and Verrazzano began to write of the Italian city-states as a culturally composite "supra-imperial power that transcended mere nationhood." The influential, allegorical poetry of Dante Alighieri (1265–1321) inspired sixteenth-century Italian writers to see Columbus as the inheritor of a great apostolic tradition—as a new Saint Paul, a *vero apostolo* appointed and guided across the ocean by God.³⁰

Italian writers such as Torquato Tasso (1544–95) and Tommaso Stigliani (1573–1651), and artists such as Giovanni Stradano (1523–

Holy Hero: In the 1580s, Giovanni Stradano designed a set of prints that fashioned Magellan, Vespucci, and Columbus as mythological heroes. Used by the De Bry family and other printmakers for centuries, the images were widely influential throughout Europe and the Americas. Of the three heroes, Columbus is styled as the Catholic crusader.

1605), thus embedded a particular version of Columbus in the culture of the Italian Peninsula, one who was Catholic, saintly, and Italian. They thereby anticipated, and helped lay the groundwork for, a series of Columbiana phenomena. One was the quest to define the new Italy in the Risorgimento period; Giuseppe Garibaldi (1807–82), the prominent unification leader and *Pater Patriae*, remarked in the 1830s, as he contemplated the possibility of a united Italy, that he felt "as Columbus might have done when he first caught sight of land." Another was the campaign to canonize Columbus—to which we are about to turn. And yet another was the appropriation of Columbus by Italian Americans. As we shall see in Life Nine, when Italian unification finally occurred in 1861, it helped prompt a half-century wave of immigration to the United States. There, mistreated immigrants seeking acceptance, if not assimilation, discovered that Anglo-America had appropriated an Italian as a Founding Father; his insertion into Italian culture as a holy hero helped those immigrants to reappropriate Columbus as theirs.[31]

In 1796, when Columbus's remains in Santo Domingo were being prepared for their transportation to Havana, the official record of that operation predictably referred to Columbus not only as "the first discoverer of this new world" but also "the first instrument that God used for the spiritual good of spreading the true religion and Holy Gospel."[32]

That perspective had thus become common and widespread, so much that it was even sewn into the fabric of histories written for children. The books of Joachim Campe, for example, were structured around a fictional father recounting history to his children (today we might view the books as young adult literature, only more violent and racist). In the German educator's trilogy *Die Entdeckung von Amerika*—first published in 1781, appearing in multiple languages in the decades that followed—every twist and turn of Columbus's life was explicable in terms of "Providence saving him" for "his great and glorious enterprise." Here was the Italian holy hero, sold to all of Christendom: "our friend," "the immortal Columbus," living and dying in a serene state of "Christian piety." He may have been underappreciated by his contemporaries, the father tells the children, but Columbus's "name will pass from mouth to mouth into the most distant posterity and will always awaken love and admiration among those who value the virtues that he embodied."[33]

THE NOTION OF CAMPAIGNING to beatify and canonize Columbus seems to have originated in conversations in Paris in the 1830s, although it had surely been aired many decades earlier. The first surviving written record of a sustained campaign begins with a series of personal letters sent between Paris and Rome in the 1840s and 1850s, leading to the first book-length argument for the idea flying off a Parisian press in 1856. The author, an eccentric clergyman named Antoine François Félix Valalette, the Comte [Count] Roselly de Lorgues (1805–97), had already written lightweight books on Catholic theology. He would devote the rest of his life to campaigning for Columbus's elevation to the pantheon of Catholic saints—a campaign later credited as his creation in the 1840s, and one viewed for the rest of the century as a primarily French effort.[34]

As outlandish as the idea of a St. Christopher Columbus may strike most people today, the Columbus invented by Roselly was far from a fringe frivolity. He titled his book *Christophe Colomb, histoire de sa vie et de ses voyages d'après des documents authentiques tires d'Espagne et d'Italie* (history of his life and his voyages, based on authentic documents taken from Spain and Italy), as if it were a contribution to the growing body of scholarly books on Columbus. On the book's strength, Pope Pius IX endorsed the canonization campaign—in writing. Indeed, it was rumored that due to his "great admiration" for Columbus, Pius IX had actually commissioned Roselly to write his book (Pius reigned from 1846 until his death in 1878). The pope's endorsement was followed by those of much of the Catholic Church hierarchy, including the heads of the Franciscan, Dominican, and Jesuit orders, twenty-three cardinals, thirty-five archbishops and bishops, and numerous other Catholic dignitaries. By 1866 almost all French bishops had made their support public. In 1876, the archbishop of Bordeaux wrote to Pius that "a large number of the Fathers of the Vatican Council have voluntarily affixed their signatures to the petition for the introduction of the cause" of canonization. The press in Europe and the United States was positive. Mexico's archbishop was in favor. Roselly's book was quickly published in Spanish, Italian, and German, and within three years a second edition came out, trumpeting how rapidly and fully the public had accepted the idea that "the history of Christopher Columbus is the glorification of Catholic genius."[35]

As with the "American" Columbus, the "genius" of Roselly's Columbus was evidenced by his "great wisdom of observation," his boldness in the face of adversity, and his recognition that he was guided by God. However, the difference between the Columbus of the early United States and Roselly's was that the latter was not a national hero, but "a hero of Catholicism." And all those who were indifferent to his genius in his own day—from the Spanish monarchs and settlers who betrayed him, to Vespucci as the thief who stole the continent's name—were secondary to the real enemy: Protestant writers denigrating Columbus in order to attack the Catholic Church.[36]

Here was another example of Columbus appropriated as a battleground, this time to wage the war begun centuries earlier (just a decade after Columbus's death) by the Protestant Reformation. The canonization campaign was fueled by the reaction to the nineteenth-century's spread of Protestantism and an anti-Catholic anticlericalism. Catholicism needed heroes, and ecclesiastical officials and Catholic writers in France and Italy (where the campaign was strongest) saw Columbus as a useful weapon in the battle to defend the traditional church. But Roselly's claim that Protestants hated Columbus failed to recognize that Protestant writers and nations—especially the United States—were more interested in appropriating Columbus as *their* hero. Their Columbus was a modern figure who had stood up to stodgy medieval Catholicism. The battle was less about attacking or defending Columbus than it was about claiming him. Indeed, Columbiana writers in the Iberian Peninsula tended to see canonization as a foolhardy distraction from their campaigns to claim Columbus as native to their region.

Roselly made another tactical error in claiming to have made a discovery that would change world perception of the First Admiral. The attention-getting gambit was signaled in the "authentic documents" phrase in his book's title. At the heart of the two-volume biography

Imagined: A completely imaginary portrait of "doña Beatriz Enríquez de Arana," included in some Spanish and other editions of Roselly de Lorgues's *Christoph Colomb*. The invention of Beatriz as noble, saintly, and married to Columbus was crucial to the campaign to make him a saint.

was the dramatic assertion that "doña Béatrix Enriquez de Cordoue" (Beatriz Enríquez de Arana, from Cordoba), the mother of Columbus's second son, was not a poor commoner who gave birth to Fernando out of wedlock—as other authors had claimed. On the contrary, declared Roselly, Beatriz was a noblewoman and "was in the eyes of the Church the wife of" the Genoese navigator—as "we have now proved."[37]

As beautiful as she was wealthy, Roselly insisted, Beatriz "belonged to the noble house of Arana, one of the most ancient families of Cordoba." She was extremely modest, as such a woman should be, and that resulted in her virtual absence from the historical record. But, Roselly assured the reader, this "Christian woman took modest pleasure in the glory of her husband." The Colón-Enríquez marriage took place "towards the end of November 1486," with Fernando born the following August. Following the logic of Roselly's storytelling, the regret expressed by Columbus in his will regarding his treatment of Beatriz did not stem from his failure to marry her, but because he loved her with such a passion ("avec la puissance de son coeur") that it pained him to accept the "sacrifice" that God called him to make: to sail to the New World without her.[38]

Roughly the last 170 pages of Roselly's 1,165-page biography were devoted to making the case that Columbus deserved a place among "the martyrs" and "the saints" ("who do not all arrive in heaven by the same route"). His character and deeds were cast by Roselly as providential, pious, and devoted in a saintly manner to a Christian mission, with the planting of the cross in the Americas—symbolizing the introduction of Catholic Christianity to millions of people—constituting the necessary miracle to seal the deal.[39]

As Spanish historian (and Roselly's contemporary) José Asensio noted, rather mockingly, Pius IX's response to Roselly's book was to declare in 1857 that because he was very busy, he had not had time to read it ("Et si ob gravissimas multiplicesque summi Nostri Pontificatus ocupationes, nihil adhuc de hoc tuo opera degustare potuermius"). Nonetheless, Pius's delicate avoidance of the issue was taken as approval, encouraging Roselly and his supporters to maintain their campaign. Finally, in 1865, Pius IX gave his official blessing. Yet efforts to bring the matter before the First Vatican Council in Rome in 1870 failed. A similar effort in 1877 also stalled. "Advocates of the beatification of Columbus" were entreated in the American Catholic magazine the *Christian World* "not to be easily disheartened." (Beatification is the first stage of approval on the path to sainthood.) The 1879 *Lives of the Catholic Heroes and Heroines of America* began with "the pioneer of a long line" of such heroes, his life "one of the most holy, heroic and wonderful on record." Drawing on Roselly's sanctifying mythmaking, "the great and holy" Columbus was presented to English-speaking Catholics as "the greatest man to whom *the world is most indebted* ... heading the list of the most illustrious men of all time."[40]

Roselly, meanwhile, continued to produce a series of books, increasingly polemical (as Asensio noted), with titles such as *Satan Against Christopher Columbus* and *A Posthumous History of Christopher Columbus*. He was joined by other French writers. Léon Bloy, that aristocrat borrowed by Alejo Carpentier, published well-illustrated books in French and English that idealized a Columbus fit for sainthood; his *Le Révélateur du Globe: Christophe Colomb et sa Béatification Future* repackaged Roselly's arguments in a diatribe that made the Genoese mariner seem like Christ himself, with those opposing the campaign condemned as ignorant infidels. Books by Corsican abbot Martin Casanova, aimed at convincing readers of Columbus's Calvi birth, portrayed his supposed compatriot as the kind of crashing bore that seemed saintlike in the late nineteenth century—a straightlaced prude, "horrified by blasphemy," "revolted by bawdy songs," "repulsed" by debauchery, and intolerant of all "excess." Charles Buet, in his 1886 *Christophe Colomb*, followed his compatriots in arguing that only true Catholics could understand Columbus and why he should be Saint Christophe.[41]

But the attacks on the campaign, and especially on Roselly, were relentless. Asensio was mildly sympathetic to the cause, in the sense that he was himself a Catholic and a great apologist for Columbus. But he couldn't help detailing how the French count was misled into "exaggerations" and "errors" by his "passion" and his "piety." The American librarian-historian Justin Winsor (1831–97), likewise writing close to the Quadricentennial, dismissed all Roselly's books as "mere disguised supplications to the Pope to order a deserved sanctification." "As contributions to the historical study" of Columbus, Roselly's tomes were "of no importance whatever." In 1870, an English translation of Roselly's *Christophe Colomb* was issued in Boston by the Catholic Publication Society, which had been founded a few years earlier by "one of the most ardent propagandists for Catholicism" in the country. But Catholic American scholars, such as John Gilmary Shea (1824–92), tended to distance themselves from a campaign they saw as French, not American, by urging conservative caution pending "calm official scrutiny of the question" by the Vatican. After all, opined Shea, the insistence by Roselly and his allies that Columbus was surrounded by enemies—both in his lifetime and now—who were "inspired by a satanic hostility to a great servant of God, is too much to ask for our belief."[42]

Someone who might have been expected to be sympathetic to Roselly's campaign was Angelo Sanguineti. A Genoese canon and admirer of "our hero" the admiral, he published a *Vita di Cristoforo Colombo* in 1846. But sympathy was one thing; supporting a campaign based on a blatant misreading of documents was another. Eventually (in 1875), Sanguineti published a refutation of Roselly's invention of the Colón-Enríquez marriage, showing how some of Roselly's sources were indeed "authentic documents"—but they proved the very opposite, that Fernando was born out of wedlock. Sanguineti's carefully argued position was then incorporated into an 1891 edition of his *Vita*, published in time for the Quadricentennial.[43]

Published in Italian only, by a small Genoese press, Sanguineti's rational defense of the historical Columbus was an umbrella in a hurricane. Although his position on Columbus as a Genoese hero but not a saint was ratified by the entire ecclesiastical hierarchy of Genoa—from its bishop to all cathedral canons to local cardinals—new books continued to promote Columbus as worthy of sainthood. The 1892 frenzy encouraged waves of Columbiana, with every claim bringing surges of counterclaims. The canonization campaign was but one of them, seemingly as unstoppable as it was unfulfillable. In 1891, a cable dispatch announced that the Vatican was investigating the virtue of the Columbus case; tellingly, it also mentioned that the question of the Columbus-Enríquez marriage was under review. American Catholics organized a petition drive to plea with Pope Leo XIII for canonization. The following year, Leo (who had succeeded Pius IX in 1878) issued the *Quarto Abeunte Saeculo* encyclical quoted above, decreeing that Columbus should be honored with "a Solemn Mass of the Most Holy Trinity." Leo XIII did not mention canonization, but his panegyrical praise of Columbus's piety helped keep the campaign alive.[44]

Sanguineti's skepticism began a Genoese tradition of focusing on defending Columbus as a local hero without ruining such efforts by going too far. Genoese senator and Columbus apologist Eugenio Broccardi

(1867–1959) described Roselly as one of the nineteenth century's most "extreme apologists," committed to making Columbus "a perfect man, free from every material or moral defect, almost invested with a divine mission, of which his name was an omen (Columba = Dove)." Later Genoese senator Paolo Taviani insisted that Columbus's myth is that

> of a genius, of a great man of history, not that of a saint. The desire to make him a saint at any cost was the compulsion of clerics during the fourth centenary of the discovery, when blatant and violent anticlericalism spawned an equally crude clericalism.[45]

A century after Roselly's book came out, an Italian Franciscan, fra Giovanni Odoardi (1916–2005), called for a full scholarly investigation into the beatification campaign, in order to end it or advance Columbus on to consideration of canonization. Odoardi noted that while both Pius IX and Leo XIII had been sympathetic to the campaign, they had not officially approved it, failing to help it leap the hurdle of Columbus's failure to marry Beatriz Enríquez. Odoardi's call fell on deaf ears. Still, as Columbus became an increasingly controversial figure in the years leading up to the Quincentennial, apologists again trumpeted his extreme piety, giving a kind of half-life to the fiction of his second marriage.

One enthusiast, in a 1982 book titled *The Nine Arguments in Defense of Christopher Columbus: The Untold Truth*, insisted Columbus was a profoundly religious figure whose "greatness" lay in "the depth of his knowledge," his bringing of civilization to "the natives," and "at all times, he spread the word of Christ." The claim that Enríquez was "Second Wife" was rather undermined by a composite illustration of Columbus's "Wives and Mistresses"—there apparently being four of the latter. *The Nine Arguments* was self-published. As myths collided in the topsy-turvy world of Columbiana at the end of the twentieth century, the canonization campaign quietly died.[46]

THERE IS SOMETHING OF an irony to the Columbus-Enríquez story being the great canonization hurdle, as will become clear when we look more closely at Beatriz Enríquez in the next Life; had Columbus been able to see centuries into the future, might he have given up social mobility in order to be elevated to sainthood?

For most of his life, surely not. But in his final years, faced with his own mortality, and delirious with the notion of his own providential importance, he might have chosen sainthood. Not according to an editorial in the *Sacramento Daily Record-Union*, however, responding to the 1877 setback in the canonization campaign—the Vatican's decision not to advance the case that year. "We have no doubt that Columbus himself would have modestly declined any such posthumous honors, for he was not a saint, and he did not pretend to be one." Furthermore, "it would have been awkward," the editor wryly added,

> to have had to call him Saint Christopher, because there is already a Saint Christopher, and supplications to either of them would have been in danger of going to the wrong address, and all sorts of confusion would have resulted.[47]

The Sacramento journalist unwittingly invoked the fact that Columbus was already a saint—in a manner of speaking. For despite the failure of the canonization campaign, Columbus had already become saintlike. Like most of the saints, the Virgin Mary, and even Christ himself, there had developed—and remain—numerous avatars or manifestations of Columbus. Like the saints, the many Columbus avatars tend to be localized, rooted in a specific place, given regional names and meanings. Like a saint, Columbus is neither divine nor the man he once was. Saint images may not be idols, but Columbus exists as a carved and craven idol, to be revered or torn down, surviving in a liminal place of undeadness, mutable yet immortal.

To believers, he is sacred, his function and meaning rooted in their faith; for devotees of faithistory, evidence-based understandings of his historical life are secondary or even irrelevant. By the same token,

Columbus's images and avatars are idolatrous to others, offensive objects of ill-conceived worship. Revered or reviled, Columbus has become more of a secular saint/idol than a strictly religious one. He may fade as a Catholic hero and candidate for sainthood, but he will surely continue to serve as a palimpsest for an embarrassment of beliefs and causes.

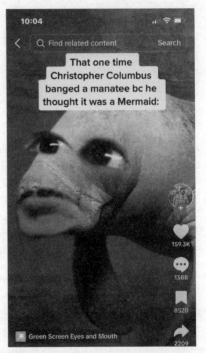

F**ker: Sample images from TikTok memes showing outrage, amusement, or comic horror in reaction to the news of Columbus's alleged predatory bestiality.

LIFE FIVE

THE LOVER

IN THE THIRD DECADE OF THIS CENTURY, AS POLITICIANS AND LAWmakers in the West argued over banning the social media platform TikTok, the app continued to spread and expand upon an old rumor: that Columbus was a "perverted psychopath," a predatory abuser of animals. His specific perversion was made explicit in video clips consistently designed to be humorous, while depicting Columbus lusting after, if not sexually assaulting, manatees. His motive was made clear in most of these memes: that he mistook manatees for mermaids. Even worse, according to some posts, he then complained "in his diary" that these mermaids were less attractive than he had been led to believe. Others insisted that he also "banged" or "boinked" llamas and goats.

These clips, numbering in the scores, were "liked," shared, and commented upon up to tens of thousands of times. It is impossible to know how many more thousands saw such memes (as of 2024 there were 150 million U.S. TikTok users, 70 percent of them under thirty-five), and how many accepted the bestiality allegations. Mock horror was the closest that most users seemed to come to outright acceptance. "OMG I didn't need this reminder," accompanied by laughing emojis, was a typical response. Terms used to describe Columbus's manatee interaction ranged from the explicitly vulgar to the romantic, with the manatee usually unable to give consent and consequently terrified or violated, often given human eyes and lips—even protesting ("please don't f@*k me"). Commenters expressed that, in response, they were "hating Columbus": "HATE Columbus," "he is sick," "he is sick for that," and so on; he "is the manatee f**ker" or simply "is a f@*ker."[1]

Why did the "fact" of his bizarre bestial desires find such traction among Americans born after the Quincentennial? Was Columbus really a lover of great appetite or a predator of indiscriminate sexual tastes?

IN VIEW OF HOW much Columbus is mocked on social media as a perverted predator, it may come as a surprise to learn that the Genoese mariner's sex and love life, prior to his marriage at the age of twenty-eight, is a blank slate. His biographies are saturated with speculation and imagination on most topics, and yet there are neither hints nor allegations of any kind of sexual activity, let alone deviancy or impropriety. One explanation is that the vast majority of biographies are laudatory, influenced by the hagiographical Colón-attributed *Historie della Vita*, and often driven by apologist leanings. But there is another explanation: there is no smoke because there was no fire.

Even later evidence, starting with Columbus's Lisbon years, is slim. Finding a place to prosper amidst the Genoese community in the Portuguese capital, Columbus acquired connections and some wealth. Enough to marry, for the first and only time, and to marry fairly well. As we learned earlier, there is little evidence of the circumstances under which the marriage occurred. Biographers have thus invented details, asserting them as facts, or echoing the imaginations of previous writers. Genoese senator-historian Paolo Taviani, for example, liked to think that Columbus had "married for love," and so he simply made it so: "he fell in love in church, as did so many Italians, Spaniards and Portuguese of the time" (and perhaps, centuries later, Taviani himself).[2]

What we do know is that Columbus's wife was Filipa Moniz—in some documents, Filipa Perestrelo e Moniz, with Bartolomeu Perestrelo as her father (by some accounts, likely wrong, her stepfather). Perestrelo's own father seems to have been a merchant from Piacenza, who, like so many fifteenth-century Italians, had settled in Lisbon, becoming wealthy enough to acquire a title of nobility and to send his son—Bartolomeu—to grow up at court as a page to the princes John and Henry (the future Henry the Navigator). It is hardly coincidental that Columbus would himself achieve the very same social ambition, albeit in neighboring Castile rather than Portugal, shedding his Italian origins to become a local aristocrat with two sons at court. His first step toward realizing such an ambition was to marry Filipa, from whom he would have heard tales of how her Italian grandfather had attained local mobility and nobility.

Filipa's father, Bartolomeu, died in 1457, long before his daughter grew up and met Columbus. But Bartolomeu had continued the family's upward trajectory, acquiring the hereditary captaincy of the island of Porto Santo in the Madeiras (a modest holding by any standard, but one which provided some income). After his first wife's death, Bartolomeu Perestrelo married into the old Iberian aristocracy; Isabel Moniz was propertied in Portugal and Madeira, tied by blood to the Portuguese and Castilian monarchies—with Queen Isabel a cousin. Columbus and Filipa Moniz's son Diego, born in 1480, was thus distantly connected to royalty.[3]

Did Columbus marry dona Filipa only for her status? We cannot know for sure. He could have been motivated by both romantic feelings *and* social ambition. Fernando Colón imagines his brother's mother as "a lady of noble blood" (una gentildonna di nobil sangre) whose habit was to look out of the window of the convent where she had been placed, thus catching the eye of Lisbon's passing Genoese. Spying Columbus, "she acquired so much time and friendship with him that she became his wife" (prese tanta prattica, et amicitia con lui, che divienne sua moglie). Harrisse spotted an element of "certain fantasy" in this. While I suspect he is right that Fernando, knowing no more than anyone else, drew upon a romantic cliché, his account may nonetheless reflect the fact that the marriage was based on friendship rather than love.[4]

I suspect that Las Casas may have likewise sensed that the record suggested ambition was more of an influence than passion. He thus imagined a further motivation, swallowed whole by future Columbus apologists: the marriage gave the future admiral access to Bartolomeu Perestrelo's (supposed but unevidenced) private collection of "maps, navigation manuals [*portolanos*], commentaries, and manuscripts." In other words, ambition was his motive, but one ennobled by intellectual curiosity and the task of discovery assigned by Divine Providence.[5]

If there is no evidence of Columbus's feelings toward doña Filipa around the time they married, what about later, around the time of her early death? In 1484, with his hopes for sponsorship from the Portuguese king finally dashed, Columbus left for Castile—and he left Filipa behind. That is the usual interpretation of Columbus's later complaint to the monarchs that, in order to pursue their patronage, he was forced "to leave wife and children." There is no evidence that he and Filipa had "children"—only one, a son named Diego—so the phrase is surely a rhetorical flourish by Columbus. Filipa may even have died by this time. Or she died in 1487, by which time Columbus had impregnated someone else. Or in 1488, when his second son was born; his brief return to Lisbon at the end of 1488 may have been connected to Filipa's death.

Either way, evidence of affection or grief is wishful thinking; "we want to believe" (wrote Harrisse) that Columbus would not have begun another relationship "until after the death of his legitimate wife." Other historians chose to believe that he genuinely regretted leaving his family in Lisbon or that he left Portugal in 1484 consumed by grief. Yet, objectively speaking, Fernández-Armesto was surely right in noting that Filipa's premature death left "Columbus free and, it seems, unencumbered by sentimental memories."[6]

WIDOWER OR NOT, COLUMBUS arrived in Spain without a wife. His marriage had produced only one child and had lasted four or five years, during which time he was often, if not mostly, at sea, plying his trade as a merchant mariner. He never remarried, and there is no evidence that for the remaining twenty-two years of his life he cohabitated for any period of time with a romantic or sexual partner. Aside from the rumors that would inevitably stem from his fame after 1492 (to which we turn momentarily), only one post-Filipa sexual relationship is substantiated.

While living in Cordoba for part of 1486 and 1487, Columbus met

Beatriz Enríquez de Arana. A native of a small village in wine country ten miles outside the city, Beatriz's immediate family—both parents, a grandmother, an aunt—had all died, one after the other, leaving her at the age of ten to live in the Cordoba household of her mother's cousin, a merchant named Rodrigo Enríquez de Arana. Enríquez de Arana knew the community of Genoese merchants and storekeepers in Cordoba, and it was probably through them, perhaps in a Genoese store, that his young ward met Columbus. In the autumn of 1487, she became pregnant by him. She was nineteen, he was thirty-six.[7]

Their son, born the following August, named Hernando, or Fernando, after the king of Aragon, was much favored by his father, and later legally legitimized by him. But efforts centuries later to prove that Columbus had married Beatriz foundered for lack of evidence. It is evident, rather, that he neither married her, nor lived with her. When she was four months pregnant, he received permission to travel back to Portugal; it is not clear why, as instead he moved to Murcia to continue pursuing sponsorship from the monarchs of Castile and Aragon. He then made his brief return trip to Portugal in December, when Fernando was a few weeks old. Beatriz does not seem to have been much of a priority. In the final codicil to his will—dictated the day before he died—he admitted that his treatment of her troubled him. He directed his son Diego

> to make provision for Beatriz Enríquez, mother of don Fernando, my son, that she may be able to live honestly, being a person to whom I am under so much obligation. And this shall be done for the discharging of my conscience because this matter weighs heavily upon my soul. The reason for which it is not fitting to write here.[8]

Five years after Fernando's birth, soon after Columbus's return from his first Atlantic crossing, he started paying Beatriz a modest annual pension, doubling it in his will of 1502—"to lessen the burden on my conscience." He also requested that Diego "take care of Beatriz Enríquez, for love of me, as attentively as you would your own mother. See that she gets 10,000 maravedís a year beyond her income from her meat business in Cordoba." The 1506 direction was thus a reiteration of that earlier order. Diego seems to have honored that commitment intermittently, for in Beatriz's final years the pension payments stopped coming.[9]

Centuries later, historians were quick to read the codicil mention of Beatriz as "the language of affection, repentance, and mystery." Certainly, much has been made of Columbus's expression of guilt, for the phrases are a rare example of him offering a glimpse into his romantic life. Or not: he left both his sons in Beatriz's care in the years leading up to and during his First Voyage, and then in 1493 he deposited them at court, all of which suited his needs but likely not hers (her stepson was thirteen and her son five when they were taken from her); those facts may have been enough to prick Columbus's conscience, with romance irrelevant. But that is a less appealing story, and generosity toward Columbus typifies the commentary by almost all historians and writers.[10]

The flip side to such generosity is a tradition of decidedly ungenerous comments on Beatriz Enríquez. Indeed, in the hagiographic world of Columbiana, commentary on Beatriz has been excessively speculative and—in efforts to spare Columbus blame of any kind—overwhelmingly misogynistic. Beatriz has been turned into a teenage temptress who lured the trusting navigator into her bed, worthy neither of his attentions nor his subsequent magnanimity. There is no evidence whatsoever to support such a scenario. In the 1930s, a Spanish historian collected and published archival records from Cordoba that documented Beatriz's family and her descendants through her son Fernando Colón; not satisfied with simply confirming that Columbus never married Beatriz, the historian suggested that Beatriz was not marriage material for such a man, that treating her merely as a lover was appropriate, and that it was likely her fault—"unfaithful" that she was—that Columbus soon lost interest in her.[11]

Whatever the truth regarding motivations and emotions, Columbus did not stay in Cordoba to be a partner to her or to help parent their son. And whatever we might make of his deathbed expression of guilt over his treatment of her, it is also clear that having acquired noble rank through his contract with Queen Isabel, Castilian law prevented him from marrying a commoner. Beatriz's relatives were carpenters, butchers, and—ironically, and perhaps part of the original attraction—weavers. Therein, of course, lies the reason why he did not take her to the altar when he found out she was bearing his child: his social ambition.[12]

The fact that social ambition trumped romantic feelings prompts this question: Did Columbus fall in love with the two women who gave him his two sons? There is no clear evidence either way (suggestive itself). On the one hand, he may have chosen them (unless, in fact, they chose him). On the other hand, he surely married Filipa at least to some extent because of her social rank and connections, while he opted *not* to marry Beatriz for the same reasons—her lack of them. Furthermore, he expressed great affection for his sons but never wrote a word expressing romantic love or sexual interest in anyone. Facing possible death at sea in 1493, Columbus wrote that "it pained him greatly that two sons that he had studying in Cordoba would be left orphans of father and mother in a foreign land." This was according to Las Casas, but it rings true: the focus on the boys and the slight to their mothers—one of whom, despite the inference, was alive and taking care of both children—is consistent with Columbus's priorities. One might well deduce, in fact, that his sole interest in women was that he needed them in order to seed a noble dynasty—a well-documented ambition.[13]

But a good story needs a romance; and the life of Columbus has, as much as any in history, been retold repeatedly as a tale of swashbuckling adventure, the kind that requires a doomed or illicit romance. His relationships with Filipa and Beatriz—brief, lacking direct evidence of passion, darkened by hints that Columbus valued the sons far more than their mothers—have consequently appealed less to storytellers than other, imagined relationships. Over the years those relationships grew in number, as writers repeated and adorned prior inventions or just made up their own. By the late twentieth century, Columbus enthusiasts had assigned him a full love life, complete with two wives, two official mistresses, and covert dalliances with two of the highest-ranking women in Spain—one being none other than Queen Isabel.[14]

The Isabel fantasy was an old one. Writers were easily hypnotized by the imaginary but irresistible notion of the queen's amorous attraction to the Genoese mariner—especially in the nineteenth century, as Romanticism, the novel, and opera, grew increasingly popular on both sides of the Atlantic. Once established, it became repeated throughout the next century too, often as history not fiction.

In fact, two invented, intertwined tales helped lend each other the ring of truth, forming a well-rooted legend: first, that Isabel was enamored with Columbus, possibly to the extent that they enjoyed an illicit romance; and second, that she pledged or pawned her jewels in order to pay for his First Voyage. The latter was included as one of the scenes in the immensely popular Columbian Series of stamps printed in the United States in 1893, and it was widely believed and repeated as fact in the decades surrounding the Quadricentennial. When the same stamp was reissued in 1992, the story was alive and well. Spanish representative to the United Nations Jaime de Piniés, speaking in support of a draft resolution to commemorate the Quincentennial, had exclaimed that Columbus would never have discovered America had Isabel not "pawned her jewels to secure the necessary funds."[15]

The Columbus-Isabel story, being fiction, has always been free from the strictures of evidence, making it all the more tempting to writers—especially those drawn to Columbus as a "novelistic character" (a *personaje novelesco*, as one Mexican literary critic put it). Popular songs,

Pledging: The first $1 stamp issued in the United States—"Isabella Pledging Her Jewels" to pay for Columbus's First Voyage—in the U.S. Postal Service's 1892 Columbian Series (reissued in 1992). Isabel appeared on seven of the set's sixteen stamps. The painting on which this image is based—fiction presented as fact—is available printed on a T-shirt, mug, phone case, and so on.

operas, plays, and novels have frequently alluded to it; Alejo Carpentier's *The Harp and the Shadow*, mentioned earlier, contains but one example. Historians have tended to dance around it, to mention it as mere innuendo or rumor, or to devote too much space to dismissing it—all responses that have the effect of lending the story some sort of mysterious credence.[16]

Less ambiguous are playwrights or novelists who have embraced the fiction fully as a plot foundation. Two vivid examples from different ends of the twentieth century are Lawton Campbell's *Immoral Isabella?* and Salman Rushdie's "Christopher Columbus and Queen Isabella of Spain Consummate Their Relationship, Santa Fé, January 1492." Campbell's play, which ran on Broadway in 1927, depicted the queen as oversexed and impetuous, full of love and lust for Columbus, her jewel-pawning crucial to America's Discovery. Rushdie's 1990 short story likewise has the Discovery completely dependent on the seduction of a flighty, mischievous Isabel by the persistent but desperate mariner.[17]

Even in academic circles the story persists. For example, literature scholar Estelle Irizarry (1937–2017), in an imaginative yet earnest way, spun one line in a letter to the queen from the Genoese mariner into a 2011 book titled *Christopher Columbus's Love Letter to Queen Isabel*. That line, written in 1501, was "the keys to my wishes I gave to you in Barcelona" (las llaves de mi voluntad yo se las di en Barcelona). For Irizarry, this was the stuff of coded eroticism, with "the keys as a tool that penetrates and permits opening."[18]

As captivating as may be the notion of Columbus using a keys/penis metaphor in a missive to the queen, it is an interpretation that can exist only within the closed confines of literary criticism. Viewed in the historical context of all we know about Columbus, about genres of notarial records at the turn of the sixteenth century, and about this particular document, the notion is risible. Columbus follows that line with "if you try a taste of my good will, you will find its scent and savor have only increased since then, and not by a little." He is not, sadly, driving home a phallic metaphor, but trying rather clumsily to deploy a "coquettish, courtly tone" (Fernández-Armesto's phrase) to shore up the queen's support and persuade her that, "without dissemblance, I am disposed with all my strength to seek your repose and contentment and to increase your lofty sway." Months earlier Columbus had returned from the Caribbean shackled and facing charges relating to the failure of the first colonies there. His efforts at courtly sycophancy are tinted by defensiveness (especially over his idea of a crusade to Jerusalem). He is anxiously trying to reassure Isabel of his political loyalty, not his love and lust.[19]

The Columbus-Isabel affair is far less popular in conservative Italian and Spanish circles, where the crusade to canonize Columbus has largely been replaced by one for Queen Isabel. Begun in 1957, the St. Isabel campaign moved slowly but steadily for many decades, with rumors periodically surfacing—most recently in 2022—that the Vatican was considering beatification. The epicenter of the campaign is the Spanish city where Columbus died, Valladolid. More than twenty volumes of documentation have supposedly been assembled there, including testimony on a miracle: in 1967, prayers to Queen Isabel by family members in Granada's cathedral allegedly cured a comatose Valladolid priest of pancreatic cancer. In 2023, Valladolid's archbishop renewed efforts to move the campaign along in Rome.[20]

The lusty queen is blatantly incompatible with the saintly one. The twentieth and current Admiral of the Ocean Sea believes that the romance story is "completely false." Why? Because the queen "had very upright morals," not because Columbus did. Instead, Columbus's affair, his living descendant has opined, was on the island of La Gomera, in the Canaries, with doña Beatriz de Bobadilla.[21]

So, who was Bobadilla? According to Spanish historian Mario Escobar, who wrote a sensationalistic book on her, doña Beatriz was a gorgeous but dangerous aristocratic widow, "a dazzling beauty" prone to "violent impulses." From the thinnest strands of rumor and conjecture, Escobar wove an entertaining but entirely imaginary love-quincunx. In his telling, Queen Isabel, intent on getting rid of Bobadilla, the lover of her husband King Fernando, had her married off to the governor of the distant island of La Gomera; but the governor was killed and Bobadilla soon met, and seduced, Isabel's "secret love," Columbus. There is evidence to support all aspects of that narrative, aside from all the parts involving Columbus. But finding Columbus absent from records of sexual intrigue at court, writers have employed a typical Columbiana technique: simply weave him into the story, turning fiction into apparent fact. The Columbus-Bobadilla romance has been less popular than the Columbus-Isabel one, especially outside Spain; the queen has the advantage of royalty and wider renown. Yet numerous historians have repeated it and even invented details—such as, for example, doña Beatriz begging Columbus to give up his life at sea to marry her.[22]

The Bobadilla story can be traced back to Michele de Cuneo (1448–1503), a merchant from Savona—the port town, adjacent and subordinate to Genoa, where Columbus himself lived for a part of his childhood. Cuneo was a boyhood acquaintance, if not friend, of Columbus's, and he participated in his Second Voyage. Recounting the journey to a friend in Savona in 1495, Cuneo claimed that when the fleet stopped at La Gomera before sailing west, there was "much parading and salvoes and fireworks, ... all done for the sake of the lady of the said place, for whom our lord admiral had once been touched with love."[23]

This "lady" was doña Beatriz de Bobadilla, an altogether (too) obvious object of alleged infatuation. Having come to the Canaries in 1481 as the wife of the islands' conquistador, Hernando Peraza, she had assumed control of the islands upon his assassination by Indigenous islanders in 1488. She crushed resistance to Spanish rule with a vengeance, slaughtering and enslaving the Indigenous population. She was, in the words of a prominent historian of the islands, "one of the cruelest and most beautiful women in Castile." Wealthy, powerful, a former lover to King Fernando of Aragon, capable of great brutality, Bobadilla was a fertile subject for gossip and rumor.[24]

The rumor of an old crush (*tincto d'amore*) is far too flimsy for anything beyond wishful thinking—and it was surely imagined. But eager to make Columbus more relatable and warm-blooded, historians turned Cuneo's vague single line of gossip into sure evidence of an actual love affair. Taviani, for example, insisted that it must be true because Bobadilla lacked a "monogamous temperament," that her husband had been unfaithful and thus "it seems only natural that she would repay him in the same coin." Furthermore, Cuneo "was a careful observer of natural phenomena and therefore of women and sexual behavior."[25]

Cuneo was, by his own admission, far more than a mere observer. Even without casting aside Taviani's flimsy arguments, we might consider the story's source. It is reasonable to speculate that Bobadilla appealed to Cuneo himself; Cuneo had fathered children by three women, only one of whom was his wife, and in his account of the transatlantic voyage he described his assault of an Indigenous girl with the kind of relish that betrays the shameless sociopathy of a serial rapist. It is worth reading in full Cuneo's account in a letter he wrote in 1495 to a friend in Savona. After detailing the killing and capture of Indigenous villagers—whom he calls *Camballi*, "Cannibals"—on the island of Guadalupe, he continues:

> The other Cannibals, together with the slaves, we later sent to Spain. While I was in the boat, I grabbed a gorgeous Cannibal woman whom the Lord Admiral gave me; when I had her in my quarters, she being naked, as is their custom, I felt a desire to sport with her. And when I tried to satisfy my desire, she, wanting none of it, treated me so with her nails that at that point I wished I had never started. Seeing this, to tell you how it all ended, I grabbed a rope and I beat her so well that she made unheard-of cries, such as you would never believe. Finally, we were of such accord that, in the act, I can tell you, she seemed to have been trained in a school of whores.[26]

Cuneo pursued and abused women; Columbus did not. He tolerated the actions of men like Cuneo, but with a distaste seemingly born of prudishness rather than higher moral standards. Columbus has taken flack for what Cuneo did, because "the Lord Admiral gave me" (il signor armirante mi donò) the victim. Columbus defenders have indignantly asserted the admiral's innocence. Partisans will defend their trenches, but the issue of Columbus's responsibility is irrelevant to a pair of points here.[27]

The first point is that the men on Columbus's voyages, especially the second crossing, were overwhelmingly Castilian conquistador-settlers, and the enslavement and rape of young Indigenous women was an elemental tool of conquest and settlement. While it is regrettably true that men can be found in any society doing what Cuneo did, he nonetheless behaved like many a conquistador-settler. Dubbing him a serial rapist is not mere rhetoric. The phrase applies even more so—in terms of the evidence available to us—to the archetypal conquistador-settler, Hernando Cortés. Just as Cuneo seemed to be unaware that his actions were wrong, convincing himself that his victim's screams signaled consent, so did Cortés's fellow conquistadors and conquest chroniclers comment casually on his sexual appetites and his particular proclivity for using sexual assault as an assertion of power—over his Indigenous victims, their families, and the Spanish men with whom he competed.[28]

The second point is that Columbus clearly did not participate in this well-evidenced rape culture. Defenders have viewed this as reflecting his upstanding morals. In contrast, one could argue that lack of interest went hand-in-hand with tacit toleration, supported by a gendered view of Indigenous islanders. The triad of women, monstrous humans, and cannibals formed "a complex metaphor for inferiority" in the mind of Columbus. Indigenous people not seen as monsters or cannibal Caribs were nonetheless inferior and suitable for commodification because they were like women. Is Columbus's misogyny and his enthusiasm for the enslavement of Indigenous peoples (especially young women, whom he referred to as "head" [cabeças] as if they were cattle) damning context enough? Probably. But the fact remains that "Columbus was not, by common standards, a sexually susceptible man." So judged his most astute modern-day biographer. I agree, especially as a strong contrast emerges when we go beyond "common standards" to compare the admiral to contemporaries such as Cuneo and Cortés.[29]

Lest those comparisons seem extreme, we might also compare him in this regard with his own sons and grandsons. Fernando seems to have been like his father. Their obsessions were different: despite accompanying his father on the Fourth Voyage, Fernando explored new worlds through his creation and cataloging of what was the largest private library in European history to date. But both men seemed indifferent to women. Fernando does not even mention his mother in the *Vita* of his father, despite Beatriz Enríquez taking care of both Fernando and his brother Diego while their father secured the approval of the monarchs for his First Voyage and crossed the Atlantic for the first time; and despite her suffering the permanent removal of the boys from their Cordoba home, upon Columbus's triumphant return, to be pages at the royal court. Not until facing death did Fernando reveal that he remembered his mother: he asked for her to be included in his funeral prayer.[30]

As for other women, ones who may have been the object of affection or attraction, there are none—save for the hint of an interaction with an innkeeper's daughter. In his will, Fernando left 3,000 maravedís to Leonor Martínez, whose father kept an inn at Lebrija (forty miles south of Seville), "for the discharge of his [Fernando's] conscience." This was the conventional phrase used to refer to indiscretions, not only sexual ones. It was used by Fernando's father "when leaving a pitiful inheritance" to Fernando's mother, and likewise used by his brother, Diego, in trying to pay off a woman he had seduced, Isabel de Gamboa. But to imagine a sexual relationship between Fernando and Leonor is to do just that: imagine. If a child had resulted, at least one that survived, there would be evidence in Fernando's will and elsewhere. And even as an exception to the deafening silence in the historical record on Fernando's sex and love life, it is the exception that proves the rule. Fernando and his father were obsessed with informational minutiae, doggedly determined, and self-absorbed—that is, absorbed in their personal convictions that they were destined to achieve a singular and supposedly impossible thing (to sail west to Asia; or to create the world's largest library and index its entire contents). That personality type (I hesitate to assign it a more modern label) left little room for sex and romance.[31]

Diego, on the other hand, was the opposite: he had impulses that his father and brother lacked, with an inability to control them. In the months after Columbus's death, his two young sons sought and found a powerful ally to help them secure the titles, offices, and income promised in their father's contracts with the Spanish Crown. That ally was the Duke of Alba, whose niece's marriage to Diego was intended to cement the alliance. But between the betrothal and the wedding, Diego managed to impregnate not one but two other women. The wedding went ahead in the spring of 1508. Both his illegitimate children were born later the same year; Diego publicly shunned their mothers while privately paying them off. But one of them, the above-mentioned Isabel de Gamboa, sued Diego, insisting that he had promised marriage in order to seduce her. A lady-in-waiting to Queen Juana of Castile, Gamboa was "the wrong damsel to distress." The suit went on for years (until Gamboa's premature death in 1516), proceeding up the legal chain to the Vatican, maintaining a cloud over the Alba-Colón alliance, reminding everyone of the sexual incontinence of don Diego, the Second Admiral of the Ocean Sea.[32]

Diego's wife, María de Toledo, had the misfortune of dealing not only with a sexually dissolute husband but also a son who was arguably even worse. Luís Colón sired an illegitimate child when he was nineteen. The following year, also in Santo Domingo, he used a marriage proposal in efforts to seduce another woman—María de Orozco, a lady-in-waiting to don Pedro de Alvarado's wife (Alvarado, a veteran conquistador of Spanish invasions in Mexico and Central America, was en route to Honduras; his wife, doña Beatriz de la Cueva, was a niece, by marriage, to one of Luís's sisters). His proposal was accepted, without parental knowledge, constituting a scandalous informal marriage; before it could be consummated (which it may have been anyway), lady-in-waiting María was taken to Honduras and married to the royal treasurer there. A few years later, Luís married a wealthy heiress in Santo Domingo, María de Mosquera, with whom he had two daughters. When he tired of his wife after five years, he tried to use his prior promised marriage as a ploy to repudiate her (not dissimilar to how Henry VIII rid himself of his Spanish wife a decade earlier). When the first María then passed through Santo Domingo on her way to Spain with her treasurer husband and their five children, don Luís had her detained—demanding that the archbishop declare her his real wife.

Following an investigation, and much scandal, Luís was foiled. And so, a year later, he adopted the simpler solution of sailing to Spain, leaving his wife and children in Santo Domingo. Settling in Valladolid, he became engaged again after a few years, claiming through a rather tortured logic that his first two marriages canceled each other out. When, after two years, the church had yet to confirm his argument's validity, Luís impatiently and secretly married the unsuspecting doña Ana de Castro anyway. When the news reached his wife in Santo Domingo, she immediately returned to Spain with their daughters. Meanwhile, Luís's new mother-in-law (the countess of Lemos) discovered that her daughter Ana's husband had not been as free to marry as he had claimed. Both María and the countess were furious, and they furiously petitioned the authorities, who in 1559 duly arrested Luís (he was now thirty-seven).[33]

Using his wealth and standing to fight both prosecution and his wife María, he discovered that she had started a relationship with a man in Toledo named Alonso de Villareal. Despite the fact that his defense rested on the invalidity of his marriage to her, Luís filed an accusation of adultery against her. Forcing her to flee with Villareal to Italy, he gained a judgment that awarded him all of Villareal's property. Luís could not avoid a 1563 conviction for triple-marriage polygamy, with a sentence of ten years in exile (five in the North African fort of Oran). But he was able to use his wealth and influence to have both first marriages annulled, with his exile delayed pending his appeal. With his third marriage declared valid, he now lived in Madrid with his wife doña Ana and his now-adult illegitimate first daughter—who was sufficiently annoyed by his ongoing womanizing that she later testified to it in detail.[34]

For example, in 1564, barely nine months after his conviction, he seduced fourteen-year-old Luisa de Carbajal. A year later, she gave birth to a son—whom Luís called Cristóbal. Astonishingly, considering his circumstances, Luís bigamously married Luisa and declared baby Cristóbal to be the legitimate heir to his namesake great-grandfather's titles and privileges. Then, realizing his folly, he tried to force Luisa to marry a servant of his as a cover. Her father was persuaded by the offer of a 3,000-ducat dowry, beating Luisa when she refused to cooperate and threatened suicide. Then Luís's appeal failed; perhaps swayed by rumors of a fourth bigamous union, the judges amended his ten-year sentence to be served entirely in Oran. Luís was transported to Africa. Weeks later, Luisa gave in and married the servant.[35]

Several witnesses would later testify as to the teenage Luisa's influence over don Luís. In particular, she persuaded him to provide for his daughters properly (he had confined them in poverty in a convent, preventing them from seeing their mother in Italy), and then to emancipate them so they could access their own property and leave the convent. Years later, one of the daughters (doña Felipa Colón y Mosquera, who would later have to marry her cousin don Diego Colón de Pravia to secure her inheritance of the Columbus legacy), lived in a small village outside Valladolid as friendly neighbors with Luisa de Carbajal (still married to that servant of Luís's). A man like Third Admiral don Luís Colón made female solidarity all the more important.

Exile in an African outpost did not change who Luís Colón was. A year or two after his arrival, a daughter was born to him in Oran. Named Petronila, she was four when her father died. She later married the governor of Melilla, another Spanish fortress on the northern coast of Africa; in 1651, when she was close to ninety, Petronila testified as a participant in the vast Colón litigation, achieving that immortality promised by the old saying.[36]

Victorian-era Columbus biographer John Boyd Thacher judged Luís to be "a thoroughly unworthy specimen of humanity." It is hard to disagree, although it might be too perverse an argument to use Luís's life as a way to cast his grandfather in a better light. Nonetheless, Columbus's son Diego and grandson Luís were shocking examples of Colón men who were "sexually susceptible"; Columbus himself was not. As Fernández-Armesto concluded, Columbus's "response was markedly prudish when his followers indulged in concubinage or promiscuity with native women in the New World. Later in life, he affected the dress and habits of a cenobite and generally appeared to prefer the company of friars to that of women."[37]

An alternative speculation regarding Columbus's sexuality is unavoidable. Indeed, Fernández-Armesto also mused that it was "surprising that no one has yet written a book claiming that Columbus was gay." That is a wry reference to how much of the Columbus literature is driven by absurd theories rather than speculation based on firm evidence. Nonetheless, Columbiana contains books spun out from less than the source material that prompted that quip. For example, Columbus scribbled 437 postils or notes in the margins of his copy of *Lives of the Noble*

Greeks and Romans by Plutarch (46–ca. 119). Almost a quarter of his notes were about omens, auguries, and visions. The next most popular topic was that of deceit or trickery by a politician or commander, a strategy that so appealed to him that he even crowed about using it himself (although in some cases the deceit lay in him making up the story). He was also interested in fatherhood, especially what we might call extreme tough love father-son parenting. And he was fascinated by the nature of heroic deaths. There were also more than a dozen markings noting all the heroes who were in some way associated with or disposed toward "the unspeakable lust" (that is, between men). Not, therefore, the topic on which he was necessarily obsessed, but nonetheless clearly one that interested or bothered him—and one that drew more attention than any aspect of women's lives. In fact, in other books he annotated, almost all the women in which he showed interest were Amazons—mythical warrior women who did not exist, but whom Columbus believed were real, and whom he insisted he had narrowly missed seeing in the Caribbean on two of his voyages.[38]

Do any of Columbus's relationships with men help us to discern homosexual curiosity rather than homophobic observation? His apparent social awkwardness was offset by a great garrulousness, and among the minority who could tolerate his verbosity were some close and loyal friends. One who stands out is Diego Deza, the Dominican friar whom Columbus met at court when fray Diego—seven years older than Columbus—was tutor to Fernando and Isabel's only surviving son, Prince Juan. The friar and the mariner were, in Fernández-Armesto's estimation, "for a long time on terms of the most extraordinary intimacy." In later letters, such as those to his son Diego, Columbus certainly expressed nostalgia for the times he had stayed with Deza, as well as confidence in the political efficacy of their "brotherly love."[39]

But in terms of reading into their friendship, I see no smoking gun. Deza rose rapidly up the ecclesiastical hierarchy, from 1487 to 1505 moving through the bishoprics of Zamora, Salamanca, Jaén, and Plasencia, to become archbishop of Seville; in 1499, he became Spain's inquisitor general (succeeding the infamous Tomás de Torquemada). He was thus a powerful patron to Columbus, who in turn was gratefully loyal, quick to credit Deza for his voyages being sponsored by the Castilian kingdom—a sharing of credit that did not come easily to him. Deza's power drew enemies, who accused him of violent cruelty, of secretly practicing Judaism, and of criminal womanizing. There are stories behind such accusations (the pope cautioned Deza for his Inquisitorial excesses, for example, and he had Jewish ancestry on his mother's side—although he was a persecutor not protector of Jews). If there had ever been a hint of "the unspeakable lust" in Deza's life, his enemies would have used it against him. The Columbus-Deza relationship was thus of mutual political value, seemingly buoyed by platonic friendship.[40]

In the twentieth century, hints of an emerging U.S. perception of Columbus as possibly homosexual proved to be hollow. "Gay Columbus Played Too Dully," ran the headline to a review of the 1932 Broadway play *Christopher Comes Across*, a comedy criticized as "too heavy" and offensively depicting Columbus as a womanizer. By the 1970s, when "gay" had taken on its current meaning, Columbus's reputation was similarly soon defended. On an episode of the CBS game show *Match Game 74*, contestants were asked for the word that finished this sentence: "One of the sailors on the Santa María said, 'just between us, I think Christopher Columbus is . . .'" Preceded by raucous laughter, the first contestant said, "gay." Amidst more laughter and applause, another contestant quipped, "How long had they been at sea?" But a third contestant indignantly offered an alternative word—"wrong"—asking the host if "this is the Santa Maria we read about in all our school books and all that swell stuff? This is *our* Mr. Columbus?" "Yes," the show's host responded, "*our* Mr. Columbus."[41]

By the time of the Quincentennial, defenses of Columbus were less common—especially among the young—than accusations of his association with genocide. A 1992 student newspaper article at the University of Michigan—where a comedy titled *The Christopher Columbus Follies* was being performed—noted that "National Coming Out Day" and "Columbus Day" were adjacent "anniversary dates" with "much in common." "Both should be occasions of mourning and of celebration," because Columbus's voyages had initiated centuries of suffering for both Native Americans and homosexuals; the AIDS epidemic and the U.S.

government's failure to act sooner, it was argued, echoed genocide and "germ warfare" against Indigenous peoples.[42]

WE LEARN SOMETHING ABOUT twentieth-century U.S. culture and the American Columbus, but very little about Columbus's actual life, in such attacks on, and defenses of, "our Columbus." For as much as the creators of histories, plays, songs, and TikTok content have willed into existence details of Columbus's intimate life—inventing for him a rich array of partners, from wives to illicit lovers to victims—the evidence reveals an existence remarkably devoid of sex and romance. He was not interested in love; his lusts were not carnal. The attention he showed to women was ephemeral; his interest in men was not about sex but about power and rank—and how other men might help him improve his own. Love and lust have inevitably been forced into his story, its gaps filled with speculation as to his passions, proclivities, and perverse penchants.

That may go a long way to explaining many of the invented details; the story just feels incomplete without them. The agenda-driven nature of Columbiana means that a heroic Columbus needs to be magnetic enough to attract a queen, while an antiheroic one needs to be predatory or perverted. But the specific phenomenon of the TikTok accusations requires similarly specific explanations.

One is the deep roots in the West of mermaid mythology, and more specifically of the manatee/mermaid dyad, which remains central to the perception of manatees. A typical children's book on the mammals, for example, begins with a chapter titled "Mermaid?" followed by "Do We Look Like Mermaids?"—explaining that "early sailors reported sighting creatures that appeared to be part fish, part woman. Sailors called these creatures sirens or, more commonly, mermaids.... What the sailors probably saw were manatees."[43]

Sirens and mermaids are not the same in the modern world—perhaps not since the success of Hans Christian Andersen's *The Little Mermaid* (1837) and certainly not since Disney's 1989 animated musical version (1989)—but in the late-medieval world mermaids *were* sirens, beautiful but deadly, either half-bird or half-fish, mythical yet believed by many to be real. European sailors looked in the seas of the Americas for sirens—

just as they sought Amazons, cannibals, dog-headed men, people with tails, and other "monsters"—and imagined they found them, passing on tales of sightings. On his First Voyage, off the coast of Hispaniola, Columbus claimed (according to Las Casas) "to see three *serenas* [sirens or mermaids] that came very high out of the sea, but they were not as beautiful as they are depicted, for in some way their faces had the appearance of a man." This disappointment lay in the manatee's failure to inspire fear rather than lust; sirens lured men into peril not pleasure, and thus an unattractive siren was no siren at all. But removed from that context, that line in the *Diario* helped seed the myth of Columbus as bestial predator.[44]

In Columbus's lifetime, and for hundreds of years to follow, stories circulated of manatees being abused—hunted, sexually assaulted, eaten—by European sailors in the Americas. The tale of the Taíno lord who rode on the back of a tamed manatee, first written down by Peter Martyr, was oft repeated and illustrated, as were other versions (López de Gómara's version has Indigenous boys riding manatees). But such playful interactions never included Spaniards, from whom manatees

Ridden: A 1621 imagining of the story told by sixteenth-century chroniclers such as Peter Martyr and Francisco López de Gómara, repeated in subsequent centuries, of manatees permitting Indigenous riders but fearful of abusive Europeans. From Honorius Philiponus, *Nova typis transacta navigatio*.

fled, recognizing the clothes of their abusers. Columbus goes unmentioned. The unnamed Europeans who terrorized manatees served as placeholders for the First Admiral to be inserted centuries later.[45]

Another factor explaining the myth of the predatory Columbus is the old claim that some of the men who returned from the First Voyage in 1493 brought to Spain a new disease. Las Casas wrote that he saw the disease spread through Seville in that year. A doctor, Ruy Díaz de Isla, claimed he treated the captain of the *Pinta* and other afflicted sailors, later stating that "its origin and birth [was] in the island which is now named Española." A deadly outbreak in Naples in 1494 and 1495 was blamed on this new disease, which we call syphilis. In the late twentieth century, modern understandings of epidemiology combined with attention given to the Columbian Exchange—as historian Alfred Crosby dubbed in 1972 the transatlantic movement of diseases, plants, animals, and so on—to make increasingly popular the notion that spreading syphilis was one of Columbus's crimes.[46]

Although scholars now believe that different species or subspecies of the syphilis bacterium *Treponema pallidum* likely existed on both sides of the pre-Columbian Atlantic, the accusation against Columbus is well rooted. It is mostly an indirect allegation (sailors on his ships spread the venereal disease) but sometimes a direct one (Columbus himself did it). In this century, both variants are easily found scattered across the internet, in news reports, online comments, and social media sites. A typical example is a 2022 *Express* posting on Columbus's burial, which mentioned in passing that "the navigator has also long been blamed for carrying the sexually transmitted infection syphilis from the Americas to Europe through his crewmen."[47]

The internet's distorting effect on logic and evidence is the third explanatory factor behind Columbus and the manatees. The "potent recommendation algorithms" and "low barrier to entry" of platforms such as TikTok have created fertile ground for "tall tales" in which "assumptions and coincidences" are easily and crudely assembled. If Columbus was bad, he must have done bad things. Social media fosters the reduction of such logic to pithy extremes, as encapsulated in the comments inspired by the Columbus/manatee memes, especially those lumping together this and his other "crimes." The abuser of manatee-mermaids becomes the transmitter of syphilis becomes the instigator of germ warfare. As one TikTok user put it, "Bro both caused a genocide and loved be@stiality."⁴⁸

Born Again: The "Christopher Columbus House" in the Corsican port town of Calvi, as photographed in 1908 for a postcard set (top). The ruin upon which people are standing is the structure claimed to be the house where Columbus was born; it looks very similar today (bottom).

LIFE SIX

THE LOCAL

IT IS EASY TO MISS. THE SMALL RUIN, TUCKED AWAY ON A SIDE street, is hardly more than a hint of an old house attached to a jumbled rock wall. Yet it is of great significance, according to tourist information offered visitors to the Corsican port town of Calvi. For this is the "Maison Natale de Christophe Colomb." An old plaque attached to the ruin reads "Ici est né, en 1441, Christophe Colomb, immortalisé par la découverte du Nouveau Monde, alors que Calvi était sous la domination génoise: mort à Valladolid le 20 de mai 1500" (Here was born, in 1441, while Calvi was under Genoise domination, Christophe Colomb, immortalized by the discovery of the New World, dying in Valladolid on May 20, 1500). Two French historians have studied the location, and they date the ruin to 1580. But it is better to ignore that fact; it spoils the fun. Take a leap of faith and drive down the Avenue Christophe Colomb to reach the great Corsican navigator's Maison Natale.

Elsewhere in Calvi there is a small monument comprising a concrete ship's bow and a bust of Columbus, this one giving him the dates of 1436–1506. Those extra years, of course, give him a Corsican childhood before he appears in the Genoese archives. As one of Columbiana's simpler slices of invented history, it is elegantly efficacious, as it obviates the need to deny all the evidence of Columbus's actual life. In a way, the people of Calvi did not need to contrive that pre-childhood; in the fifteenth century, the port was an integral part of the Genoese state, even more so than implied by the plaque's phrase "sous la domination génoise." The year before Columbus was born (in Genoa), the Genoese Bank of San Giorgio took over administration of the island. Columbus would have known natives of Calvi when he was growing up in Genoa. There's no evidence that he went there in his early days at sea, but it is very possible. Calvi sailors were part of the crews on his second and third transatlantic voyages. Calvi is a part of the Columbus story. It's just not where he was from.[1]

Locals chuckle at the Corsican claim to this imaginary Colomb. But some also will admit, when pressed, that they believe in it too. In other words, they view it the way they may view the detailed claims of the Catholic religion in which they were raised. Intellectual skepticism, if not downright dismissal, does not undermine their faith. People are capable of believing two contradictory things at the same time, or of *knowing* one thing while *believing* another that contradicts it—a kind of interior dialectic that helps make Columbiana possible.

"How could the wildest flights of fantasy have conjured up a Greek Columbus, an English Columbus, a Corsican Columbus, a Swiss Columbus, and no less than three Columbuses from France and Portugal respectively?" indignantly demanded Taviani, the great defender of Columbus's Genoese origins. Taviani's question is not merely rhetorical; he has an answer, arguing that "the stature of the Navigator himself," "the grandeur of his enterprise" and its "historical consequences," are sufficient to explain "such outrageous claims."[2]

How, indeed, can there be so many claims on Columbus? Growing steadily over the centuries, their number now approaches thirty (a dozen cities in Italy, and more than twenty in other countries)? I counted almost one hundred fifty books, published between Henry Harrisse's *L'origine de Christoph Colomb* of 1855 and Carlos Machado Roma's *A Nacionalidade Portuguesa e o nome de Cristóbal Colón* of 1934, devoted specifically to Columbus's name, birthplace, and origins. To add pamphlets and books published before and since likely doubles that total.

It is almost as if such books self-reproduce. Which, of course, they do. Taviani was right, but his answer was too wrapped up in his apologist agenda; his focus was always on defending Genoa's claim, not in understanding the others. The more indignant the protests from Italians—especially the Genoese—the more the shouting match tempted others to join the clamor.[3]

In Calvi's case, therefore, it was French claims in the nineteenth century that prompted Italian responses; those stirred the French in return. The Calvi claim played a central role in that contest, kept alive through the middle decades of the century by claims in publications in Paris and Corsica, then spurred by a series of articles in Corsican newspapers in the 1870s, followed by a book by a Corsican abbot named Martin Casanova de Pioggiola in 1880. Then, in 1882, a decree from the French president charged the minister of the interior to execute an old 1816 decree that had approved the erection of a Columbus statue in Calvi—taken by claimants and detractors as tacit French governmental approval of the Calvi tradition. In 1886, both a statue and the plaque on his "birth house" were put up and ceremoniously unveiled.[4]

Abbot Casanova's position was based on the assertion that the Calvi claim drew upon local, oral tradition, which trumped archival evidence; "even better than written texts, it is the result of the real data of history, and, of all the proofs, it is the strongest." Harrisse responded with a brief, resounding dismantling of such "astonishing" logic, noting that both Casanova and Corsican officials had "again proved nothing," resorting merely to "audacious assertions." Their declaration that Columbus "was a compatriot of Napoleon" was blatantly mendacious, jingoistic boasting, while the claim that the Corsican archives contained records of Columbus's birth in Calvi in 1441 was a falsehood that could not be proved. Italian commentators and historians also protested the Calvi claim. But, as with so many Columbiana cases, the delivery of public addresses and the printing of pamphlets and books on the matter only gave it further attention.[5]

The proliferation of birth claims created a kind of catch-22 for historians seeking to present the evidence rather than prove a partisan or patriotic theory. Such claims and theories were a patently "inane" distraction from the well-evidenced truth (as Harrisse put it); yet their repetition demanded refutation, which in turn served to further air the theories, regardless of how thoroughly they might be disproved. Furthermore, the focus on Columbus's life (real or imagined) rather than local context made the viral phenomenon of birth claims a battleground rather than something to be understood. Corsica, for example, threw off Genoese control in the eighteenth century, but its short-lived republic was soon conquered by France, which still rules the island. A modest independence movement persists, as does Corsica's complex cultural relationship with France and Italy, circumstances that help make sense of the island's otherwise baffling birth claim to "Christophe Colomb."[6]

Through the twentieth century, it had become common to argue that the very existence of such claims proved that "it is impossible to know, definitively, where he was born." It was not enough to simply point out that this is not the case. "There must be reasons," as partisan claimants declared, "why other historians doubt his origins in Genoa." Indeed. But the obvious reasons—patriotic, micropatriotic, and localist agendas—were ignored by such claimants, willfully or unwittingly, as such reasons would have doomed their own arguments.[7]

CALVI IS NOT FAR from Genoa, roughly one hundred twenty miles almost due south across the Ligurian Sea. But the phenomenon of localist competition over Columbus's birthplace began even closer to his real hometown. Cuccaro is about fifty miles east of Genoa, in the Ligurian mountains. There, the late-sixteenth-century claim on the Columbian legacy by Baldassare Colombo and his family became the first written challenge in writing to Columbus's Genoese roots.

Then, in the early seventeenth century, other northern Italian towns made claims. They were encouraged in part by the *Historie della Vita*, that Columbus biography attributed to his son Fernando, first published in Venice in 1571. Regardless of who wrote and edited its opening pages, the *Vita*'s blatant purpose was to disguise Columbus's humble origins by inventing a vague, implied noble ancestry (with that vagueness justified as being Christlike). Fernando and his brother, Diego, both enjoyed status as Spanish noblemen, thanks to the contract their father had signed with the monarchs. But because Fernando had been born illegitimate, and of commoner parents, he tried to bolster his acquired noble status by suggesting a different family past. The *Vita* even avoided mention of his father's birthplace (although he let it slip that his uncle Bartolomeo was a native Genoese).[8]

The *Vita*-encouraged claims from Cuccaro and other towns in northern Italy were soon reported in influential places. Spanish royal historian Antonio de Herrera noted in 1601 that, while "it is important to know, based on the best information" that Columbus was Genoese, "some would wish it" that he was from Plasencia (that is, Piacenza), Cugureo (Cogoleto, near Genoa), or Cuccaro. In devoting more space to such fanciful wishes than to Columbus's actual Genoese origins, as well as speculating on Columbus's link to non-Genoese Colombo noblemen back in 940, Herrera indulges in a dubious habit that would come to characterize Columbiana: airing unsubstantiated rumors and claims in detail, thereby undermining their dismissal. His remark that the matter "will be determined" with certainty by the Council of the Indies, "where it is being litigated," served only to confirm the notion that the doubts regarding Columbus's origins justified litigation.[9]

In fact, they did not. The early Italian claims went nowhere. They also prompted the Genoese to start publishing early in the seventeenth century the extant documentation on the Colombos and Columbus's early life. For centuries, therefore, the phenomenon spread slowly, chronologically, and geographically. Late in the seventeenth century, it crossed into southern France. In 1697, a French lawyer from Digne (one hundred eighty miles west of Genoa) named Jean Colomb claimed he was descended directly from Columbus—and that Columbus himself had in fact been French. Yet this remained fringe stuff. In seventeenth- and eighteenth-century histories of the early voyages of discovery, in various languages, Columbus was incontrovertibly Genoese—just as Amerigo Vespucci was Venetian and Fernão de Magalhães (Magellan) was Portuguese. Rumormongering similar to Herrera's was not uncommon, but it remained understood as such. Not until the turn of the nineteenth century did the pace of fabulations pick up.[10]

The claims of those old lawsuits then resurfaced in a series of Italian publications debating the precise details of the Colombo family's Italian origins. Various attempts were also made to link Columbus to French families called Colomb or Coullon in Burgundy, Bordeaux, and Savoy. By 1814, there were enough claims to prompt a Genoese senator, the marquis Domenico Franzoni, to write a book refuting them. His targets were still overwhelmingly from northern Italy and southern France (he also had a go at Vespucci as a false "discoverer" promoted by Venetians). But that soon changed.[11]

The speculations drew renewed attention to the ambiguities in the *Vita*, some of which seeped into Washington Irving's best-selling Columbus biography of 1828. Localist appropriation of Columbus's birthplace had fomented a mythistory, and the more that speculation, invention, and imagination were applied to the legend, the further Columbiana spread. The 1890s saw a feeding frenzy on the manufactured mystery of Columbus's origins. By the 1930s, the Genoese government was protesting how much evidence had been invented over the previous century by unscrupulous foreigners, "wishing to obtain notoriety by attaching themselves to the eminent figure of Christopher Columbus." Sometimes claims were so personal as to be risible—such as a Swiss man named Colomb claiming, purely on the basis of his name, to be descended directly from Columbus, who had therefore himself been a native of Geneva (an error allegedly based on one tiny spelling mistake, as the Genoese call their city Genova). Historians such as Ballesteros were not alone in denouncing such claims as "ridiculous in the extreme."[12]

Within that Spanish historian's "ridiculous" category was also the assertion that Columbus had in reality been a Greek pirate. The minor mythistory of Columbus's piratical origins has for long periodically surfaced, primarily to underpin either French or Greek claims to Columbus. Corsairs and privateers were far from being an amusing topic in the late-medieval Mediterranean and early modern Caribbean. But the successful assertion of a monopoly on seaborne violence in the Atlantic world by the great imperial trio of the eighteenth century—Britain, France, and Spain—permitted a very different vision of piracy; it flourished in the era of Romanticism and persists to this day.[13]

The Greek claim drew upon the grasping of a handful of straws. One was the pirate fable originating in the *Vita*, whereby Columbus was related to a famous fifteenth-century French corsair named Guillaume de Casenove (aka Casseneuve, Casenovo, Colombo, Colombo the Elder, or Coullon). That was part of a tangled web of invented coincidences, a circular argument in which imaginative speculations are turned into "facts" that "prove" other speculations. For example, Columbus was imagined as related to various pirates, such as Colombo el Mozo ("the Younger"), who may also have been a Greek pirate named Georgi Bissipat; the *Vita* appropriated one of Bissipat's adventures for Columbus's life, and that seemed to substantiate Columbus's claim that "I am not the first admiral in my family." Yet that oft-quoted line is found in only one place: the *Vita's* summary of a letter Columbus wrote to Prince Juan's nurse. Although the boast sounds like a fabulist exaggeration of the kind Columbus would make, it is absent from both versions of the letter that seem to have been made by Columbus himself—meaning it is either apocryphal or a lie.[14]

Columbus as a descendant of noble pirates is an obvious dead end. The Genoese scholars who a century ago published the archival evidence of Columbus's origins dismissed the pirate story as a well-known invention; it "is not pardonable today to perpetuate the error." But the Greek claim grasped other straws. Columbus once mentioned that he visited Chios, a Greek island that was a Genoese colony in the fifteenth century; was this not evidence that he was actually born there? It also makes sense, Greek claimants insisted, that a Greek discovered America: Plato, Homer, and other ancient authorities showed that Greeks had discovered Atlantis, which was part of America, and hence the Greek Columbus was really *rediscovering* America. As for its name, that was nothing to do with Vespucci: "the Indians called the first Europeans in America Homeric because they spoke the Homeric language and from it called the new world Homerica (America)."[15]

Like most non-Genoese claims on Columbus, the Greek claims also fixated on his coded signature. That signature was not intended to be cryptic or mysterious; the very transparency of its code was essential to its purpose as a promotion of Columbus's status. But a code it was, and

The Greek: Christopher Columbus was a Greek prince, according to a 1937 book of that title by a Greek American. Illustrations from the book connect this noble Greek discoverer to the modern United States.

it has thus proved irresistible to those with imagination and a partisan agenda. For Spyros Cateras, for example, the letters X.N.Y. clearly represented Chios (or Xios), Nicolaos (his "true" first name, and the origin of the name Colón), and Ypshilantis (his "true" surname). The *Vita*'s description of Columbus's *naso aquilino* (aquiline nose) was further evidence he was an Ypshilantis, insisted Cateras, because that Greek family's men "have hawk noses and long faces."[16]

The farther from Genoa that the alleged Columbian birthplaces become, the more our credibility is stretched. That has not stopped northern European claimants, largely due to the parallel but far older history of Scandinavian voyages to the northern shores of the Americas. As an Associated Press story put it in 1991, "Norwegians have been miffed for centuries about Christopher Columbus stealing credit from Leif Ericsson for discovering the New World." The story was prompted by the release that year of *Christopher Columbus—en europeer fra Norge?* (a European from Norway?), written by a maritime writer called Tor

Borch Sannes, dubbed "Norway's funniest cult archaeologist" by the country's Research Council newspaper. But Sannes (1940–2007) was in earnest. And he was building on precedent: a "self-educated" Norwegian named Svein-Magnus Grodys had suggested in 1969 that Columbus might have been related to an old Norwegian noble—and sometime royal—family called Bonde.

For Sannes, the emblem in Columbus's coat of arms denoting paternal lineage was that of the Bonde dynasty. He spent years weaving that coincidence into an insistent theory. After all, Sannes noted, some biographers described Columbus as tall, blond, and blue-eyed—a very Scandinavian look. Bonde must have fled persecution to Italy, where he changed his name to Columbus. His discovery of America? That had already happened in 1477, when he reached Canada; fifteen years later, Christopher (né Bonde) Columbus *rediscovered* America.[17]

The trope of rediscovery has surfaced in other ways beyond the Greek and Norwegian claims. In 1476 a Danish expedition sailed to Greenland, led by a pair of Danish admirals and guided by a pilot named Joannes Scolvus. This expedition is evidenced by surviving correspondence between Denmark's King Christian III and Portugal's King Afonso V. Whether the Danes went beyond Greenland to visit the North American coast is unclear. Likewise, we know little about Scolvus, encouraging Columbiana to appropriate him as a mysterious figure, and Poland to join the fray by claiming Scolvus was a Pole who discovered America fifteen years before Columbus. A similar contrivance has the Danes as the true discoverers.[18]

As you may have guessed, enterprising minds have played with Scolvus's name—was it Scolnus, thus Colonus, thus Colón or Columbus?—turning him into a double-discoverer of the New World, first under his real name and then under his assumed name. No wonder he was so persistent in pursuing his enterprise at the Portuguese and Spanish courts: he had already discovered America.[19]

There are revealing patterns to these claims. They tend to be built upon a common set of weak argumentative contentions: that the variety of names Columbus used (or we use for him, including "Columbus") means that there exist "doubts about his real name"; that the absence of a Genoese birth certificate casts doubt on his birth in that city; that the lack of documents written by him in Italian means he cannot have been Italian; and that all this mystery can only be solved by the discovery of a cypher or previously hidden solution, typically something coded, often deliberately so.

As birth outside Italy would have meant a different birth name, these "doubts" have often been imagined as a mystery to be solved by the discovery of his "real" name. Once the writer has asserted that Columbus's name and birthplace are mysteries, he can be revealed to be almost anybody from anywhere. But the "doubts" are, in fact, simple ignorance of European naming patterns in the fifteenth and sixteenth centuries. Personal names were not fixed, socially or legally, as they are in the modern world. It was common for a name to be spelled in various ways (none being "correct"), as it was for a person's name to change as a result of mobility of any kind—from social mobility, through marriage or business connections or wealth acquisition, to migration to places where one's name might be localized. Within a system this fluid, Columbus's name actually underwent *less* change than we might expect, a relatively straightforward translation, along with his home-base moves, from Genoese to Portuguese to Castilian—with the all-important "don" title added upon his return from the First Voyage.[20]

Likewise undermining non-Genoese claims is ignorance of the history of Italy, which did not exist as a single country or unified state until four centuries after Columbus's birth. No birth certificate for him exists because in his lifetime neither the Genoese state nor the Catholic Church had yet implemented such a bureaucratic requirement; birth records were not yet required anywhere in Christendom. Columbus "never wrote or spoke a word of Italian" because, as a Genoese, he spoke the Ligurian dialect, not the Tuscan dialect that was the core of what we call Italian today. Legal records in the Genoa of his day were written in Latin, which he did not learn (and then not well) until after he had left Genoa; in fact, evidence strongly suggests that he did not become literate in any language until after leaving Genoa.[21]

Another weak argumentative contention deploys a visual or linguistic coincidence recast as "evidence" to the exclusion of all other lines of evidence. The Norwegian coat of arms argument is one example. Anticipating a technique used by twenty-first-century conspiracy theorists, writers have taken arguments used to debunk their claims and pointed them back at the debunkers in order to dismiss evidence inconvenient to such claims. For example, scholars have shown how a number of Columbiana documents are fraudulent. Rooted in the centuries of lawsuits over the Colón legacy, that tradition has been stimulated since the early nineteenth century by the growing monetary value of such materials and the competition for localist birthplace claims. Consequently, claimants have tended to use accusations of fakeness to dismiss evidence of Columbus's Genoese origins; as one Spanish historian said of another, "when a document is troublesome to him, he declares it a fake, and by this trick the obstacle disappears."[22]

These contentions and lines of argument are all willful, driven by agendas whose motive is crucial for us to make sense of them. That motive is almost always localist, be it personal (I am descended from Columbus) or patriotic (Columbus belongs to our nation) or the many levels of regionalism and micropatriotism in between. Within the overarching localist motive are two motivating motors. By asserting an association with an internationally renowned historical figure, an otherwise marginal place is moved closer to the center of the world stage—with potentially real and measurable benefits, such as a boost in tourism revenues. Often closely related is the motor of regional or national competition, whereby each claim stimulates rival responses—be it within northern Europe, between northern Italy and southern France, or among Mediterranean islands and Iberian regions.

WHEN I WAS A boy, I thought that there was an island in the Atlantic Ocean called M'Dear, devoted entirely to English holidaymakers. An elderly couple, family friends, regularly vacationed there and raved about its climate and friendly people. They seemed to find the island's name—M'Dear—amusing. Confusingly to me, there was also "M'Dear" cake and a wine of the same name, prompting winks and nudges and, after a few glasses, guffaws. Not for many years did I acquire the crucial missing link of knowledge: a 1957 song by Flanders and Swann, titled "Have Some Madeira M'Dear." The song's darkly comic lyrics—by today's standards, too dark to be comic—center on the attempted use of Madeira wine by an old letch to seduce a teenage girl.[23]

Not for decades did I grasp the deep roots of the association in the English mind between the Portuguese Atlantic archipelago of Madeira and seduction (predatory or not), pleasure, and fantasy. Treasuring the island as a holiday destination, the nineteenth-century English wrote dozens of books "extolling its beauty and salubrious airs," repeating in such books the invention that Columbus had once lived there. The invention was likely not local, but a figment of visitor imagination, as "Madeira cared little about Columbus until the English came along in the seventeenth century."[24]

Considering how far across Europe the Columbiana virus of birthplace claiming has spread, it should not surprise us to find it in England itself. Still, the English claim never amounted to much of a tradition, existing as a snippet of legend or mythistory originating in the seventeenth century. A frequently (mis)cited example is in Charles Molloy's *De Jure Maritimo et Navali*, first published in London in 1676. Molloy (1646–90) praised "a discontented Native of this Isle, the Famous *Columbus* who prompted by that Genius that naturally follows a Native wise man, discovered a New World"—adding in the margin the clarifying note "Born in *England*, but resident at *Genoa*."[25]

In Molloy's day, the English already had their eye on Madeira—and not just on its wine, most of which was drunk by the English from the seventeenth century on. In 1662, King Charles II asked for the island, in vain, as part of Catarina de Bragança's dowry; he was disappointed when offered Bombay and Tangiers instead. That did not stop the British establishing a tradition of élite visitors, then tourists. And when the opportunity arose, as it did in 1801–2 and again in 1807–14, the Royal Navy seized and occupied the island. Such moments were the exceptions, however, as Britain's alliance with Portugal had provided the navy with privileged access to Madeira as a base, just as British merchants enjoyed such access to the wine trade. That context explains why incipient attempts to claim Columbus in England soon developed into the British cultivation of a Columbiana Madeira myth. Its roots seem to have been Britons asking whether and where Columbus lived on the island; to which helpful Madeirans responded by pointing out ancient houses or crumbling buildings, some no longer standing. The number of candidate buildings grew during the nineteenth century to five, as British and then American curiosity increased, even as Madeirans did not hesitate to raze such buildings. In time, as visits by privileged foreigners evolved into the modern tourism industry, Madeirans would pivot to promote the myth themselves.[26]

Amidst the travel guides and diaries repeating that this or that house "had claimed for it the honor of having been inhabited by the discoverer of America" (as the 1860 edition of White and Johnson's *Madeira: Its Climate and Scenery* put it) is a passing mention in 1863 by the famous English explorer Sir Richard Burton:

> My last pilgrimage was to the spot where Christopher Columbus is supposed to have lived when he resided, probably for health, at Funchal, during the intervals of his trading voyages. In 1851 his house near the Carmo was, like Shakespeare's tree, impiously destroyed; two other localities have claimed the honour, but hitherto with little effect.[27]

The phrase "for health" is revealing. Britons and Americans traveled to Madeira for a climate reputed to be beneficial to one's health, a motive here projected by Burton back onto Columbus. Typical of such references is Burton's skepticism regarding the claims, combined with acceptance of the tale's essence ("his house").

As the nineteenth century wore on, and the Quadricentennial approached, apparent written evidence of Columbus's residency on Madeira and its sibling island, Porto Santo, was added to these repetitions of supposed local folk traditions. In fact, Columbus twice visited the islands, the first time in 1479, when he was sent to Madeira to buy sugar. Bungling the trade, he returned to Genoa in debt. Las Casas applied his imagination to those bare facts, asserting that in 1479–80

Columbus lived on Madeira with his new Portuguese wife Filipa. Later writers changed the location to Porto Santo, where Filipa's late father had once been governor, and added that Diego was born on one or other of the islands. There is no evidence for any of that. The Genoese controlled four-fifths of Madeira's sugar trade at the time, so Columbus possibly made other brief, undocumented visits to the island. But to turn that into a narrative of domestic residency is wishful thinking.

The second Columbus visit for which we have evidence was in 1498. He stopped in at both Porto Santo and Madeira on his way to cross the Atlantic Ocean for the third time. The day that the fleet dropped anchor off Porto Santo, the town was deserted; having heard rumors of French pirates in the vicinity, perhaps taking Columbus's fleet to be such pirates, its residents were hiding in the hills. The ships stayed just a few hours, long enough to load water and wood, and then, taking the rumors seriously, sailed on to Funchal—where they "remained several days to take on provisions, sailing again on Saturday afternoon." As one Madeiran historian concluded in 1956, noting the local development of Columbiana legend, "we have only one document to prove that Columbus was in Porto Santo and even then, only for six hours, in addition to a longer visit of six days in Madeira."[28]

So, neither in 1479 nor 1498 did Columbus—save by wild stretches of the imagination—live or reside on the islands. No matter. If a Columbus residence never existed, one needed to be invented. As Columbus's Madeira myth slowly grew during the nineteenth century, one old house after another in the capital of Funchal was claimed as his, but then—if any of it was still standing—demolished. One of the more popular such structures was called the Esmeraldo House, after a Flemish merchant named Jean d'Esmenaut who settled there in or soon after 1480, changing his name to João Esmeraldo (in accordance with the era's naming culture). Columbus could not therefore have stayed in 1479 with d'Esmenaut/Esmeraldo, who had yet to settle on the island, nor is there a shred of evidence that he and the Flemish merchant were ever acquainted, let alone friends. We also know that Columbus was in Funchal only for less than a week in 1498. But, in the world of Columbiana, such details become irrelevant.[29]

The Esmeraldo House was heavily damaged by a 1748 earthquake (legend has locals stealing debris as Columbus mementos), then badly hit by an 1803 tidal wave. Unfit to be a residence, it survived as a storehouse for grain until being demolished in 1877. By this time, the local Columbus myth had taken root, and a Funchal historian named Agostinho de Ornelas—who had written a series of brief essays transforming Columbus's brief visits into detailed residencies—claimed and paid for a memento from the building before its destruction (he had invented the earlier memento legend). A double-window frame was found with a date on it, supposedly 1494. The frame's style suggests it is not that old, dating to one of the house's much later renovations. But the Ornelas family kept what they called "the Columbus Window" for generations.[30]

Madeira did little to mark the Quadricentennial. But the frequent mention of the islands in Columbiana writings and in commemorations from Genoa to Chicago had an impact. Four times between 1907 and 1913 the archipelago's daily, the *Heraldo da Madeira*, gave the Columbus-Madeira connection front-page attention. A historical novel set on the islands in Columbus's time—with him a resident—proved popular, as did performances of its theatrical version. Portuguese and English-speaking visitors seemed increasingly thirsty for local Columbiana. The notion of rebuilding the Esmeraldo House as the Columbus House began to circulate.[31]

Meanwhile, a young local named Mário Barbeito, whose passion for Madeira history matched his ambition to be a vintner, began to amass documents and books on Columbus. While his Vinhos Barbeito (established 1946) flourished, his collection grew to more than three thousand items by the time he died in 1985. The city acquired the collection, inaugurating the Museu Biblioteca Mário Barbeito de Vasconcelos in 1989 with considerable fanfare, a roundtable of scholars, and the U.S. ambassador to Portugal. The museum-library can be visited any weekday, located beneath Diogos Wine Shop.[32]

The seed for Barbeito's Columbus obsession was planted by his father, a native of Porto Santo—where a parallel interest in the local legend had developed. In 1848, the city council annals recorded a formal statement by one of its members, João de Santana e Vasconcelos, that Columbus had married and then "lived for some time" on the small island, and that it was from his father-in-law's "inherited manuscripts" that he "got his ideas for the great discovery of the new world." A few imaginative lines from Las Casas's pen were thus transmuted into "facts" that placed obscure Porto Santo firmly in the middle of history. Without funds to capitalize on the legend, it simmered and circulated for generations, with nothing concrete done—until modern tourism came to the archipelago and the Columbus Quincentennial approached.[33]

Not to be outdone by Madeira, Porto Santo picked the ruins of an old governor's house and declared it to be where the Genoese explorer had lived. Some local historians protested that such a building dated to long after Columbus's death, and that the legend "of a Columbus residence in Porto Santo is wrong, since there is no indication nor any document to prove such a thing." João Carlos Abreu, secretary of the archipelago's Tourism and Culture Department, told a local journalist: "Proof or no proof, the work of reconstruction of the Columbus House on Porto Santo will go forward."[34]

Opening to visitors in 1989, it still features on the list of key sites for tourists to consider. As does Madeira's new Praça de Colombo (Columbus Square), built in Funchal near to where the Esmeraldo House once stood. That house has been rebuilt—or rather, a building has been constructed on the square, beside the recently built Sugar Museum, and promoted as the place where Columbus "stayed when he visited the island, the first time in 1478" (as one tourist website puts it), or "stayed where he visited Madeira in 1498" (as another does, adding that "the house known as Casa de Colombo was actually owned by a friend of his, João Esmeraldo"). Not far away is Ornelas's old "Columbus Window," "restored" and erected in the Quinta Palmeira botanical gardens once owned by sugar magnate Harry Hinton, where tourists are told it is "all that is left of the house where Columbus once lived."

All this is intended to give a veneer of authenticity, without going so far as to expose the contrivance. Artifacts dug up by archaeologists in Funchal, displayed for decades, comprise small pieces of ceramic, glass, wood, bones (human and animal), and coins of various centuries (including the fifteenth). None of the objects connect to Columbus—or even to Esmeraldo. But they give the *impression* that local historical tradition is supported by scientific evidence. They also add a physicality to the legend, just as the Columbus Museum, Columbus House, and Columbus Window lend a (literal) concrete presence to Columbiana myth. The Madeira-built replica of Columbus's flagship from his 1492 voyage, dubbed the *Santa María de Colombo*—on which tourists can explore, voyage, and sample Madeira wine—has the same alchemic effect of making fiction seem like fact.

The British holidaymaking tradition in Madeira continues unabated. By the 2020s, a third of a million Brits visited the islands every year—more than any other nationality after Portuguese mainlanders. The Columbus connection is not loudly proclaimed, but it is gently prompted in every tourist guide and online site, and hard to

We Are Sailing: Tourists visiting Madeira's capital of Funchal (seen in the background) can spend a few hours on this replica of Columbus's 1492 ship, the *Santa María*. That the original ship never sailed near Madeira, and that Columbus never lived there, does not detract from the holidaymaking fun.

miss, especially in Funchal. It is as prominent as in Genoa, more so than in other claimant places such as Calvi. Secretary Abreu noted in 1987 that "today in the United States it is a reality that Columbus lived in Porto Santo" and that "from the cultural and touristic point of view, we have to exploit that reality." That local reality is now well established for tourists to observe and absorb. Believe it or not, m'dear, Columbus is also Madeiran.[35]

SENATOR CRISTÓBAL COLÓN, THE Sixteenth Admiral of the Ocean Sea, shared with the French embassy in Madrid his thoughts on Casanova's 1880 book on Columbus's Corsican origins. The Corsican abbot, wrote Colón, deserves "sincere praise for the meticulous care with which he looks for arguments favorable to his thesis, and if he does not prove by irrefutable arguments" that the First Admiral was born in Calvi, "he invokes truly important traditional testimonials."

The aristocratic politician's true opinion was apparent. But Casanova ignored the letter's dismissiveness and latched onto its compliments, quoting it at the start of a new edition of his book, declaring that the endorsement of the current admiral sealed the deal. "The question is resolved and doubt no longer possible," he crowed; "the Discoverer of the New World was born in Calvi."

Four Columbus generations later, the mantle of tactful responses to such claims sits as heavy as ever on the admiral's shoulders. "I am very respectful towards all those people who have investigated the origin of Columbus in order to arrive at all those theories," responded the Twentieth Admiral, when recently asked by a Colombian journalist about the birth claims from Greece, various Mediterranean islands, and parts of Spain. "That is a subject that I deliberately leave to historians. I cannot say if he is from here or there." The admiral's diplomacy was admirable. But it was understandably stretched thin. The plethora of "theories" demand an opinion, and today's don Cristóbal Colón cannot resist bluntly hinting at his: "However, we should not lose sight of the principal theory of his origins—the Genoese Columbus—which is the one held by historians."[36]

The Twentieth Admiral might have said "firm proof" instead of "principal theory." But his politic prevarication reflects the reality, which he presumably recognized, that there is no winning a debate in which belief trumps evidence. Better to concede that claiming Columbus is a universal right. Better to accept the logic used by Secretary Abreu in 1987: "Proof or no proof, Christopher Columbus lived" here.[37]

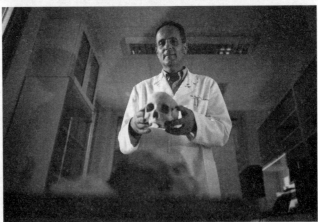

It Is Believed: In the early decades of the twentieth century, postcards were sold in Galicia of a ruined building in Poio "where it is believed that Christopher Columbus was born." A century later, that belief is bolstered by the building's renovation as part of a Casa Museo de Colón. Meanwhile, the University of Granada's José Antonio Lorente continues his decades-long campaign to prove with DNA evidence his belief that Spain has Columbus's bones—and that Columbus was a Spaniard.

LIFE SEVEN

THE IBERIAN

THE TOMB IN THE SMALL CHURCH OF SAN MARTIÑO DE SOBRÁN HAD not been opened since 1496. That autumn, months after Columbus's return from his Second Voyage to the Caribbean, the body of a Gallego priest named Xohán Mariño de Soutomaior was sealed into the stone sepulcher. There is no evidence that anybody believed that the two men were connected—until now. For in the autumn of 2022, after 526 years of rest, the bones of Xohán (Juan, in Spanish) were pulled from his tomb under bright lights and cameras. The sanctioned tomb raiders insisted that the sepulcher's mortar and stone wedges had not been moved since 1496. That mattered because the justification for their grave disturbance was the claim that it was "a key piece to confirm the investigations that point to the Gallego origin of Christopher Columbus."[1]

The somewhat circumspect language of the Spanish news reports that covered the tomb opening reflected both the uncertain science and the root rationale behind the deed. For that rationale sprouted from a belief, which had blossomed into a theory, which could now be proved by new evidence—the classic Columbiana sequence of a smoking-gun solution to an invented mystery. The belief was that the Gallego nobleman Pedro "Madruga" Álvarez de Soutomaior had in fact been the one and only Columbus. The theory was that because Madruga disappeared to become Columbus, Madruga's bones could not be tested. Or rather, they could, because—by the logic of this faithistory—Columbus's remains in Seville are really Madruga's. Any close Soutomaior relative buried in Galicia was thus also a relative of Columbus's. The proof was in the bones.

In an old cemetery in Poio, some seventeen miles south of Sobrán, more graves were dug up. Poio, an old town across the river from Pontevedra, is the epicenter of the Colón Gallego movement. On its Plaza de Cristóbal Colón sits the Casa Museo de Colón, a museum dedicated to the Galician Columbus theory (*hipótesis*), in a building that claims to be a reconstruction (*una construcción de nueva creación*) of Columbus's local *casa natal*. Where faithistory meets history, the hedging terms of circumspection meet the bolder language of asserted reality. The Casa Museo is one manifestation of governmental support for local tourism, but the exhumation of graves is driven by the Asociación Colón Gallego, in collaboration with the University of Granada's José Antonio Lorente (earlier we met him and his decades-long campaign to prove with DNA that Seville holds Columbus's real bones). That effort was taken a step further by the Galician grave-digging, designed to prove not only that Columbus's remains are in Spain, but that they are therefore home—as he was a Spaniard, more specifically a Gallego.[2]

Leveraging science to revive old theories and resuscitate old faithistory is a clever strategy. Historians debating centuries-old documents, with Columbiana arguments dissected in lengthy books, can seem ambiguous and outmoded, if not dull and confusing. But scientific findings can be made clear and precise, tailor-made for the succinct and sensationalist "news" upon which online sites thrive. DNA—with its reputation as relatively new, infallible, able to overturn old errors with its revelatory conclusiveness ("it's a match!")—is perfect for claims that Columbus's Genoese origins are an old error. In numerous corners of the world, cold cases are being reheated, and wrongful convictions overturned, by the new science of DNA. By asserting that Columbus is a misunderstood mystery, the historical equivalent of a cold case, the regional claimants and believers can tap into the public's expectation that DNA can correct the past errors of professionals. It is insider science turned into a tool for outsiders.

Not surprisingly, therefore, the twenty-first-century strategy used by the Colón Gallego movement and its allies was simultaneously deployed in Portugal. In 2017, a cluster of bones was found in an ancient wooden box in the church of a village near Lisbon called Quinta de Santo António de Castanheira. The box had been opened before, but its label indicated that the bones belonged to António de Ataíde, the first count of Castanheira. Antonio's cousin, Pedro, was killed in a 1476 naval battle off the coast of Portugal. But what if he had in fact survived by swimming ashore and changing his name? What if he had borrowed the name of a Frenchman who had died in the battle, a Culon, remaking himself as Colom? You'll remember that Columbus settled in Lisbon after swimming ashore in flight from a sea battle. Surely that is no coincidence, but instead evidence that Columbus came not from Genoa, but from the sea as surviving Portuguese nobleman Pedro de Ataíde. Except (as you'll also remember) the Columbus sea-battle story is apocryphal, and there is no evidence that Pedro survived the actual battle. But those facts are irrelevant to faithistory's method of sequencing speculation, imagination, and circumstance—all given a veneer of truth by the capstone sequencing, that of DNA. So, in Quinta de Santo António de Castanheira, up came the remains of Pedro's cousin, to be compared with the DNA taken from Fernando Colón's already disturbed and tested bones in Seville. (As with all tests from the bones in Seville, the initial tests proved inconclusive, but the campaign continues.)[3]

We live in an era when the disturbance and display of human remains has come under heavy scrutiny, although the change in attitudes has been slow. In the United States, more than three decades after Congress passed NAGPRA (the Native American Grave Protection and Repatriation Act), at least 100,000 ancestors of Native Americans and Hawaiians remain in the hands of U.S. museums, universities, and federal agencies. It is easy to find skulls, skeletons, and mummies to gawk at or even study—all in the name of science and education. Still, it becomes harder every year; in 2023, the Smithsonian removed its human bones and the like from display and research access, pending the implementation of a new policy for "the appropriate care, shared stewardship or ethical return of human remains."[4]

Why, then, have ethical questions not been raised over the opening of graves in Spain? How is it that the bones of human beings, whose descendants are identifiable today, can be removed, handled, scraped, photographed, and displayed, all as possible solutions to problems formulated or mysteries invented centuries after their ceremonial burial? After all, it is precisely such treatment of human remains that accompanied the colonial origins of anthropology, today seen as invasive and disrespectful, even sacrilegious.

The difference, of course, is that mummies are Egyptian or Andean, not European; and Europeans created modern anthropology. The mass collection of human remains was a by-product of European imperialism and colonialism, and thus the shift in attitudes toward such collections is a symptom of changing attitudes toward the colonial past. Fifteenth-century Gallego priests and noblemen are not viewed as victims, either of colonialism while they lived or of neocolonialism when pulled from their graves. On the contrary, their alleged connection to Columbus places them, in some vague sense, on the other side of that line of sensibility.

But more specifically and significantly, such present-day disturbing of graves falls under the protective umbrella of Columbiana. The dust and bones of Columbus have been moved and quarreled over for so many centuries that they have ceased to be perceived as actual human remains. Rather, they have become representative artifacts, along with all the statues and monuments to him, and the documents written by and about him, all to be denounced and destroyed, or claimed and deployed, as war matériel in battles that ultimately have little to do with him.

As we saw in the previous chapter, those battles have often been waged below the level of nationalism—they began, after all, at the personal or family level, and have persisted at personal, local, and regional levels. But imperialism, and its pugnacious offspring nationalism, came increasingly to lurk beneath the surface of such disputes in the nineteenth century—the era when imperialism peaked, and modern nationalism was born. Eventually, nationalist claims came to play dominant roles in Columbiana conflicts, spreading beyond the birthplace battles to the battles over bones and even into the study of European discoveries (with patriotic scholars arguing for the primacy of explorers from their own nations). That complex dynamic of nationalist and regionalist approaches to Columbus, including an ambiguous mix of both, has been most evident in Spain and Portugal. Spanish and Portuguese citizens have been primed for many generations to be receptive to a belief in a Columbus from somewhere in the Iberian Peninsula, repeatedly given in books, television shows, and online assertions such as "Cristóbal Colón died... leaving an enigmatic historical legacy regarding his true origin."[5]

When Columbus's two sons were children, he deposited them at court to be raised as royal pages—a privilege of nobility in the Iberian kingdoms. They grew up as Spaniards, founders of the Colón dynasty, never once setting eyes on their Genoese grandparents. It is therefore not surprising that the hispanization of the family would, almost from the start, extend into an imagining that Columbus himself was Spanish all along.

However, such claims have an internal tension to them: they seek both to make him Spanish and to root him in a particular region—usually one historically subjugated by Castile-centered Spain. Viewed as a pan-Iberian phenomenon, the regions that historically endured efforts to center the nation on Madrid are the autonomous communities of Catalonia and Galicia and the nation of Portugal. All three have used Columbus as a way to re-center the Iberian narrative on them. Micropatriotism is, by definition, both rebellious against the center and locally competitive. Thus, just as birthplace claims by Corsica and places in southern France competed with claims from northwest Italy (Genoa included), so have claims by Portugal and Galicia—adjacent, sibling regions of Iberia with a deep history of rivalry—stimulated each other in a to-and-fro that still flourishes today. As with the battle over Columbus's remains, advances in DNA technology have encouraged regional campaigns to open graves and publicize "the hidden truth." But nationalism also plays a role. Portuguese claims are both regionalist and nationalist, while Spanish claims that glorify the national and imperial past have often been forceful responses to regionalist claims (even if they are paradoxically sometimes tied up with them, especially those asserting that he was not a Genoese Christian but a Judeo-Spaniard).

EVA CANEL WAS A Spanish novelist, playwright, and journalist who spent much of her adult life in Cuba and South America. In 1913, she gave a lecture at the university in Lima on the origins or "cradle" of Columbus (*La Cuna de Colón*). Insisting on her impartiality and commitment to Italian-Spanish harmony in all things, she nonetheless leaned heavily on the same prejudice that underpins pseudoscience and pseudohistory in this century; that is, the belief that disagreement and debate among specialists (in Canel's case, nineteenth-century historians) is symptomatic of collective ignorance and bias, and that such specialists seek to protect their territory by denying and blocking truth-finding discoveries by those outside their closed shop.

Canel was persuaded by a young Argentine journalist and politician, Juan Solari, that the documents proving Columbus's Genoese origins had been forged by the Colombos in Cuccaro; that Columbus was really from a town in the mountains behind Genoa (where Solari also had ancestral roots); and that the leading Columbus historians of the day secretly doubted the Genoese "story." Thereby converted to belief in an unsolved mystery, Canel fell for the arguments made by Celso García de la Riega, a government official from Pontevedra whose passions were journalism and painting. "Don Celso"—who "had found in Galicia reliable data and evidence to prove that Columbus was Galician"—became the hero of Canel's own journey of discovery.[6]

Around the time of the Quadricentennial, García de la Riega had noticed that toponyms around his hometown were often similar or identical to those assigned to places in the Caribbean by Columbus. From that observation, he began to construct an argument that Columbus was born in Pontevedra, and that the Colón family were all Gallego. Facing skepticism, he boosted his claim in 1898 by "discovering" local documents that mentioned all three Colón brothers—Cristóbal included—as well as their mother. Canel took great pride in this outsider trumping the experts. She didn't directly compare him to Columbus, but the implication was there: like the navigator struggling to convince the sages at Salamanca, García de la Riega faced disbelief and mockery ("all doubted and many smiled"), until his discovery redeemed him.[7]

The documents were quickly certified as forgeries by Spanish paleography experts, and the Gallego claim, along with all non-Genoese claims, was systematically demolished by Spanish historians. This did little to dampen local belief in the veracity of their Columbus; after all, one would expect Spaniards from outside Galicia to deny and attack Gallego reality. The document scandal slowed but did not end the Colón Gallego campaign, as its enthusiasts took comfort in the belief that García de la Riega had merely "retouched" the documents as an act of clarification that did not render them "falsified" or "adulterated."[8]

They also found a simpler way to "discover" archival evidence. The surname "Colón" was not uncommon, certainly not unique to the Colombos who settled in Spain, yet any document mentioning a Colón in Galicia was taken as evidence. Such notions, combined with ongoing debate over García de la Riega's claims and rival Portuguese theories, fueled a steady flow of pro-Galicia articles and books for generations. Earlier writings tended to be blatantly micropatriotic (the "truth" was a "vindication of the Gallego people") and vitriolic (Columbus's Genoese origins were "a false and stupid legend"); in this century, Colón Gallego proponents are less strident, offering a more diplomatic enticement to visit a friendly region.[9]

Modern tourism is also compatible with the local belief, originating with García de la Riega, that Columbus gave Gallego names to places in the Americas—in a covert nod to his youth in Galicia. By discounting the facts that toponyms such as Punta Santa and Santa María were not rare outside Galicia, and that most place names given on Columbus's voyages lacked the Galicia connection, local believers can enjoy the resonance of a direct line between two places named, for example, Porto Santo. Arguments made by claimants from other regions were also appropriated by the Colón Gallego faithful. The lack of a Genoese birth certificate, for instance, and the fact that Columbus wrote only in Latin and Spanish (not in Italian), are easily and incontrovertibly explained (as we've seen). But explanations based on specialized knowledge have weak appeal for those who want to believe and who find simple discoveries—however false (no birth certificate!)—to reassuringly bolster such belief.[10]

"The question as to whether he belonged to one place or another is very important for a nation or a region," says Eduardo Esteban, the Asociación Colón Gallego's president, clarifying the significance of his campaign. "Columbus's achievements marked the beginning of the modern world," he explains. "It's like a debate over where Jesus was born." With stakes that high, Esteban is quick to throw shade on the Italians, accusing them of making a "deliberate mistake" in claiming that a Genoese man named Colombo became don Cristóbal Colón, the First Admiral. He goes easier on rival claimants within Spain, some of whom are collaborators in ongoing efforts to use DNA to prove that the real Columbus was from the Iberian Peninsula.

Nonetheless, "the Galician theory doesn't need a genetic analysis or DNA verification to be confirmed," he cleverly concedes, understanding that it is "Galician theory visibility" that ultimately matters. Whether DNA results confirm or disprove the Galician or any regional Spanish claim, or continue to be inconclusive (the most likely outcome), the Colón Gallego campaign has surely already succeeded: its ideas continue to spread virally online, mostly presented as facts, being especially prominent on tourist websites; and visitors to Galicia are less and less likely to miss local *lugares colombinos*—regardless of whether they take in Pedro Madruga's castle in Soutomaior or the Casa Museo Colón in Poio.[11]

"After many centuries no historical question has been treated with more passion and dishonesty than the problem of the birth of Christopher Columbus," wrote one Portuguese historian in 1937. Yet, despite a rational dismissal of ten non-Genoese birthplace claims, the historian still deemed the "question" to be a "problem," titling his essay "The Mystery of Columbus," and going on to trumpet arguments by his historian brother that Portuguese navigators discovered America before Columbus. There is the dispassionate appraisal of evidence, and then there are family ties and the pull of patriotism.[12]

Like Spanish claims, the Colombo Português tradition was encouraged by the fact of Columbus's ten-year residency in that country, as well as by regional rivalry. Portuguese and Galician claims were reactions to each other, starting when the Galician claim that Columbus was born in Pontevedra catalyzed the earliest rival Portuguese counterclaim. Patrocínio Ribeiro (1882–1923), a soldier, librarian, and clerk in Lisbon's municipal archives, heard Enrique Arribas present the Pontevedra claim in a Lisbon lecture hall. It seemed to Ribeiro that the Gallego arguments regarding toponyms in Galicia and the Caribbean applied just as well, if not more so, to Portugal—where there were also many places named after the same saints and geographic features. Furthermore, there was a village in southeast Portugal named Colos; Ribeiro decided it had to be Columbus's birthplace. He also fixated on Columbus's acronymic signature, concluding that it was a cryptic code signaling Columbus's true identity as a Portuguese prince—a member of the Zarco family and a grandson of King Duarte (who reigned 1433–38).[13]

With neither access to the full array of historical sources on Columbus's life nor training to interpret them, early proponents of the idea that Columbus was a coded mystery to be deciphered focused on a handful of examples of that S./S.A.S./XMY signature. No academic degree or title was needed, just intuition and inspiration. Colombo-as-Zarco was pure historical fiction (or, in Taviani's blunt judgment, an "absurd and unfounded claim"). But many Portuguese readers were happy to suspend their disbelief and accept the Zarco story as revealed truth.[14]

And who can blame them? Columbus's alternative Portuguese biographies are great fun—with their violent conspiracies, spy plots, cabalistic codes, disguises, illicit journeys between Portugal and Castile and Madeira and Genoa, and multiple discoveries of America. Ribeiro was followed by a former finance minister, known as Pestana Junior, and by three members of the Ferreira family, who all wrote Colombo-as-Zarco books. With the foundation of the tradition well laid, new books came every decade (but clustered in the 1920s–1950s and 1980s–1990s). Each cited previous books as authoritative sources, constructing the illusion of scholarly revelation—even while adding ever more fantastical details and claiming authorial authenticity by being outside academia's supposedly closed shop.[15]

Aside from the Portuguese names that were claimed to be Columbus's "real" ones—mostly but not all Zarco, with varying first names—the core features of the Colombo Português literature were those found elsewhere in Iberia or in Europe. For example, Columbus was typically imagined as a nobleman, closely related to royalty, and often a secret agent working for the Portuguese Crown. He was often a covert Jew. The eccentricities of his Castilian were taken as proof that he was Portuguese by birth (or Galician or Catalan). Mythistorical figures from rival claims—such as the Danish pilot Joannes Scolvus—were appropriated.

The Portuguese tradition was carried through the Quincentennial and into this century by Portuguese anthropologist Augusto Mascarenhas Barreto (1923–2017) and by Azorean-American historian Manuel Rosa. Barreto reached for popular audiences while claiming scholarly credibility—his *"Colombo" Português* is a two-volume, 830-page set of Colombo-as-Zarco documents. Historians pointed out that Barreto's alternative reality was "absurd," based "more on passion and faith than on a scientific basis." Overwhelming audiences with irrelevant and invented details, he constructed arguments "with a complete lack of concern for proof or historical methodology." Yet Barreto "gathered about him a considerable following among a lay, patriotic public." Rosa—whose books have titles such as *O Mistério Colombo Revelado* (The mystery of Columbus revealed)—likewise claims his many decades of research support the "secret agent" story. He is less willing than was Barreto to accuse the Genoese (and professional historians worldwide) of conspiratorial fraud, arguing that there were two Columbuses: a Genoese one, who discovered nothing; and a Portuguese one, who discovered America. Simply imagine that this book's Life One and Life Two recount the history of two different men. For readers—especially Portuguese ones—unfamiliar with the larger mass of historical evidence, and willing to accept the assertion that Columbus is a complex mystery, that solution has an obvious Occam's-razor appeal.[16]

The years surrounding the Quadricentennial also saw the rise of a new Catalan micropatriotism—or nationalism, as "nation" or *patria* was appropriated by the Catalan patriots, in counterdistinction to the "state" that was centered on Madrid—articulated in such writings as

Enric Prat de la Riba's landmark manifesto of 1906, *Catalan Nationality*. Catalan nationalists such as Prat generated a somewhat paradoxical position on Columbus. On the one hand, Spain's Golden Age—from Columbus's First Voyage to the Spanish-American War of 1898—was turned upside down and seen as a dark age of Castilian tyranny. On the other hand, Columbus was also appropriated by Catalans, with Prat and others insisting that it was King Fernando that financed that voyage; and it was not Castile, but Aragon and Catalonia's maritime merchant empire that underpinned transatlantic expansion. Atop the 131-foot column erected in Barcelona in 1888 stands a 24-foot bronze Columbus. His arm is raised, pointing neither toward Genoa nor toward the Americas, but into the Mediterranean, where the might of Aragon and Catalonia made his voyages possible.[17]

Pointed: A bronze Columbus atop a 131-foot column in Barcelona. The Mirador de Colom was designed by the Catalan Gaietà Buigas i Monravà and built in 1882–88 to coincide with the Exposición Universal de Barcelona—Spain's first World's Fair, anticipating the Esposizione Italo-Americana in Genoa and the World's Columbian Exposition in Chicago by four and five years, respectively.

Catalan claims to Columbus, coming in the wake of the Columbus competition between Galicia and Portugal, tended to deploy the same arguments. Book titles announced the revelation inside; Luís Ulloa's 1927 *Cristòfor Colom Fou Català* is a good example. Archives were scoured for anyone named Colón or Colom, their existence "proof" of the admiral's local origins. Columbus wrote poor Castilian because he was a Catalan. Some, like Ulloa, even latched onto Scolvus, making him a Dane turned Catalan turned Colón the Discoverer. The fact that Columbus announced the findings of his First Voyage to the monarchs in Barcelona, with his 1493 letter to Santángel published in Catalan, was used as further evidence (the monarchs happened to be in that city at the time, and the letter was published in multiple languages).[18]

As Catalan claims proliferated and multiplied, they birthed microregional claims. These focused on other Aragonese provinces, including Mediterranean islands such as Majorca and Ibiza—situated between Aragon and Genoa, part of a self-governing vassal kingdom subject to Aragon. A typical such variant made Columbus the son of a Majorcan prince, and thus a nephew to monarchs Fernando and Isabel. The Majorcan claim (or *La tesis mallorquina de Cristóbal Colón*, as one of its many books was titled) originated in the 1960s and is still going strong, its elements taken from the older Catalan tradition and that of other Iberian regions. Gabriel Verd, who calls Columbus's Genoese history "a cock-and-bull story," started promoting the Majorcan tradition in the 1980s; his Columbus is the lovechild (from illicit sex in an abandoned castle) of a Spanish prince and a local Jewish girl. Having heard as a teenager that Columbus was from the Majorcan town of Felanitx, Verd, now president of a local Asociación Cultural Cristóbal Colón, has spent a lifetime trying to prove it. Lorente's DNA project gave him—like many other regionalist claimants—renewed hope. He exhibits the resentment-tinged passion of the outsider. "In Spain, if you don't have a degree, they don't pay you much attention," he says, having worked in tourism and hotel public relations. "Even if you have proof, as I do."[19]

AS THE QUINCENTENNIAL APPROACHED, plans to commemorate it in Spain and the United States did not go smoothly. The Jubilee Commission in the United States was so divided between its Italian American and Hispanic/Latino members that a rival organization emerged—the National Hispanic Quincentennial Commission. People of Spanish descent on both sides of the Atlantic were far from united on the issue of Columbus. Still, most agreed that the "Genoese genius" emphasis was too much—as one Spanish historian complained, pointing across a 1988 Santo Domingo conference room at Paolo Taviani. Without the Pinzón brothers and all the other Spaniards who made the Discovery possible, Columbus "would have ended up as shark food, hung from the back of his ship."[20]

Amidst such squabbling, the Spanish embassy in Washington sent a copy of a Columbus biography to every member of the Jubilee Commission. From a vast pile of possibilities, they chose one that was a half-century old, but it was by a Spaniard—and it argued that Columbus had himself been a Spaniard. Its author, Salvador de Madariaga (1886–1978), was a diplomat (including a brief stint as Spanish ambassador in Washington) and a government minister before the Spanish Civil War, subsequently becoming an exiled man of letters in Oxford. The choice of *Christopher Columbus, Being the Life of the Very Magnificent Lord Don Cristobal Colón* was thus an obvious and resonant political act, considering the post-Franco timing of the Quincentennial, the nature of Madariaga's career, and his claiming of Columbus for Spain. But it also reflected—and helped to perpetuate—a more complex tradition within Columbiana.

Spain's Columbiana tradition comprised two intertwined threads—one nationalist, one regionalist. The nationalist one was fed from across the Atlantic, with books by U.S. historians such as Washington Irving (1783–1859) and William Prescott (1796–1859) promoting a highly nationalist narrative of a period when Spain did not yet exist as a nation-state—one that mattered even more in the wake of the 1820s loss to Madrid of the vast swathe of mainland Spanish American colonies. The late-nineteenth-century explosion of regionalist claims on Columbus further stimulated nationalist Spanish claims. In the 1880s, the conservative government of Antonio Cánovas del Castillo attempted to persuade all Hispanic nations to adopt October 12 as an international

Columbus holiday. That initiative failed, but the nations of the Americas did gradually adopt their own versions of Columbus Day, while Spain itself marked the day as the Día de la Hispanidad from the 1920s until after the 1975 death of dictator Francisco Franco.[21]

The Civil War (1936–39) and Franco dictatorship (1939–75) "indelibly marked the study of the Americas"—Columbus included. Histories of Spain's past tended to have an epic, nationalistic tone. Columbus could be appraised critically, but not in a way that opened the door to criticizing the newly idolized *Reyes Católicos*, the Catholic monarchs Fernando and Isabel (she headed, some hoped, for official sainthood). Madariaga's *Life of the Very Magnificent Lord*, written during the war, came out within months of the war's conclusion; an immediate hit, it sold well for decades in multiple editions and languages.[22]

The book succeeded because Madariaga managed to have his cake and eat it. He mixed moments of sound scholarship based on evidence with wild speculations and pure inventions. He appealed to the general reader by claiming to privilege "the real lives of men and women" over "dusty documents," while also showing more discerning readers that he knew how to read such documents. The book seemed erudite, yet its writing was engaging and often flighty. Madariaga argued that Columbus was really a Spaniard—although not a Gallego, despite the fact that Madariaga himself grew up in Galicia (a small Madariaga statue stands in La Coruña, but he was an anti-Fascist nationalist, not a regionalist). He avoided the problem of the evidence of Columbus's Genoese youth by asserting that "the Colombo family were Spanish Jews settled in Genoa," continuing to speak Castilian and nurture their Judaism. Avoiding regionalist rant, the book appealed to Spanish patriots of varying political stripes. Its nationalism was relatively subtle (Madariaga was, after all, a republican exiled in England), but it was easily read as a hispanocentric treatise *contra* Morison's biography (published the very same year) of a Genoese Columbus.[23]

But the oddest cake-and-eat-it feature of Madariaga's book was his claim that Columbus was a crypto-Jew, secretly hiding his Hebraic ancestry. There is no case to be made, merely a repeated assertion delivered "in default—and sometimes defiance—of evidence" (as Fernández-

Armesto noted). It was not the book's core argument, as it would be to later books with titles such as *Christopher Columbus's Jewish Roots*, even if covert Jewishness was the important and imaginary plot device—used by many before and after Madariaga—that explains why Columbus, as the argument goes, only *pretends* to be Genoese, and half-heartedly at that. Madariaga invented neither the Spanish Columbus nor the Jewish Columbus, let alone their pairing; decades earlier, Columbus had been made a Gallego of Jewish origins, and a member of an old Extremadura converso family. The odd part, then, is less the claim itself, but this: Madariaga rests his argument on a series of blatant and blunt stereotypes about Jews.[24]

There were large and prosperous Genoese communities in Lisbon and Seville when Columbus lived there, some of whose members were transplants, some second or third generation or more. But for Madariaga, Columbus's arrival and departure was done "with typically Jewish adaptability" and "typically Jewish mobility." Indeed, there was "a hint of Jewishness" in the "extreme mobility" of both Cristóbal and Bartolomé, confirmed by "their special indifference to Genoa." Columbus's piety was made to work both ways: lapses were proof of Judaism; expressions of it were attempts to hide Judaism. Even Las Casas is somehow part of this conspiracy, his references to "the great devotion" of the First Admiral were cover-up attempts by "the man who knew him best" (there is no evidence they ever met). It gets worse. If Columbus "was as good a Christian as he was keen to show himself in his devotions, why was he so fond of the company of Moors and Jews, and why had he become entangled in a love-affair with a Córdoban girl, of whom he had a bastard"? Not only was it clear that Columbus was therefore Jewish, but so was the "Córdoban girl" Beatriz Enríquez, as "the sexual morality of Jews was of course different from that of the Christians." As for gold, Columbus "cannot resist" its "glittering" attraction, "so typically Jewish" (the origin, he notes, of "the curious subconscious pun on *Jew* and *jewelry*").[25]

Madariaga's book was therefore pro-Jewish and anti-Semitic all at the same time. Although his argument rested on prejudicial anti-Semitic stereotypes, and he asserted that "anti-semitism was always a democratic, and pro-semitism an aristocratic, attitude in Spain," the *goal* of his argument, to promote Columbus as a Spanish hero, was not anti-Semitic. Furthermore, the prominent British Jewish historian Cecil Roth (1899–1970), who taught at Oxford for most of the decades when Madariaga lived there, was immediately taken by the theory—even if he noted that Madariaga tended "to exceed the boundaries of historical license," and months before his death Roth came close to a retraction, declaring it "impossible to exclude or to confirm the hypothesis that [Columbus] was descended from a Jewish or ex-Jewish family."[26]

Madariaga's Jewish claim was thoroughly debunked by historians—including Spaniards such as Ballesteros, who as early as 1945 explained why the so-called theory was *una pura fantasia*. Yet it persisted as local legend, fed by micropatriotic impulses and the publicity given to other Iberian claims, while also serving to popularize the idea in and outside Spain. By the same token, the claim helped Madariaga's *Life of the Very Magnificent Lord* to sell well in multiple languages for decades, including in the United States, where it became a classic long before the Spanish embassy gave copies to the Jubilee Commission.[27]

The notion that Columbus was Jewish had begun to surface in the United States in the late nineteenth century, with the peak of such claims being the particularly anti-Semitic decades of the late nineteenth to mid-twentieth centuries. It was thus gratifying for Jews on both sides of the Atlantic to believe that a man revered as a foundational, civilizational hero—for whom there was a sustained international campaign during the same period to canonize him as a saint—had been a covert Jewish survivor during another anti-Semitic age, outwitting the Spanish Inquisition to achieve fame, wealth, and social status. The fact that Jews were expelled from Spain in the same year that Columbus began his First Voyage was seen as meaningfully symbolic, with a Jewish Columbus personifying "the transference of Jewish cultural power directly from Sephardic Spain to English-speaking America" (as a scholar of Roth recently put it). Just as Italian immigrants in the United States had leveraged his memory into supporting their claims to being American, so did American Jews use a Jewish Columbus to give deep historical reinforcement to their claims of New World belonging.[28]

During 1892's Quadricentennial celebrations, New York City's chief rabbi, Jacob Joseph, published a special prayer in Hebrew that credited Columbus for the two Jews that supposedly accompanied him on the First Voyage and for America's subsequent flourishing as a safe Jewish haven. He stopped short of claiming that the First Admiral was Jewish himself, as did Budapest's Rabbi Kayserling (whose work was sponsored by the American Jewish Straus family). But the rabbis' argument that Jewish participation in Columbus's voyages was crucial to subsequent religious toleration and democracy was influential on both sides of the Atlantic. Thus, while García de la Riega's core claim that Columbus was Galician had limited impact outside Galicia and Portugal, its secondary claim that the Colón Gallego was secretly Jewish had a broader impact. By 1914, a Rear Admiral Baird of the U.S. Navy could write to the superintendent of the U.S. Capitol's building and grounds, praising its artworks, especially sculptures of Columbus—"a Jew" and "a great and nervy man with the courage of his convictions: the other side of his life was that of a kidnapper, smuggler, freebooter etc. as many another great man has been."[29]

In subsequent decades, the anti-immigration movement and accompanying racist sentiment in the United States, combined with anti-Semitism and the Holocaust in Europe, all served to draw increasing attention to the notion of a Jewish Columbus—and provided Madariaga's book with fertile ground for the reception of both his intertwined Jewish and Spanish Columbus arguments. Prominent proponents of the Jewish theory included Holocaust survivor and Nazi hunter Simon Wiesenthal (1908–2005), whose 1968 book arguing the case was a best seller. Wiesenthal borrowed most of Madariaga's made-up facts, adding a new one: his Columbus was not only secretly Jewish, but motivated by his Judaism to find a New World where his fellow Jews could find refuge; furthermore, that world wasn't new to Jews, having been discovered by the Lost Ten Tribes of Israel (a Jewish twist to the double-discovery trope found in regionalist claims). Taviani had famously dismissed Madariaga's biography as "a novel," and indeed Wiesenthal's book likewise read more like historical fiction than anything else—despite the back-cover claim that it was "COMPLETELY FACTUAL."[30]

The old myth that "history silences or deliberately falsifies the keys we need to solve" Columbiana's "mysteries" gave the Jewish Columbus momentum through the twentieth and into the twenty-first centuries. It was also propelled by confirmation bias, the repetition of a series of contrivances: Columbus's reluctance to discuss his Genoese origins hid his Judaic roots (it was in fact "the self-defense of an insecure social climber"); his preference for Jewish bankers reflected his own Jewishness (most merchants, including all Europe's royal houses, were clients of Jewish or ex-Jewish bankers); he wrote mostly in Spanish, never in Italian (as already explained, his first language was Ligurian, not Tuscan-derived Italian, and he did not become literate until settling in Portugal, only starting to write in Castilian or Spanish after several years living in Spain); the Portuguese elements in his Spanish were elements of Ladino or Judeo-Spanish (they were, obviously, Portuguese); Columbus taking a thirteen-year-old Fernando on his Fourth Voyage was a giveaway, as that is when Jewish boys come of age (a reasoning unworthy of even the most generous of responses); his millenarianism was Judaic in nature, and his pious tendencies were a cover (by that logic, the entire Franciscan order, and much of the Catholic Church, was secretly Jewish; Columbus's increasing immersion in the Old Testament in his final years, which exposed him to its Hebraic thinking, made him more of a Christian and ideologically closer to the Franciscans—not the crypto-Jews); his acronymic signature was a crypto-Jewish code left for us to crack (such signatures were a medieval Christian tradition, designed not to hide covert Jewish or Muslim identities, but to highlight unambiguous Christian identity and high rank conferred by the Christian God and His monarchs); and finally, Columbus looked Jewish (aside from this being as obviously problematic as the claims that he looked Norwegian or Greek, descriptions of him vary, and—as already mentioned—not a single extant portrait of him was made from life).[31]

The blurring of the line between history and fiction, central to Columbiana, is particularly blatant in the Jewish Columbus tradition. For example, the Jewish Columbus of Abel Posse's *Dogs of Paradise* is a web-footed mystic more imaginatively fictional than historic, while the equivalent figure in Stephen Marlowe's *Memoirs of Christopher Colum-*

bus is drawn heavily from Madariaga; his Jewish mother wants to name him Moses Maimonides Colón, settling on "Cristóbal" out of fear, her baby born on a ship bound for Genoa as the family flees anti-Semitic persecution in Catalonia. Unsuspecting readers could easily read this as a based-on-a-true-story novel, "history at its swab-decking, swashbuckling, bawdy best" (as a *Miami Herald* review gushed). Readers might similarly take Edward Kritzler's *Jewish Pirates of the Caribbean* as a work of history (as it purports to be) that happens to be written like a novel; to privilege plot, Kritzler "forfeits argumentation," as one literature scholar put it, "and foregrounds emotion." A final example is *The Columbus Affair*, a historical novel that clearly draws on Wiesenthal and Kritzler. Steve Berry's Columbus is duplicitous and violent, but he is nonetheless a Jewish hero. Like Kritzler, Berry focuses on Jamaica, leveraging its early colonial history as a Colón dynasty possession into a site where crypto-Jews, surviving Taínos, and escaped enslaved Africans (Maroons) find common ground as defiant resisters of colonialism. Berry uses his novelist's license to sidestep or invert Columbus's controversial nature today, even while leaning on that canard, "what is clear is that virtually nothing is known of Columbus."[32]

The late-twentieth-century shift in the perception of Columbus, with his historical life buried beneath the battleground over what he represents today, partially slowed the momentum of Jewish Columbus claims. A book like *Columbus's Jewish Roots*—published a century after Kayserling began to publicize his arguments, and thus in a position to repeat all the claims from García de la Riega to Madariaga to Wiesenthal—nonetheless came to a carefully measured conclusion: while "Columbus opened a New World to which the Jews could flee from the religious persecution of the Old," nonetheless "his Jewish roots are yet to be established with certainty." In fact, they had not been established at all, and nor have they yet, despite periodic efforts to deploy DNA "proof." Yet those roots, as Columbiana myth, are deep. Scattered across the decades, writers and scholars alike have tossed off lines such as "the widely accepted hypothesis that Columbus was of Jewish birth" and arguments for a Judeo-Spanish Columbus "strongly point to the conclusion that Columbus was of Jewish descent." Most recently, even a usually careful writer such as Marie Arana can casually mention that "very likely, the forebears of Christopher Columbus were Jews."[33]

The Jewish claim was always secondary to the regionalist and nationalist claims in Spain. And yet it is those same claimants who gave the Jewish Columbus life. In this century, it may be the Jewish Columbus who keeps the Iberian Columbuses alive—on both sides of the Atlantic.

"FIVE HUNDRED YEARS: FIVE Hundred Programs," trumpeted official publications in the buildup to the Quincentennial celebrations in Spain, the nation that devoted by far the greatest effort and resources into 1992 (spending more money "than probably all other nations combined"). All seventeen autonomous regions and all fifty provinces, as well as numerous town councils and city halls, museums and universities, and institutions and organizations of every kind, laid on events and programs. With regionalist claims on Columbus now a century-old tradition, 1992 was surely a boon to those in Iberian regions seeking to seize him for themselves, once and for all?[34]

Not quite. First, such programs were seldom about Columbus, let alone focused on his regional connections. Although the remnants of Fernando Colón's library, the Biblioteca Colombiana in Seville, received funds, it was just one of sixty archaeological restoration projects, half in Spain and half in Latin America (at sites such as Tikal, the ancient Maya city in Guatemala). Columbus documents were a drop in the ocean of manuscripts to be digitized in the Archivo General de las Indias (the archives of Spain's imperial centuries), a project that still continues at a snail's pace today. The "floating university" of thousands of students sailing between Spain and Latin America, the hundreds of fellowships supporting U.S. high school teachers studying in Toledo, and the local cultural enrichment programs that reached hundreds of thousands of Spanish schoolchildren were all about Spain and Latin America—and their shared national culture and history.

Second, local programs were overshadowed by three massive national endeavors: the Olympic Games in Barcelona; Seville's Universal Exposition, or Expo '92; and Madrid's Cultural Capital of Europe programs. The billions spent on all three used the Quincentennial as a pretext for a massive *national* project to promote Spain in the world. The timing was perfect. After decades of isolation under Franco, the country managed a transition to democracy, unfolding between the dictator's death in 1975 and King Juan Carlos I's dramatic facing down of a Guardia Civil revolt in 1981. That year, the first of a nexus of Quincentennial national commissions was created. The following year, the Socialist Party won the elections, and its leader Felipe González presided over Spain's campaign to rejoin Europe (joining NATO in 1982 and the European Union in 1986). The massive infrastructure projects for 1992—from the renovated airports to the Seville-Madrid high-speed train—were part of transforming Spain into a modern European nation.

They were also designed to bring Europe to Spain. For example, more than fifteen million people came to Expo '92 (including multiple attendances, some forty-two million visits). If regional Columbus campaigns are effectively *local* tourism drives—as we've seen in places ranging from Galicia to Mediterranean and Atlantic islands—then the Quincentennial was a massive *national* campaign to boost tourism. Did it work? Although many factors explain the tourism boom, Spain has seen a steady increase in visitors since the 1970s, especially in and since 1992. By the 2020s, it was the second-most visited nation on Earth, annually bringing in 100 million visitors and more than 100 billion euros in tourist revenue. The Spanish public came to see the Quincentennial as an excessive, corrupt governmental boondoggle. But there is no doubt that, as a leveraging of 1992 into a national infrastructure and tourism project, it was a huge success. Furthermore, as much as Columbus was used at the time as mere pretext, the core vision of him as an Iberian icon—national or regional—was confirmed.

As a result, a contrast emerged between the Columbuses in Iberia and in the Americas. Spain's main 1992 event in the United States, the Texas-to-New-York tour of replicas of the caravels from the First Voyage, was a weak echo of a similar event in 1892—making Spain's vision of the Quincentennial seem as old-fashioned as its U.S. embassy's gift of a Columbus biography from 1940. And while the president of Spain's commission insisted that "all of Ibero-America is as much a protagonist of that date as is Spain," tone-deaf Spanish officials were oblivious to the debates in the Americas over Indigenous rights and the use of "discovery" as a term that symbolized blindly neocolonialist views.[35]

The invader-invaded dichotomy may strike us as stunningly obvious, just as it was predictable that most Latin Americans would see a *celebration* of Columbus as an offensive neocolonial public-relations exercise by Spanish conservatives. The Dominican Republic was the rule-proving exception. With its unique claims to being the first Spanish colony in the Americas and the holder of the bones of the Discoverer, the republic was the only one among the dozen or so Latin American nations that formed Quincentennial commissions to embrace Spain's language of "celebration" and "discovery." It even included "evangelization" in its commission's title—the tone having been set by Pope John Paul II's 1984 visit to the nation on October 12.

In contrast, Mexicans took issue with *descubrimiento* (discovery). Their commission head, Miguel León-Portilla, pushed for *Encuentro de Dos Mundos* (Encounter of Two Worlds), adopted in 1985 in the titles both of Mexico's and the Organization of American States' commissions and programs. Not only was this in contrast to Spain's (and the Dominican Republic's) use of terms, it was rumored that when King Juan Carlos visited Mexico in 1987, he complained about the highly *indigenista* (pro-Indigenous) stance of the Mexican commission. León-Portilla resigned; his successor, the more conservative Leopoldo Zea, tried to impose a compromise term, *encubrimiento* (covering over); *his* successor, Enrique Florescano, argued that *indigenismo* had caused a national identity crisis. Meanwhile, influential Mexican novelist Carlos Fuentes proclaimed the Quincentennial to be the perfect pretext for celebrating Spain's cultural significance in the New World.[36]

The apparent triumph of Spain's vision of the Quincentennial killed enthusiasm for it in Mexico. Official plans stalled, while incensed younger Mexicans marched in protest on October 12, attacking statues of Spanish colonists. In the capital, stones and eggs and other projectiles were thrown at the Columbus monument, which marchers tried to pull down with ropes attached to a public bus. They failed but tried again on the same day the following year. "We are here to repudiate the celebration of the so-called Encounter of Two Worlds," declared a student protester.

Such attacks continued on and off for decades. When Columbus statues in the United States came under attack, the monument to him in Mexico City was finally removed—this time by the government, who took it down on October 10, 2020, two days before another threatened attack on it by protestors. Would it be restored or relocated or destroyed? Years of political squabbles ensued. Protesters meanwhile placed their own statue on the plinth, calling it an *antimonumenta*, dedicated to *Las Mujeres que Lucha* (Women who Fight). In 2023, the city's mayor, Claudia Sheinbaum (elected president of Mexico the following year), unveiled the replacement, a stark modern rendering of an Aztec-era statue of an Indigenous woman, dubbed *La Joven de Amajac* (The Young Woman of Amajac). Nearby bus stops were renamed accordingly. As debate over the sculpture's artistic and political worth smoldered, the ghost of Columbus hovered over the plinth where for a century and a half he had been admired and then reviled.[37]

The Mexican story throws the Spanish one into relief. The same year that Columbus was removed in Mexico City, hundreds took to Barcelona's streets to protest the city's famous Columbus statue as a monument to racism. In 1888, it had been erected as a symbol of regional historic identity. A century later, the sculpted Columbus stared out over the 1.5 million attendees to an Olympic Games whose hosting had clearly (if not officially) been awarded to the city because of its Quincentennial connection. Another thirty years later, he had become something of an embarrassment, a symbol of colonial oppression, whom the Candidatura d'Unitat Popular (a *Catalunya* independence party) repeatedly demanded be pulled off his high pedestal. Has Columbus's great height above Barcelona's streets so far spared him being vandalized or toppled? Perhaps. But Madariaga must be chuckling in his grave, knowing the claim that Columbus was really a Catalan—which Madariaga had not invented, but made internationally popular—along with all the other nationalist and regionalist claims on an Iberian Columbus, from Majorca to Galicia, would surely outlive monuments like the one in Barcelona, and might even save them.[38]

Capitol Columbuses: A small Columbus-with-globe (on the left, outlined by a treetop) is at the back of Horatio Greenough's *George Washington* (removed in 1964). In the distance, to the left of the steps to the U.S. Capitol's *Columbus Doors* (still there), sits the large Columbus-with-globe of Luigi Persico's *Discovery of America* (removed in 1958). This albumen print is an 1863 Andrew Russell photograph.

LIFE EIGHT

THE ADAM

WASHINGTON WAS THE NATION'S FATHER, COLUMBUS ITS GRANDfather. The capital city of the new United States of America was named after one, its surrounding district named after the other. Visitors to Washington, D.C.—especially from the mid-nineteenth to late twentieth centuries—would have been struck by the prominence of the Genoese-mariner-turned-American-hero. Had they arrived by train after 1912, they would have stepped out of Union Station to see the monumental *Columbus Fountain*, centered on a fifteen-foot sculpture of the man "WHOSE HIGH FAITH AND INDOMITABLE COURAGE GAVE TO MANKIND A NEW WORLD"—as its dedication declared. That "New World" was symbolized by the smaller sculpted figure of a fearful, naked "Indian maiden." The marble Columbus gazed—as he still does today—toward the seat of Congress, the U.S. Capitol, a third of a mile away.

Prior to 1964, visitors approaching the Capitol's east central entrance would have walked past a large sculpture of a seated George Washington staring at the building. Observant visitors might have noticed two small figures that the artist, Horatio Greenough, had added to the back of Washington's throne. One was a stereotypical melancholy "Indian" (meditating on the imminent extinction of his race, a common theme in the Capitol's art), the other a long-bearded Columbus, looking like a biblical prophet but identifiable by the small globe he was holding and contemplating (musing upon how his deeds led to the structure at which Washington stares and gestures).[1]

Walking on from *George Washington* toward the Capitol, visitors would have encountered (until 1958) an oversized Columbus beside the entrance—bearded and armored and poised dramatically on the cheek block to the left of the stairs, holding what looked like a large ball. At the stairs, visitors would realize the ball was another globe, just as they'd see that the cowering figure beside Columbus was another naked woman representing "the Indians." (We shall return later to this *Discovery of America*.)

Entering the Capitol building, visitors were again presented with Columbus. Echoing his "entry" into America, visitors to the Capitol entered the building through bronze doors, each nineteen feet high and nine feet wide, depicting nine scenes from Columbus's life. Created in Rome by an American artist named Randolph Rogers and installed inside the Capitol in 1863 (moved to the east façade, to create that symbolic entrance, in 1871), the doors told a version of Columbus's life that would have been familiar to American visitors—an exemplary triumph over adversity, from the mythologized rejection in Salamanca to the First Landing to the triumphal entry into Barcelona to his arrest in the Caribbean and return to Spain in chains. The doors were framed and trimmed with allegorical representations of the four continents, as well as with sixteen small statues of historical figures of the day—from European monarchs such as Henry VIII and Queen Isabel to conquistadors such as Cortés and Pizarro to Columbian protagonists such as brother Bartolomeo (a likeness of Rogers himself) to imaginary romantic interest Beatriz de Bobadilla (a likeness of Rogers's wife). The impression given was that Columbus's life placed him at the very center of history—of the history that gave birth to the United States.[2]

From the 1870s to 1960s, visitors passing through Rogers's *Columbus Doors* would step almost immediately into the great Rotunda. Taking in its gorgeous dome one hundred eighty feet above the floor and its

Arrival: John Vanderlyn's (1775–1852) 12 × 18 foot *Landing of Columbus*, completed for the U.S. Capitol Rotunda in 1847, has been viewed by many millions of visitors, also serving as the basis for numerous representations of the First Landing—including on the 2-cent stamp in the Quadricentennial and Quincentennial series.

art-crowded curved walls, visitors would have seen among the massive mural paintings a depiction of the *Landing of Columbus*. The pose of the explorer—bearded, a sword in one hand and a territory-claiming banner in the other—was like that of the marble Columbus that visitors passed on the steps outside. John Vanderlyn had won the "discovery of America" commission in 1837, delivering it ten years later, drawing his inspiration from Washington Irving's best-selling *History of the Life and Voyages of Christopher Columbus*. The image is profoundly and militantly imperialist, Columbus and his men bringing the light of civilization to the "tawny savages" (as the Guide to the Capitol's art told nineteenth-century visitors). Following Irving, and again echoing "the Indian maiden" outside the building's entrance, "the Indians" are naked and afraid, cowering in the darkness, terrified but awed by these "beings of a higher order" (in the Guide's phrase).³

It was often (and still is) criticized as a work of art. In 1892, the Cuban American author Néstor Ponce de León deemed it tolerable as a creative work but "historically it is worthless." He was right, but Vanderlyn was not aiming for historical accuracy, seeking rather to capture the symbolism of the American Columbus, to give him immortality in the Capitol. In that, he succeeded. When U.S. vice president Levi Morton opened Chicago's World Columbian Exposition in 1892, he specifically cited Vanderlyn's *Landing of Columbus*, its hero inspiring "the homage of mankind." On Columbus Day seventy years later, a newspaper article praising the painting for showing Americans "why we honor Columbus by setting aside a special day for him" typified popular opinion.[4]

Above the paintings such as Vanderlyn's are panels, into which are carved small bust portraits of a selection of early explorers. Although created by Italian artists, the two Genoese explorers are identified by their English names—Christopher Columbus and John Cabot—thereby contributing to the illusion of their places in the pantheon of Anglo-American founding fathers. And above the panels, encircling the base of the dome almost sixty feet from the ground, is yet more artwork that includes Columbus. Constantino Brumidi's *Frieze of American History* is in grisaille (painted to look like it is sculpted in stone), presenting the nation's history in a series of scenes, beginning with the *Landing of Columbus*. Although Brumidi's Columbus is less of a conqueror than Vanderlyn's, and the Indigenous people are more primitive than savage, more welcoming than fearful, the Discoverer is still imperious and imperial, a figure commanding obeisance, one familiar to Capitol visitors.[5]

Welcome: Another "Landing of Columbus" scene in the U.S. Capitol Rotunda, this one part of Constantino Brumidi's (1878–1953) *Frieze of American History*.

Security measures in today's Capitol deny visitors the luxury of lingering long before the art, let alone the freedom to wander the building's corridors to see the hundreds of other paintings and sculptures. But for most of the building's life, visitors wandered and thereby encountered more renditions of Columbus—including two others by Brumidi in the Senate wing, one of them *Columbus and the Indian Maiden*, yet another interpretation of the Discovery as gendered. For past visitors, Columbus was simply everywhere. Even today, with the east entrance permanently closed, the *Discovery of America* gone, and Rogers's *Columbus Doors* visible only with some difficulty from inside the building, the presence of Columbus is a powerful one. How did the nation's Capitol become a Columbian shrine?[6]

AS INESCAPABLE AS HE became as a U.S. nationalist icon, Columbus played no such role in the English-speaking world for centuries after his death. Certainly, he was praised, even adulated, in histories of the sixteenth through early eighteenth centuries—from those translated into English, such as the *Decades* of Pietro Martire (known as Peter Martyr in England, where he was read starting in 1555), to those written by Englishmen such as Richard Hakluyt and Samuel Purchase. But the main concern of English writers was religion, as it was to Anglo-Americans such as Cotton Mather, who in his *Magnalia Christi Americana* of 1702 took a simple Lascasian position on Columbus: he was part of God's long-term plan to make the Americas Christian (and, in Mather's version, English). Other Anglo-American writers of early histories of the colonies gave Columbus similarly positive but brief coverage. Even those for whom his Catholicism and connection to Spain was too unfortunate to be completely excused nonetheless sympathized with the ingratitude the Discoverer suffered from both "the Heathens" and "the *Spaniards*" (to quote the New Jersey judge who published as Sylvanus Americanus).[7]

A slight but crucial shift occurred at the very dawn of the war that led to U.S. independence. Three books published around the same time (1776, 1777, and 1778) were popular in the new nation, and they would prove to be crucial influences on the book that would replace them as *the* Columbus biography of the nineteenth century, Washington Irving's *History of the Life and Voyages of Christopher Columbus*. One was by a French Jesuit, the Abbé Raynal, the others by Scotsmen named William (both dying in 1793)—the Presbyterian minister Robertson and the writer Russell. These books helped lay the foundation for an American Columbus in two key ways. First, Columbus's emotional life and mental state were fully imagined, turning him into a flesh-and-blood hero, allowing future writers to make the Discoverer believable and real—even though they were repeating invention as fact, myth as history. Second, these three books borrowed a trick from Las Casas and created a dramatic, moral separation of Columbus from the Spaniards.[8]

Russell's Columbus is more flawed than Raynal's and Robertson's—he is far from the St. Christopher of later invention. But the legendary tales of his genius, his fortitude and "composure," and his providential destiny were all included, waiting to be further developed by Irving. Robertson's treatment of Columbus was much longer than Russell's, published in time for Russell to draw upon (and even paraphrase) it, but more unabashedly hagiographic. By leaning heavily into the stereotype of a noble hero, underappreciated in his time but guided by Divine Providence, Robertson strengthened the foundation laid by Las Casas for the construction of a saintlike Columbus. Columbus was a man of "perseverance, spirit, and courage," of "genius and industry" (Raynal); he was "courteous in his deportment" and "irreproachable in his morals" (Robertson), the very picture, anachronistically, of a "respectable" gentleman of the eighteenth century's age of civilized (that is, non-Spanish) empires. By contrast, the Spaniards were "treacherous," ungrateful, and violent, barely less savage than the "Heathen Indians," the court of their capricious monarchs "haughty and turbulent."

Russell and Robertson were scathing about almost all Spaniards, from monarchs ("ungrateful") to conquistadors ("avaricious" and "violent") to colonial officials ("jealous" and "cruel"). This anti-Spanish tradition among Protestant writers, rooted in the late sixteenth century and to be dubbed "The Black Legend" in the very early twentieth century, runs thick in Robertson's and Russell's accounts, and would also be deployed by writers in the nineteenth-century United States such as Irving and William Prescott. War between Protestant Anglophone nations (Britain and the United States) and Catholic Hispanophone ones (Spain and Mexico) helped nurture Black Legend depictions of the age of Columbus and the conquistadors. One such war (that of 1775–83) began while Russell and Robertson wrote, others were the Mexican-American (1846–48) and Spanish-American (1898). Black Legend tradition created distance between Columbus and Spain—a nation, empire, and even "accursed race" that could easily be depicted as ungratefully pushing aside the Genoese genius. The more tainted the Spaniards, the more wholesome was Columbus; the deeper and darker the Black Legend became, the more Columbus shone in comparison. Thus, spared blame for the deplorable treatment of "the natives," the lone, noble Columbus could die

> with a composure of mind suitable to the magnanimity which distinguished his character, and with sentiments of piety becoming that supreme respect for religion, which he manifested in every occurrence of his life.[9]

Would such histories have been so successful had they also tarred Columbus with the Black Legend brush? Probably not. During the same decades, two American poets portrayed Columbus in contrasting light. Philip Morin Freneau (1752–1832) put into verse dark critiques of empire, of the masses doomed as "dupes to a few," of a psychologically complex Columbus who served as a flawed "second Adam" who should have initiated a glorious future for the New World—but failed to do so. By contrast, Joel Barlow (1754–1812) whitewashed Columbus, offering him as a pious national symbol of the noble "march of man" that had resulted in America as promised land. Using "the paradigmatic literary form of empire," epic poetry, to compose his 1807 hit *The Columbiad*, Barlow anticipated the Columbus of an expanding nineteenth-century United States, and thereby contributed to its development. As one scholar has put it, Freneau, "certainly the best American poet of the eighteenth century," penned the age's "best depiction of Columbus," but as "it did not fit the patriotic stereotype," instead "it was Barlow's vision that prevailed."[10]

An admirer and friend of Barlow's, who shared his vision of Columbus as an Adam to America, was Bostonian clergyman Jeremy Belknap. When, in October of 1792—the third centennial of "the Discovery of America" and "the seventeenth year of the Independence of the United States of America"—the Massachusetts Historical Society met to hear an address on "the active and enterprising genius" of Columbus, the author who presented was Belknap. Here was the semi-fictionalized larger-than-life hero supposedly created by Columbus's son Fernando, familiar to Belknap's audience as a quintessential American hero. This "American" Columbus was "an enterprising adventurer," a bold, brave agent of God. Even his erroneous grasp of world geography is couched as necessary to God's plan, twisted around to reflect well on the Genoese navigator; "for great minds overlook intermediate obstacles." His triumph is gilded with other obstacles: "the envy of his rivals," the naming of the continent "after an inferior adventurer," and "the jealousy of his sovereign"—one that resonated powerfully with those who had survived the War of Independence.

Belknap, preaching to the converted, proposed to Massachusetts men that Columbus had advanced "science" and "the history of man"; that Columbus had thereby promoted true liberty, serving as a force against the pernicious powers of Europe, especially England and the empire of "the avaricious Spaniards." He denigrated both, England for promoting distorted and "imperfect notions of liberty," Spain for creating the "detestable species of traffic" that was the slave trade (blamed not on Columbus). Here was history seen through the prism of what Americans needed from it—and specifically from Columbus—in 1792.[11]

Following Belknap's address in Boston was the singing by an organ-accompanied choir of an "Ode," the words to which were also published in Belknap's book. The choir proclaimed a vision of the pre-Columbian world as darkened on both sides of the Atlantic by war and "Black *Superstition's* dismal night," with freedom lost and "nations trampled" by the alliance of "*Crown* and *Mitre*." The continent of the Americas was dubbed "Columbia"; "Hail! Great Columbia! favour'd soil," began the chorus. Columbia "lay conceal'd" until the "daring" voyage of Columbus, "guided by th' Almighty Hand," brought happiness, freedom and religion, health and wealth. The structure of the story is patently Christian, with Columbus as the agent of redemption.[12]

An oration and an ode in Boston may not seem like much of a Columbus commemoration, and compared with 1892 or even 1992 they were insignificant. The only other U.S. city that put on a similar celebration was New York. The modern American Columbus was still in its genesis by 1792. But many of the elements that would mark not only 1892, but eventually annual Columbus Day celebrations across the nation, were present in 1792: in addition to speeches, odes, and toasts, there was a parade in New York, and monuments were commissioned there and in Boston. The monuments were obelisks, not statues, and both disappeared not long after (New York's went to P. T. Barnum and thence oblivion). One 1792 Columbus obelisk does still exist, but it was not connected to a city commemoration event. However, in 1792 there already existed the idea that Columbus could be commemorated in October, the month of the First Landing, with certain public rituals being patriotic celebrations of both America and America's Adam, connected to a permanent monument of some kind.

Sadly, one element of the New York 1792 celebration did not stick. It had been organized by The Society of Tammany, or Columbian Order—Tammany being the seventeenth-century Lenni-Lenape chief Tamanend. The inclusion of a long-dead Indigenous leader, with no Indigenous people involved, is obviously not how it would be done today. But its effort to recognize that the America of the day was founded by both Columbus *and* "the Indians" was an anticipation of a paired Columbus Day–Indigenous Peoples' Day that for two centuries failed to emerge—and in the present century is far more often a battle rather than a pairing. (The Tammany Society, meanwhile, added a "St." to "Tammany," or a "Hall" at the end, with the "Columbian" fading away, as the society became the famous political machine, with a particular focus on Irish American politicians from the 1850s to 1960s.)[13]

Meanwhile, variants on Columbus's English name—Columba, Columbina, the United Columbian States, but mostly just Columbia—

emerged in opposition, or as alternatives, to Britannia (the empire from which the new nation had broken) and America (Vespucci being denied heroic reinvention, even derided as a reputation thief). As Britannia was the British Empire's poetic monicker, so Columbia, with the same rhythm and syllable count, tended also to be used in poetic contexts—beginning in the 1760s but with increasing popularity in the century's closing decades, with hundreds of examples across dozens of poems; this was an age of popular poetry. The poetic association, combined with the name pairing of British Empire–Britannia, likely helped fix in American minds a similar United States–Columbia name pairing.

Besides, the name was soon prominently used to name things other than the nation. The poetic association became broadly literary and cultural, with "Columbian" ubiquitous in the early nineteenth cen-

Progenitor: An 1876 Centennial lithograph printed in New York, with Washington at center, showing Columbus three times: seated with sword; in a medallion portrait with the same beard and hat; and in miniature at the "Landing of Columbus."

tury in the names of magazines, textbooks, music guides, museums, schools, historical and literary societies, and then—as the century wore on—everything from hotels and gas stations to insurance companies, ships, and, of course, towns and counties. New York's King's College, closed during the War of Independence, reopened in 1784 as Columbia College (then University). Columbus and Washington were frequently paired. The commissioners who oversaw the carving out of ten square miles from Virginia and Maryland to be the new nation's capital announced in 1791 that its title would be "the City of Washington, in the Territory of Columbia." That echoed the frequent tying together of the two patriotic icons—as one historian later put it, in his book's dedication, "Our Country: Discovered by Columbus, Liberated by Washington." That surely also reduced the likelihood that the United States would become Columbian in name. Early in the nineteenth century the historical societies of all northeastern states mounted a campaign for a new national name. "America" (with its advantage of already being the nation's name) beat out "Washington" (New York's choice), "Allegania" (Maryland's pick), and "Columbia" (popular but second). Similarly, the chorus to a patriotic song increasingly popular in the first U.S. decades, based on an English drinking tune retitled "Hail Columbia," ran "The sons of Columbia forever shall be / From oppression secure, and from anarchy free." But new lyrics, penned in 1814, omitted the nation's name and soon dominated the version that became a national anthem dubbed "The Star-Spangled Banner."[14]

In fact, a new nation in the Americas *was* named after Columbus—northern South America became Gran Colombia in 1819, dissolving by 1831 into separate republics, Colombia included—giving a misleading impression of toponymic patterns across the hemisphere. That example is something of an exception proving the rule that Columbus became far more important to national identity in the United States than in the former Spanish American colonies. Furthermore, the United States not only dominated the pattern, but it did so in a way that reflected a localism at the heart of the nation's claim to Columbus. Viewed at the macroregional level, the distribution of Columbiana place names is loosely, evenly hemispheric: one each in Canada (the province of British

Columbia), in the United States (the federal District of Columbia), in Panama (the province of Colón), and in Colombia (the nation itself). But at the microregional or more local level, the phenomenon is overwhelmingly a U.S. one. Of the roughly eighty counties, cities, towns, villages, and neighborhoods in the Americas named after the Genoese navigator, some sixty-four (or 80 percent) are in the United States, scattered across all forty-eight contiguous states (Ohio has the most, but the greatest concentration is in the Northeast).[15]

The few Latin American examples tended to be nineteenth-century name changes; for example, the only Mexican municipality named Colón was created in 1882 from two adjacent colonial-era towns. Toponym-changing throughout Latin America tended overwhelmingly to refer to heroes of the independence struggles *against* imperial Spain, not to the founding colonists. In contrast, U.S. Americans appropriated Columbus as an Adam of their nation by bypassing the years between Columbus and Washington to make Columbus theirs from the start. Indeed, the majority, perhaps all, of the U.S. examples were named after Columbus at the moment when settlers founded that place. Often that name choice symbolized the settler replacement of Indigenous peoples.

In Ohio, for example, Columbus and its sister settlement of Circleville were founded in 1812 and 1813 on top of—and making destructive use of—ancient earthworks, mounds, and stonework, including a massive circular mound and structure after which one of the new towns was named. The settlers seemed blithely unaware of the direct link between the builders of the site they had taken and the living peoples being displaced. They wondered "what cause could have so annihilated such a numerous race, that not a vestige of them can be found except their works," as the new Columbus paper, *Western Intelligencer*, put it in 1814—barely a year after the death of Tecumtheth (Tecumseh) and the U.S. Army's defeat of Shawnee forces in Ohio. These were the same Shawnee whose villages around what was now Columbus had been razed as recently as the decade of U.S. Independence. Thus, the founding of Columbus during Tecumseh's War "amounted to a civilian front in the US war of occupation" (as one historian recently observed).

Whether the settlers admitted it, or even knew it, their name choice was loaded with the symbolism of invasion, occupation, and racial displacement.¹⁶

WASHINGTON IRVING'S FAT, READABLE biography of Columbus thus fell on highly fertile ground when it first came off the presses in 1828. Irving did not pretend to be a historian in today's sense. He relished telling stories, be they satirical fictions of his own making ("Rip Van Winkle," 1819), stories based on older fiction (*Tales of the Alhambra*, 1832), or—like his *History of the Life and Voyages of Christopher Columbus*—biographical stories based on historical characters and sources. As a consummate storyteller, he knew that Columbus's tale was already associated with mystery, and that the success of his *History of the Life and Voyages* depended on a claim to veracity and the solving of mysteries. That faulty premise has characterized Columbiana ever since, to some extent due to Irving. He prefaced his two-volume account with a little story about being invited to Madrid to meet Martín Fernández de Navarrete, then secretary of the Real Academia de la Historia, Spain's Royal Academy of History. Shown Navarrete's new publication of Columbus documents, Irving was inspired to assemble them into a narrative for readers of English—aided by access to such additional sources as "the royal library of Madrid" and the personal family archives of "the Duke of Veraguas [*sic*], the descendant and present representative of Columbus." That archival evidence allowed Irving "to construct this history" in an impartial manner without "indulging in mere speculations or general reflections."¹⁷

Irving's claims were a smokescreen. While some of his descriptions and interpretations pass modern smell tests, especially those reflecting his careful reading of documents published by Navarrete, many others show him not only indulging in speculations but going further into outright invention. He had access to Las Casas's reworking of the *Diario* and to the imaginative hagiography attributed to Fernando Colón, both of which he treated as authoritative sources, liberally using his own imagination to fill gaps and reconcile contradictions—all presented as facts, rendered persuasive by Irving's narrative skill. After all, he insisted, "the general reader" wanted "clear and continued narrative," not the "disconnected papers and official documents" that he was apt to find "repulsive."[18]

This was a variation on what British novelist William Boyd would call, two centuries later, "fictography." Whereas Boyd's "whole life" novels are structured as biographies, in which fictional characters interact with historical characters and events, Columbiana books like Irving's make the traditional truth claims of biographies, while blending semi-fictionalized historical characters with both historical and fictional events. One of the joys of a Boyd novel is not being sure if certain lesser characters or events are his invention or more-or-less accurate historical portrayals. The danger of a Columbus biography is that the reader often cannot be sure how much of it is well evidenced, how much invented, how much of it mythistory repeated by an author unwilling or unable to drill down to the original source—in other words, how much of it is "fictographic."[19]

Irving had access to, and could read, both the primary sources assembled by Navarrete and other relatively dispassionate accounts of Columbus's life—such as Juan Bautista Muñoz's 1779 *Historia del Nuevo-Mundo*. But he found such careful treatments of evidence to be dry and dull, preferring the lively versions of boyhood favorites such as *The World Display'd* (English pocket-sized volumes, first published in 1759, that stressed the "daring" of "the greatest man of the age"). He fully adopted the strategy of Russell and Robertson, based partly on Las Casas, of separating Columbus from the Spaniards. That allowed him to credit Columbus for bringing Christianity as the pious Discoverer, to blame the colonists for the violence that soon swept the New World, and to present Columbus as a martyr to the ingratitude of Spain and its monarchs. Irving embraced every piece of Columbiana mythistory invented to date, adding some more from his own imagination, giving Columbus a swashbuckling yet studious youth and an irresistible charm and courageous magnanimity. To point out how much of this was hogwash was to indulge in "pernicious erudition"; history was about great men, and America needed to know about this one.[20]

Irving's book was so successful and influential that its details became permanently lodged in the American perception of Columbus, reiterated in paintings and sculpture, poems and plays, and hundreds of books of all kinds—from the art in the U.S. Capitol to children's books to the immensely popular 1892 (and 1992) postage stamps. Indeed, Irving imagined the life of Columbus punctuated by epic history-changing moments that could be captured in simple images—such as Queen Isabel pledging her jewels, the First Landing, the reception at Barcelona, the great man unjustly in chains, all depicted on those stamps. Although Irving stated that only some of the "clerical sages" of the "Council at Salamanca" were flat earthers, he depicted them all as inept, superstitious buffoons, embodiments of "the errors and prejudices, the mingled ignorance and erudition, and the pedantic bigotry, with which Columbus had to contend." The notion of Columbus as a brilliant visionary, ahead of his time, the wise man of action who knew better than all the men of learning, was given a great boost by Irving's book; it appeared in hundreds of ways through the nineteenth and twentieth centuries. Irving's fictionalized version of Columbus at Salamanca supported, and was in turn supported by, the nineteenth-century Protestant American invention of a so-called medieval Dark Ages mired in Catholic superstition and ignorance—an era of fanatical flat-earther pedants. Reflecting the linking of the Flat Earth myth to Columbus, it was more widely believed in the decades surrounding the Quadricentennial than before or since—although confirmation bias and the persistence of Columbiana's mythistory allowed the myth of Salamanca to survive all through the twentieth century.[21]

It wasn't just small yet resonant pieces of Columbiana mythistory that flourished in the century after Irving, it was the notion of history as comprising the instructive biographies of "great men." From Robertson decades before Irving, to William Prescott decades after him, historians sought history's heroes, whose greatness could be thrown into relief by flawed heroes and villains. In the four books Irving wrote in Spain, published in 1828–32, all on Columbus or on Granada, he developed a negative portrayal of "King Ferdinand" of Aragon—"that perfidious monarch." In contrast, Granada's last sultan, Boabdil, emerged as a flawed and long-suffering hero, "alternately the foe, the prisoner, the friend, and always the dupe of Ferdinand." In such company, Columbus shined all the more.[22]

History Men: Christian Schussele's *Washington Irving and His Literary Friends at Sunnyside* (1864) depicted a rather smug Irving in the center, appraising us, with William Prescott in gray to his right.

That technique was also employed by Prescott, whose history of Fernando and Isabel permitted a hagiographic vision of Columbus that both reflected and boosted prevailing opinion. Loyally devoted to the Spanish sovereigns, guided in all things "by the nicest principles of honour and justice," Columbus suffered not "a single blemish in his moral character." Coming close to the language of the imminent canonization campaign, Prescott concluded that

> There are some men in whom rare virtues have been closely allied, if not to positive vice, to degrading weakness. Columbus's character presented no such humiliating incongruity. Whether we contemplate it in its public or private relations, in all its features it wears the same noble aspect. It was in perfect harmony with the grandeur of his plans and their results, more stupendous than those which heaven has permitted any other mortal to achieve.[23]

There were, meanwhile, counternarratives and critical studies. Henry Harrisse (1829–1910), for example, a Parisian who spent much of his long adult life as a lawyer and historian in the United States, dived deeper than anyone before him into the primary sources, read everything in its original languages (as good scholars do today, but few did then), and applied to them a penetrating nonpartisan mind. As a result, he saw through much of Columbiana mythistory, making his studies among the few of the nineteenth century still worth consulting. Likewise, Justin Winsor (1831–97), an American librarian and historian, was ahead of his time in publishing an 1891 biography that took Irving to task and sought to present a balanced portrait of the historic Columbus, not the hagiographic version that ruled the era of the Quadricentennial.[24]

But contrary or critical or just more balanced narratives, from Freneau to Winsor, had little impact on popular, public perceptions. This was partly due to how Columbus was presented in public parades and art; even mildly contrary views had little hope against the steady building of the Capitol as a Columbian shrine and the Columbiana frenzy of 1892. The public preferred Columbus not as a complex, flawed, egomaniac, but as a simple American hero—"underdog, individualist, pathfinder, and Pilgrim-like agent of Christianity." That's the phrasing of a recent writer, who also quoted another: "Creation myths identify heroes and sacred places, while evolutionary stories provide no palpable, particular thing as a symbol for reverence, worship, or patriotism."[25]

But also, Columbus's image in children's literature was relentlessly positive. The genre was still in its infancy in the late eighteenth century, but series such as *A World Display'd* (which a young Irving carried around in his pocket) and that of Joachim Campe offered up a heroic Columbus whom later writers—now influenced by Irving—could turn into a role model for American schoolchildren. Dozens of histories for schools written by the Goodrich brothers and others went into hundreds of editions. Books like *Stories of Great Men* (being Columbus, Washington, Penn, and Franklin) and *History of the United States of America, from the Discovery of the Continent by Christopher Columbus to the Present Time*, promoted the "great man" and his "character so grand and elevated" as a moral example for schoolboys. He was a model of imperial man fit for emulation by young Americans. "Don't you think our country should have been named for him, instead of being called America?" an 1895 book asked its young readers. "I am always glad when I hear it called COLUMBIA." American schools ("long the custodians of the Columbian legend," as one scholar noted) were specifically tasked in 1892 by President Benjamin Harrison to acknowledge the day that became Columbus Day as a day for patriotic rituals, evolving state by state (and, in 1934, at the federal level) into a patriotic holiday. The reciting of the Pledge of Allegiance is a school ritual, along with the schoolhouse flag, that originated with the Quadricentennial. Americans learned, from elementary school on, that honoring Columbus was a dutiful way to celebrate America. Respect for the American Columbus became a firm part of how American schoolchildren were socialized as good patriots.[26]

NO SOONER HAD THE dust settled on Philadelphia's Centennial Exposition of 1876—a $7 million event that drew 10 million visitors (the U.S. population was only 45 million)—when plans began to be discussed for an 1892 Columbian fair. This was a chance for America to show the world how far it had come since the days of the Adam that had founded it and the Adam that had freed it. A competition gradually emerged, with Chicago and New York as the clear finalists by the time the third Paris Nationale opened in 1889. In an era when imperial nations jostled to display their economic, political, and cultural might through vast, lavish "world's fairs," the French threw down a gauntlet with the most elegantly spectacular fair to date (including the unveiling of the Eiffel Tower). Early in 1890, the U.S. Congress voted to give Chicago the honor of picking up that gauntlet and hosting "the World's Columbian Exposition"—destined to be the most extensive and expensive world's fair ever created.[27]

Why Chicago? By 1890, the American Columbus was not just an Adam figure, but a success symbol, a role model for Americans who believed that—like Columbus—with courage and daring they could build a new future in the Promised Land. Chicago was an exploding new metropolis, having doubled its population in just ten years, quadrupled it in twenty. Columbus was an ideal icon to appropriate in the service of promoting the city and the United States overall as humankind's ultimate examples of progress.

For the exposition, Chicago built an entire city on the edge of Lake Michigan, larger and with more buildings than the 1876 Philadelphia and 1889 Paris fairs combined. Half its $38 million cost went into twenty-eight imposing glass-and-white-stucco buildings, one of them the largest building in the world. The fair's scale was so vast, it could not be completed in time for its official opening in October of 1892; visitors were admitted the following January, and more than 27 million would come during the 184 days of the fair's life. It was then, and remains to this day, "arguably the greatest of American world fairs." Columbus made prominent appearances—astride a Barge of State within a Columbian Fountain, for example, and driving a four-horse chariot atop a triumphal arch—and he was ubiquitous in the imagery and nomenclature of the whole sprawling extravaganza. But visitors learned not a thing about him beyond his apparent aptness as an icon of empire, a symbol of American power, prosperity, and progress.[28]

In other cities, too, there were Quadricentennial celebrations, none even remotely comparable with the fair in Chicago, although the triumphal arches and days of parading in Boston and Baltimore, London and Madrid, anticipated an emerging Columbus Day culture that would flourish in the twentieth century on both sides of the Atlantic. Of those celebrations, none was more lavish than New York City's—which was bedecked in patriotic decorations on every building, attracting a million visitors just on the last of its five continuous days of parades. Columbus was again the excuse, the American invention rather than the historic figure, conceived to celebrate the nation and—above all—the city. Columbus was thus "the pioneer of modern civilization," but New York was its ultimate "living, bustling, instructive monument." Taking advantage of its Atlantic location, the city also welcomed a massive parade of ships representing many nations, the United States most prominently, with replicas of the *Niña*, *Pinta*, and *Santa María* towed by Spanish warships (apparently not trusted to cross the Atlantic under their own sails).[29]

The ships sailed again—or rather, were towed again—a century later, as part of a Quincentennial that at first promised to be a transatlantic celebration as extensive as the Quadricentennial. There were commissions established throughout the Americas and southern Europe. In the United States, there was a spike in the building of Columbus monuments not seen since 1892, while the Post Office reissued the hugely successful Quadricentennial Columbus stamps with only one tiny update—an 8 changed to a 9. But the Spanish caravel tour "went bust"; the ships drew meager crowds in New York, where they became stranded as of January 1993, eventually being sold to a museum in Corpus Christi, Texas. The lack of imagination behind the caravel idea, the assumption that 1992 could repeat 1892 and be just as successful, and the resulting disappointment, all summarized the Quincentennial's failure. In 1992, empires were no longer being celebrated; on the contrary, there was more interest in uncovering the stories of those peoples who had suffered empire, and Columbus could all too easily be turned into a symbol of that suffering. The U.S. commissions, undercut by political indifference and "ethnic infighting," struggled to accommodate the apparent clamor for an Anti-Columbus Quincentennial. Chicago's effort to reproduce 1892's exposition with a 1992 world's fair collapsed, even before a site had been selected, under the combined weight of local activism and conflicting notions of patriotism. In retrospect, might 1992 have been styled as a patriotic festival that virtually ignored Columbus, as it was evident in Spain and Genoa? For as much as celebrating empire had fallen out of fashion, celebrating America (as in, the United States) was as in vogue as ever. But 1892's white American consensus on what that meant was crumbling by 1992.[30]

"WE FIND THAT WE do 'pity ol' Chris,' and would like to give him a home, if we could," pleaded George Mathews in a letter written in October 1977, addressed simply to "National Archives, Washington, DC." "Indeed," added Mathews, who ran Columbus Foundries, Inc., he hoped to "fit him into our quarters in Columbus, Georgia." The letter eventually reached the desk of George White, Architect of the Capitol (and thus responsible, with his curator, for all its art). He responded that the fate of "the statue" in question was mired in multiple discussions, including an effort by the mayor of Beaufort, South Carolina, supported by that state's U.S. senator, Strom Thurmond (1902–2003), to place the statue in a new waterfront park. Mathews gently pushed back. He suggested to White that Beaufort's planned park was more for "military artifacts." In contrast, Columbus was "now Georgia's second largest city," and the city would happily "pay a fair price for this sculpture" or bear the costs of borrowing the statue. Columbus would display "Columbus prominently."[31]

Columbus, Georgia, and Beaufort, South Carolina, were not the only cities to "pity ol' Chris"—a reference to the title of an *Atlanta Constitution* article seen by Mathews in the Columbus Day 1977 issue. Furthermore, interest in this particular sculpture was widespread, amounting to a virtual competition to wrest it from governmental hands in the nation's capital. What was this Columbus statue, and why were cities fighting over it?

In the middle decades of the nineteenth century, as the United States dramatically expanded, so too did its Capitol, along with the building's burgeoning art program. The content and messaging of that art reflected the ideology of a former colony that had become itself an aggressive colonizer. Less obvious, but equally important, was the flip side to that coin: the Capitol's art offered "an instructive lesson" (as one congressman put it in 1845) to elected politicians, their staff, and visitors, thereby helping to shape policy and practice.[32]

Twenty years before he became U.S. president, and a decade before he helped James Polk, as Polk's secretary of state, to invade Mexico, James Buchanan was a U.S. senator for Pennsylvania. It was Senator Buchanan who in 1836 initiated the process of commissioning two sculptures to sit on the stone cheeks either side of the U.S. Capitol's main exterior staircase. And he had just the sculptor in mind: Luigi Persico, an artist from Naples who had come to the United States in his twenties, living for seven years in Lancaster, Pennsylvania. Buchanan also lived in Lancaster (it had been the 1799–1812 state capital), and the two men, the exact same age, had become close friends.[33]

Senators William C. Preston and John C. Calhoun, both of South

Carolina, argued that the commission should go to "an accomplished citizen of our country." They had in mind Horatio Greenough, the young Bostonian already at work on the Congress-commissioned statue of George Washington that introduced this chapter. Persico was also a veteran of Capitol commissions, having completed three sculptures on the east façade in the 1820s, and Buchanan countered the flag-waving of the South Carolina senators with the argument that Persico may not have been "a native, but he intends to spend his days among us, for he loves liberty with all the enthusiasm of genius."[34]

In the end, the commission was split between the two sculptors. By 1839 they were in Florence together to compare design notes and order marble. Greenough's work, not installed until 1853 (he had died suddenly the year before), was titled *The Rescue*. It was an unabashed statement on racial superiority, a representation of the prevailing view that America was where civilization triumphed over barbarism. *The Rescue* comprised a large, fully clothed white settler restraining a smaller tomahawk-wielding "Indian" man, wearing only a loincloth, while a small baby-holding Indigenous woman recoiled. There was also a dog. Persico worked faster, producing a less cluttered version of the same vision. His *Discovery of America* was installed in 1844, comprising a pair of figures: a large Columbus in masculine pose, garbed like a conquistador, holding aloft a globe; and a small, cowering female "Indian," almost completely naked.[35]

The significance of *Discovery of America* lay largely in its prominent placement by the Capitol's entrance, an outside reification of the composite depiction of Manifest Destiny that was being assembled, piece by piece, inside the building. And its timing was perfect. That phrase—styled as "the right of our manifest destiny to spread over this whole continent"—was first used in Congress in 1846. In that year, the United States invaded Mexico and seized Oregon, having annexed Texas the year before; by 1848 the United States had taken half of Mexico's national territory, including California.[36]

Columbus was a frequent justification for this rapid imperial expansion. The Founding Fathers were merely taking their cue from Columbus, himself "the agent Heaven employed to place us in possession," as

Nine-Pins: In 1844, the year that Luigi Persico's *Discovery of America* was placed at the east façade to the U.S. Capitol, Senator Charles Sumner of Massachusetts quipped that Columbus seemed to be "getting ready to play a game of nine-pins." The sculpture's future has remained uncertain since it was moved into storage in 1958.

Representative William Sawyer of Ohio put it in 1846; and thus "we have the right to every inch" of the continent. Persico's Columbus was specifically cited as embodying "the march of civilization and Christianity," with the sculpture capturing "the destiny of the New Continent" and its "providential guidance." Everyone entering or leaving the Capitol could be reminded that Columbus was evidence of God's intention for America.[37]

The same was true of the crowds who gathered for presidential inaugurations. As had all presidents since Andrew Jackson in 1829, James Polk was in 1845 sworn in on the eastern portico of the Capitol, adjacent to the cheek block on which Columbus stood. His speech echoed his platform; he vowed "to extend the dominions" of the United States across "*our* territory that lies beyond the Rocky Mountains" (emphasis mine). Polk declared that "the title of numerous Indian tribes to vast tracts of country has been extinguished." How? Through the genocidal

A Small World: The globe-holding Columbus of Persico's *Discovery of America* overshadows James Polk as he is sworn in as U.S. president on the steps of the Capitol. From the *Illustrated London News*.

violence that followed the Indian Removal Act of 1830, supported by a vision of national destiny as "a struggle for supremacy between savages and civilized men" (as one senator put it).

Persico's sculpture loomed as an icon to illustrate the vision embraced by Polk and most of Congress. A couple of months before the inauguration, Representative Belser of Alabama had used Persico's "two figures" as that "instructive lesson" on the nation's future: his Columbus with submissive "Indian" illustrated how "freedom's pure and heavenly light would continue to burn, with increasing brightness, till it had illuminated this entire continent." That was inevitable because of "the power of civilization" embodied in Persico's Columbus. "Gentlemen might laugh at the nudity of one of" the other figures, smirked Belser, "but the artist, when he made Columbus the superior of the Indian princess in every respect, knew what he was doing." Twelve years later, Buchanan himself was inaugurated as president on the eastern portico, flanked by *Rescue* and *Discovery*, whose existence he had initiated and whose iconography promoted the racist ideology that underpinned the imperialism he supported—and which dominated the century's national politics.[38]

But the next century brought change, as the world wars and their decolonial aftermath swept away the old empires. The imperial United States emerged stronger than ever, but its methods and ideology were obliged to adapt. That included attitudes toward Indigenous peoples— even if such changes were painfully slow after centuries of dispossession and slaughter. Citizenship was extended to "Indians" in 1924, and 1934's Indian Reorganization Act rejected "the erroneous, yet tragic, assumption that the Indians were a dying race—to be liquidated." *Rescue* and *Discovery* looked increasingly outdated. In 1939, Representative Clark Burdick of Rhode Island introduced a joint resolution that the sculptures be "ground into dust, and scattered to the four winds," for they were "a constant reminder of ill-will toward the American Indian, who has now become a part of this Nation."[39]

The resolution did not pass. But two years later, Representative James O'Connor of Montana submitted that the "statuary group is an atrocious distortion of the facts of American history and a gratuitous insult to the great race." His resolution proposed that the statues be replaced by "one of the great leaders famous in American history." O'Connor was ahead of his time. Such substitutions were almost a century away, in the United States and elsewhere. Nonetheless, objections to the sculptures gradually mounted, and by the 1950s there was a steady stream of letters to Congress and to the Architect of the Capitol—who tended to respond that only an act of Congress could remove the statues. Then, in 1958, all of a sudden, cranes were placed in front of the eastern portico, and the offending sculptures were removed. Even as he remained prominent inside the Capitol, had the campaign to banish Columbus from the façade and entrance finally succeeded? [40]

The removal of *Discovery* and *Rescue* was actually neither sudden nor a result of opposition to their existence. A massive expansion of the eastern front of the Capitol had long been planned. As it involved moving the cheek blocks forward—along with hundreds of tons of stonework to be moved, rebuilt, or replaced—in 1958 the two sculptures were lifted, crated, and deposited in the Capitol Power Plant coal yard on nearby Virginia Avenue. There they sat, their future "placement" status "in limbo." When the power plant was renovated eighteen years later, the artworks were again moved, this time to the Smithsonian Institution's storage site in Silver Hill, Maryland. Both were damaged in the move, and continued to deteriorate, Smithsonian-held but still Capitol-owned. A mid-1980s memo internal to the Architect of the Capitol's office detailed "the ravages of freezing and thawing" and of pollution to "these <u>MAJOR</u> sculptures," calling the Smithsonian's neglect "a disgrace."[41]

Public interest in the vanished statues, and the persistent prevarication that such interest encountered on Capitol Hill, had begun before the extension work started and only increased after its 1962 completion. A stream of queries, small but persisting for decades, reached the offices of the architect and the curator. More curious than critical, almost all hoped for the sculpture's return to display. In 1958, a Mrs. Carla Weinwurm of Chicago sent Speaker of the House Sam Rayburn a photograph she had taken of *Discovery*, asking whether it would be moved "with the extension." A Harry Sachtler of Portland, Oregon, asked Architect of the Capitol George Stewart in 1964 when we might have "these fine

Half-Lives: The two halves of Persico's *Discovery of America*, separated, trapped, and slowly deteriorating in outdoor storage for many decades—symbolic of Columbus's half-life as both reviled and celebrated, with images from the heyday of his apotheosis suitable neither for public display nor complete destruction.

statues" back? Stewart reassured him that the statues were "in protected storage," and "no decision had been reached" on their return, as "the Commission in charge desired to observe the building for a period of time." His successor, George White, repeated that rationale in letters to concerned citizens in 1971, adding that the sculptures' condition meant "they would have to be reproduced in new marble."[42]

Newspapers periodically reported on the limbo status of *Discovery*. One article went into syndication in 1977, appearing in papers all over the country, titled "Stone Columbus Stands in Exile with a Maiden." Stating that Congress remained undecided on the sculpture's fate, the article gave the impression that *Discovery* had always been controversial mostly due to its artistic flaws, not its content (repeating the old joke, attributed to Senator Charles Sumner of Massachusetts in 1844, that Columbus seemed to be "getting ready to play a game of nine-pins"). In 1980, the *Washington Star* made a similar report, including photographs of the half-crated, deteriorating figures.[43]

As word of the exiled neglect of Persico's Columbus spread, so did efforts to rescue it—especially following the spike in patriotism brought on by the U.S. Bicentennial. Columbus, Georgia, and Beaufort, South Carolina, found themselves in competition with Columbus, Ohio. The Smithsonian passed their request to the Architect of the Capitol, who responded that he had just become aware of a "specific restriction" whereby the sculpture must stay within the city of Washington—in a site to be determined by "mutual decision" of the Capitol and the Smithsonian. Senator Bob Dole made a similar request, in the hope of "securing this statue" for Columbus, Kansas. As did the City of New Orleans, whose Mayor Moon Landrieu sent his executive assistant, Anthony Gagliano, up to Washington in a vain attempt to circumvent capital stonewalling.[44]

As requests continued to reach the Smithsonian in 1978—Kansas City, Missouri, also sent a representative with letter in hand, for example—a new brush-off was added to the rejections: the restoration and relocation in Washington of *Discovery* was now being planned for the 1992 Quincentennial, and "it would indeed be unfortunate if this important element of Washington's sculptural history, and one so replete with symbolism, were not available for display at such a significant time." When Columbus, Kansas, tried again in 1981, this time with Representative Bob Whittaker taking up the cause, the same reasons for rejection were given. Then, in 1989, Representative Robert Walker of Pennsylvania made the case for Lancaster, citing "increased public concern for this artwork" and hoping to see "the statue returned to the artist's hometown." Capitol Architect White gratuitously corrected that hometown claim, and dismissively noted that "numerous cities" have been denied the Columbus statue "in hopes that funds may become available for conservation" and display in Washington. One can almost hear him sigh as he types the lie.[45]

For, of course, there were no such hopes or plans. *Discovery* had been controversial all along, and that was even more the case as the Quincentennial dawned. A 1991 plan to dust off and patch together just the Columbus statue from *Discovery* for a Quincentennial display in the Capitol was abandoned. As C. Douglas Lewis, curator of sculpture at the National Gallery, told Barbara Wolanin, the Capitol's curator, in a memo that year, there is "no way to argue that as a work of art it is of any importance." What's more, its depiction of Columbus is "simply not politically correct." That phrase would go out of fashion, but its view of *Discovery* remained pertinent in the next century, as the Washington buck-passing continued. In 2005, the Architect of the Capitol asked the Smithsonian to re-crate both *Discovery* and *Rescue* (they didn't). In 2006, the National Museum of the American Indian (opened two years before) was asked to take the sculptures (they declined, citing "lack of relevance"). In 2007, the Smithsonian demanded that the Capitol take them back by 2008 (they didn't).[46]

As *Discovery* and *Rescue* approached their seventh decade in storage, plans to display fragments—Columbus's severed globe-holding hand, for example—were hatched and shelved. Inquiries continued to drift in "about the sculptures' whereabouts and condition," as the curator put it in 2021, admitting that "their marble surfaces are eroded and crusted with gypsum. Hands and fingers are missing, feet and heads severed, and torsos disconnected from legs."[47]

Among the various images of Columbus, carved and painted, on Capitol Hill, the tale of Persico's is particularly "instructive." Its birth in 1844 coincided with a moment in the history of U.S. expansionism that was particularly brazen in both deed and ideology—and the sculpture matched that mood. The use of Columbus by *Discovery* and its advocates to promote white supremacy underscores the sculpture's survival, openly and prominently until 1958, and covertly since then. A spike in interest in the sculpture in the wake of the 1976 U.S. Bicentennial showed that patriotic sentiment quickly attached itself to Columbus. But while his promotion as a nationalist icon was two centuries old, the Quincentennial revealed a fissure in public perceptions of Columbus. Just as the flurry of Columbiana literature in the 1990s was more critical than apologist, especially among scholars, so did requests for *Discovery* come from towns, not universities.

Conservatives' attitudes toward such icons remained positive in the years between the Bicentennial and the 1992 Quincentennial, while the story of the government's removal, hiding, and endless misdirection regarding the statue's future tapped into localist suspicion of federal authority. One can sense—between the lines of polite letters asking the Capitol and the Smithsonian to loan or sell a sculpture they clearly don't want—a quiet, befuddled indignation. After all, there has always been a tension, in all Columbus-claiming countries, between his use to promote nation or empire and his localist appropriation. Thus, the numerous U.S. cities that claimed him as theirs—including cities named after him and the dozen or so that in the late twentieth century sought specifically to claim Persico's Columbus—were the equivalent to those European islands, towns, and provinces who claimed Columbus as theirs by birth, and have continued to do so with as much vigor as ever.

If Persico's Columbus was a contentious figure in the twentieth century, he is even more so in this century. *Discovery* can neither be destroyed nor displayed, and the liminal space in which it has existed, its half-life between revealed restoration and being "ground to dust," remains the only apparent solution, as necessary now as in any of the decades since its removal from the Capitol.

On a Pedestal: A 1909 stereograph card of New York City's Columbus Circle, with Gaetano Russo's marble sculpture of Columbus atop a granite column decorated with three bronze galleys representing the First Voyage's caravels. Despite recent efforts to have it torn down, the monument still stands, surrounded by traffic and dwarfed by skyscrapers.

LIFE NINE

THE ITALIAN

IN 1892, A MARBLE STATUE OF CHRISTOPHER COLUMBUS WAS placed atop a seventy-foot granite column, located at the southwest corner of New York City's Central Park, in the middle of a traffic roundabout whose name was thereby changed from The Circle to Columbus Circle. That much is not surprising: that was, after all, the peak year for the hero worship of the American Columbus. But there's a twist. The monument included a dedication, in the form of a small poem: "To CHRISTOPHER COLUMBUS / The Italians Resident in America, / Scoffed at Before, / During the Voyage, Menaced, / After It, Chained, / As Generous As Oppressed, / To the World He Gave a World." The poem was carved into the Baveno granite base in Italian as well as in English. It continued, "Joy and Glory / Never Uttered a More Thrilling Call / Than That Which Resounded / From the Conquered Ocean / In Sight of the First American Island, / Land! Land!"

In light of the nature of the vast majority of the thirty or so Columbus monuments built in the United States up to 1892 (and more in Latin American nations)—with their emphasis on the triumph of civilization over barbarism, Christianity over heathenism—the Columbus Circle monument's focus on migration, sacrifice, and Italian identity *is* surprising. Admittedly, there were Indigenous figures included in one of the two bronze bas-reliefs set into the base. But they were crammed into a fifth of one bas-relief, mere heads in feathered headdresses, hiding in the bushes as part of the flora and fauna at the First Landing, a colonialist yet modest echo of their inclusion in Vanderlyn's iconic *Landing* painting. The other bas-relief depicted the caravels at sea, at the moment when land was first sighted. In other words, these images have very little to do with Manifest Destiny or justifying the creation and expansion of Anglo-America. The bas-reliefs, like the column and statue and dedication, are about crossing and landing, about migration, about the experience of the Italians who witnessed the monument's unveiling on October 12, 1892. When they saw *"Terra! Terra!"* at the foot of the inscription, those Italians surely thought of their own Atlantic crossing more than they did the Genoese mariner's of four hundred years earlier.

At the grand unveiling, President Benjamin Harrison made a brief speech, as did Governor Roswell Flower, prior to the blessing and consecration of the monument by Catholic archbishop Corrigan. But the Italian connection was predominant. The statue had been created in Rome by two Sicilians: the sculptor Gaetano Russo, who was present and bowed to much applause; and the poet and artist Ugo Fleres, who composed the inscription. The monument had been paid for by subscriptions from Italians in the United States—and some in Canada, Mexico, Guatemala, and Italy itself, as well as the Italian government. The campaign, which took only three years from initiation to its Columbus Day unveiling, was started by Charles (Carlo) Barsotti, publisher of the *Progresso Italo-Americano*, one of the leading Italian-language newspapers in the city. Italian diplomats, officers of visiting Italian warships, and dignitaries from all the city's Italian American societies paraded in to applause. The packed crowd—ten thousand in the stands and thousands more in the circle, streets, and Central Park—"greeted vociferously by cheers" the arrival of the local *prominenti*, "the prominent residents of the Italian colony in New-York," as the *New York Times* put it the next day.

Annie Barsotti, the publisher's daughter, "attired in the Columbus colors" (perhaps *il Tricolore* of the Italian flag), pulled the unveiling rope after a speech exclaiming that "as a child of Italian parentage and of American birth I can ask nothing greater than this: God bless Italy! God bless America!" Her words were echoed by Baron Francesco Saverio Fava, Italy's first ambassador to the United States, who enthused in Italian about "the bond of brotherhood that Columbus created between Italy and America"—which made no sense historically, yet deftly captured the spirit of this alternative Columbus, this new Italian American Columbus.[1]

The greatest challenges of the Fava ambassadorship of the 1880s and 1890s were the regular attacks on Italian immigrants across the country—including one of the largest mass lynchings in U.S. history, the mutilation and killing of eleven Sicilians by a racist mob in New Orleans just eighteen months before the Columbus Circle Quadricentennial ceremony. That had been *the* most covered story in U.S. newspapers in 1891 (as the World's Columbian Exposition would be in 1893). Lynching in the 1870s and 1880s overwhelmingly targeted Black Americans in the Deep South, with the press and politicians nationwide tending to view victims as criminals receiving local justice. The New Orleans killings prompted a shift in that portrayal, as its victims and their families had advantages that Blacks did not: diplomatic pressure, in the form of Baron Fava's public complaints; and international pressure, a shaming of the United States in the eyes of the "civilized world" for permitting a "barbarous" practice—against Europeans, no less.[2]

Three campaigns vied to create an 1892 Columbus monument in New York City. A group of Spanish Americans proposed a Columbus Fountain to grace the entrance to Central Park at Fifth Avenue and East Fifty-Ninth Street—the same spot requested by Barsotti's group. Supported by the Spanish government, the fountain scheme had a design (Columbus standing on a huge globe, with a Pinzón brother on either side of him), a dedication ("*A COLON y los PINZONES / Los Españoles / E Hispaño-Americanos / De Nueva York*"), and a plan to melt down old Spanish cannons for the statues (one from Honduras had already been acquired). Such a monument would have offered a fascinating contrast between the two Columbuses, Italian and Spanish, half a mile apart, both at corners of the great park. But when the Italian Americans were offered The Circle, and the Spanish Americans were offered 120th Street and Fifth Avenue, a then-isolated spot four miles to the north of The Circle, the latter refused such a "second-class place"—and their project faded away.[3]

All along, the wealthy (non-Italian, non-Hispanic) élite had been raising funds for their own monument. Contributor names included Astor, Belmont, Fish, Gould, Grant, Morgan, and Vanderbilt. With such élite support, they were bound to succeed, and indeed their statue—a bronze replica of a marble effigy in Madrid created by Jerónimo Suñol—was placed on the Mall in Central Park, where it still stands. Cross in hand, staring up at the heavens, a globe beside him, this is a low-key version of the American Columbus, without the cowering, half-naked "Indian" of depictions in the Capitol and elsewhere. But in terms of location, scale, and impact, the white élite's Columbus-in-the-Park could not compete then, or now, with the Columbus Circle monument. Its unveiling was also an anticlimax: the king of Spain had agreed to do the honors in the summer of 1892, thus to beat the rival Italian monument to the post, but he suddenly died; don Cristóbal Colón (the Duke of Veragua and Sixteenth Admiral of the Ocean Sea), agreed to step in, but the shipping of the statue from Spain was delayed, and the duke went on to the World's Columbian Exposition in Chicago; not until May of 1894 was the Suñol statue installed. Nonetheless, the local press response to the two monuments reflected the prevailing anti-Italian sentiment of the time. "New York has reason for satisfaction that at last she has what, to her shame as the chief city of the American Continent, she has long lacked," enthused the *Tribune* in 1894, snubbing the Italian monument in favor of the new one on the Mall: "a creditable statue of Columbus." As for that Italian monument, it had fast become "a sort of mecca" (according to the *New York Times*, a few months after the unveiling), a "new shrine for Italians," where "troops" of the "swarthy sons of the Sunny South wander about the bit of marble"—a dig at the southern Italian origin of many of the city's tens of thousands of immigrants.[4]

A century and a quarter later, disdain for the Columbus Circle monument surfaced again. On Columbus Day, 2017, thousands of Italian Americans paraded, as was their custom going back to the Quadricentennial, only to encounter protesters at the monument—holding signs calling Columbus a murderer, enslaver, and rapist. Was this an expression of the same old prejudice toward Italian immigrants? Some took it as such. But it was far more complex than that. One state assemblyman, Charles Barron, who agreed with the protesters that no U.S. holiday should be "named for an enslaver or glorifying a murderer," suggested that Columbus Day should be renamed Italian American Day—with the very same parades and celebrations, only minus Columbus. The problem wasn't Italians or Italian Americans, or even the Italian American Columbus, but the twenty-first century perception of the American Columbus—the one developed in the nineteenth century to glorify the United States as it was seen by its white political rulers and cultural élite. Calls to pull down the Columbus statue in Columbus Circle were not expressions of anti-Italian racism; rather they stemmed from the perception of Columbus as a symbol of racism—against Black Americans, Indigenous Americans, any Americans on racism's receiving end (which, historically, included Italian immigrants).

By the end of 2017, the Columbus Circle monument was added to Mayor Bill de Blasio's list of possible "symbols of hate" in New York City. The resulting outcry in the Italian American community prompted Governor Andrew Cuomo (whose grandparents were all born in southern Italy) to nominate the monument to the National Register of Historic Places, and by the end of 2018 it had fallen under federal protection. Leaders of Italian American organizations in the city understood the problem well. As the president of one organization noted,

> For many people, including some Italian Americans, the celebration of Columbus is viewed as belittling the suffering of indigenous peoples at the hand of Europeans. But for countless people in my community, Columbus, and Columbus Day, represent an opportunity to celebrate our contributions to this country.

Explained another: "Columbus Day and the Columbus monument have played a vital role in Italian American acceptance and the celebration of Italian culture." Their Columbus is a symbol of their twentieth-century assimilation in America—a separate icon from the other Columbuses, the historic fifteenth-century mariner and the nineteenth-century American hero.[5]

Another Columbus monument in a city in New York State, likewise in a space called Columbus Circle, also came under threat. Some two hundred fifty miles northwest of New York City, the City of Syracuse (named after the city in Sicily) had attracted Italian migrants to work on the railroad in the late nineteenth century. By 1909, when the state declared October 12 to be an annual Columbus Day holiday, with Syracuse's population having tripled since 1880, the city's Italians were sufficiently established to begin a crusade for their own Columbus statue. After a long campaign, given impetus by a fund-raising Columbus Monument Corporation, by the leadership of the *prominenti* of the local Pietrafesa family, by a donation from Benito Mussolini, and finally by the support of Mayor Rolland Marvin, the monument was unveiled in 1934. Its Columbus faced the Cathedral of the Immaculate Conception, his back to the county courthouse. In standing atop a square-pedestaled neoclassical column, with a dedication and bas-relief bronze scenes at the base, he echoed his New York City sibling.

Unlike the monument in New York City, however, the Syracuse one featured four bas-relief scenes, expanding the story of the First Voyage by adding (at the start) Columbus petitioning the Spanish monarchs for support, and (at the end) his presentation to them of "Indians" taken from the Caribbean. The iconography was more that of the Capitol Columbus and its depiction of the nineteenth-century self-image of the United States as a bastion of civilization against barbarism. By depicting the First Voyage as a cycle of imperialist triumph, rather than a simple journey to America, the monument shifted away from the "Citizens of Italian Origin of Onondaga County" (as the dedication called them). The mayor had insisted on Dwight James Baum as architect. Neither were of Italian descent. Baum drew upon Irving's 1828 biography, on Rogers's *Columbus Doors* in the Capitol, and on a stereotype of all Indigenous Americans as wearing the feathered war bonnets of the Plains Indians. He added four more generic Indian heads, larger than those in the bas-reliefs, disembodied and attached to the column below Columbus's feet.[6]

Five Heads: The monument to Columbus in Syracuse, New York, has for decades been the site of protests and rallies. In the 2020s, Columbus Day events both against and in favor of the statue's survival were organized annually—as evidenced by this Columbus Monument Corporation flier.

Soon after its 1934 unveiling, it became clear that the bronze heads were staining the granite pillar, and they were unbolted and placed in storage until a solution was found. Paralleling the long wait of Persico's stored *Discovery* figures, the heads sat untouched for half a century, until two Syracuse men stole them and sold them at auction in Atlanta in 1986. The buyer, insisting he was unaware that he paid $32,400 for "hot heads," planned to display them in his restaurant in Orlando. But before they had a chance to stare down at Floridian diners, the heads were recovered by Syracuse officials and replaced on the restored monument in preparation for the Quincentennial. The heads' well-publicized 1989 recovery drew attention to the monument, whose bas-relief scenes were painted red during 1991 Columbus Day protests. Calls for the monument's removal were opposed by the Columbus Monument Corporation (which still existed, still supported by the Pietrafesa family, as it is today), and it survived to face another round of attacks in 2021. Mayor Ben Walsh proposed at various points to remove the whole monument, or just to remove the offending heads and bas-relief scenes; a 2022 state Supreme Court ruling determined that he lacked the authority to pull Columbus down.[7]

Syracuse is built on "hallowed ground," the former "capital of the Haudenosaunee confederacy" (which comprised the Mohawk, Oneida, Cayuga, Seneca, and Onondaga—after whom the county is named), and thus it is "burdensome for the people of Onondaga to see Christopher Columbus memorialized with a statue," as Chief Sidney Hill of the Onondaga Nation put it in 2020. For Indigenous protestors, the American Columbus was conceived as an icon of imperial triumph over Indigenous America, and the Syracuse Columbus has one foot firmly in that tradition. As the other foot is in the Italian American Columbus tradition, the Columbus Monument Corporation supported the mayor's 2020 suggestion to rename the circle "Heritage Park"—keeping Columbus to represent Italian immigrants while adding statues to represent other interest groups. These are clear positions, but the battleground is complex. As one scholar recently observed, the Black Lives Matter (BLM) movement "draws attention to racial violence in public space." It is, in other words, overtly iconoclastic. Columbus monuments have thus attracted BLM-related protests across the country—protests that resulted in the removal, either by protestors themselves or by pressured local officials, of almost fifty Columbus statues in the first half of the 2020s. That Columbus was involved in the Indigenous, not African, slave trade, is superfluous detail; such monuments are targeted as symbols of racial violence and slavery in a general sense.[8]

A further complication is that not all Italian Americans see the issue the same way. The assertion that "the violence of Columbus and those who came after him" was "rooted in white supremacy" (as one protestor told the crowd attending a 2021 Indigenous Peoples' Day rally in Syracuse) is confusing and offensive to some Italian Americans—many of whom know that the Klu Klux Klan and other white supremacist groups had violently discriminated against Italian immigrants, including stopping them erecting Columbus statues. But to third- or fourth-generation Americans of Italian descent, the historical context that led to the creation of the Italian American Columbus is now a very distant one. "This monument does not represent *my* heritage," one Syracuse native of Italian descent declared at the same 2021 protest, asking for "the removal of this false idol based on genocide."[9]

Such debates echo the contradictions of Columbiana's mythistories, prompting these questions: How and why did an Italian American Columbus, separate from the American Columbus as Adam, come to be developed? And why is he caught up in the battles over Columbus and his monuments in this century?

ITALIAN MOVEMENT TO THE United States in the late nineteenth and early twentieth centuries was the greatest migration moment in human history—in terms of how many people moved from one place to another in a concentrated period. A slow trickle in the half century before 1850 rose to almost 10,000 per decade until the 1870s, when the flow increased to 50,000, then 300,000 in the 1880s, more than 600,000 in the 1890s, and more than 2 million in the first decade of the twentieth century. Aside from a dip during the First World War, numbers were strong through 1921—when in that year alone a quarter of a million immigrants arrived from Italy. But the Emergency Immigration Act of 1921 established a quota system that lowered numbers from Italy—and 1924's Immigration Act lowered them further. Migration continued through the century, but under tight controls (the quotas were not repealed until 1965). Of the 5.3 million Italians who came to the United States in the nineteenth and twentieth centuries, four million arrived between 1880 and 1920.[10]

To give those numbers some comparative context, only 420,000 Spaniards migrated to their colonies in the Americas between Columbus's time and 1820—that's fewer migrants in more than three centuries than Italians who came to the United States in 1905–6 alone. From 1800 to the outbreak of the First World War, the European population more than doubled. But the United States population increased 21-fold, from 5 million to a whopping 106 million—a staggering growth achieved largely by immigration.

The majority of Italians migrating to the United States came

Genoa–New York: An 1899 poster advertising regular shipping service between Genoa and New York. Although Sicilians and other southern Italians would come to dominate the great 1880–1920 wave of Italian migration to the United States, the initial origin regions were in the north—such as Tuscany, Lombardy, and Liguria— making Genoa a source of migrants as well as a major port of departure.

through New York City, where many remained. Between 1880 and 1920, more than fifty thousand Italians settled in Manhattan's South Village alone, and tens of thousands more in nearby Little Italy. But they also fanned out across the Northeast and West, first to major cities— Boston, Philadelphia, Chicago, Denver, San Francisco—and then to smaller cities and to the countryside. Their cheap labor made possible the U.S. economy's unprecedented expansion, while their culture "fascinated, disturbed, and influenced" American culture—an influence that in the twentieth century became so profound that the U.S. today is culturally inconceivable without its elements of Italian and Italian American origin, from politics to religion, food to clothing, sports to music.[11]

That impression of a nation opening its arms to Italian migrants is misleading, however. Back in 1889, President Harrison warned in his inaugural address that because the United States did "not cease to be hos-

Italy in New York: Italian immigrants filled Mulberry Street in Little Italy, New York City, in 1900.

pitable to immigration," it was vital to avoid becoming "careless as to the character of it. There are men of all races whose coming is necessarily a burden upon our public revenues, or a threat to social order." Irish immigration had peaked in 1851, German in 1882, but not until 1907 would Italian arrivals peak, and Harrison's audience knew he had the "dagoes" in mind. In an age of open racism, Italians were publicly feared and smeared as "ignorant peasants," "filthy," and "racially degraded," derided both for taking "Negro jobs" and for mixing with Black Americans, assumed to be low-life thieves, rapists, and murderers, as well as members of international terrorist organizations. Internal Italian racism by northerners against southerners merged with white-on-Black American prejudice, resulting in "guinea" being added to "dago" and "wop" as quotidian slurs. If Italian men weren't Black, as "guineas" or "nigger wops" they weren't white either; they were sub-white or "dark white," recorded upon entry as Italian by "race," "white" in "color" but "dark" in "complexion." Kept poor by low-paying industrial jobs, Italian immigrants faced segregation in, or exclusion from, certain schools, theaters, labor unions, clubs, churches, and numerous other organizations.[12]

"The Italians are spreading, spreading, spreading," warned the New York Zoological Society's president in 1913, "If you are without them to-day, to-morrow they will be around you." He was incensed by the tendency of these "bird-killing foreigners" to hunt songbirds for food, as poor Italian migrants did throughout Pennsylvania and the Northeast out of necessity. Likewise, the tendency of immigrants to join unions and socialist organizations, even anarchist groups—all an understandable response to suffering low-paid, dangerous work in mines, sweatshops, mills, railroads, and construction sites, and being denigrated for it to boot—made them all seem criminally radical. The 1927 execution of Nicola Sacco and Bartolomeo Vanzetti, for a 1920 robbery in which two men died, was widely viewed outside the United States (and by many within it) as the unjust sacrifice of innocent men on the altar of anti-Italianism.[13]

Forced into crowded tenements in cities such as New York, "unsanitary" immigrants were then blamed for the conditions there, and for deadly outbreaks of disease throughout cities. In the late nineteenth century, *cholera infantum* and other diarrheal diseases took 20 percent of children who died before turning two in the United States; when infants in wealthy households died, multiple causes were touted, but poor children dying in tenement housing were seen as the victims of poor habits by their immigrant parents (such as allegedly inadequate bathing and keeping windows closed). Nurses who went into tenements to educate immigrants were well meaning and often self-sacrificing, but prevailing prejudices instilled in them a kind of condescension. "Many of these mothers were a little flattered to have an American lady take all that trouble about little Giovanni," said one in 1908. Two years earlier, a Quaker doctor named George Newman had published his *Infant Mortality: A Social Problem*, a widely read and influential book that effectively blamed mothers for their children's deaths. Poverty itself was not the cause, but "the social life of the people"—the parental culture of certain (especially Italian immigrant) communities.[14]

But the anti-immigrant racism of the era soon suggested an alternative solution: to deploy eugenic policies to prevent babies being born in such communities in the beginning, to "eliminate the unfit by birth, not by death" (as one prominent New York pediatrician put it). Most states had by 1907 passed laws permitting forced sterilization. The success of the 1916 best seller *The Passing of the Great Race*, which warned that Italians and other southern and eastern Europeans were causing racial "dilution" and disease, helped spur the 1921 law that dramatically restricted Italian immigration. Its 1924 sequel (the Johnson-Reed Act or National Origins Act) was overtly designed to stop "racial degeneracy," as manifested in the inherently diseased and criminal nature of Italians, its revised formula further restricting their incoming numbers written by the Eugenics Committee of the U.S. Congress's Committee on Selective Immigration.[15]

Attitudes would shift, but slowly and not because racist intolerance of immigrants declined. The end of the sixty-year wave of Italian immigration, the peaking of the eugenics movement in the 1920s, and the development of treatments such as antibiotics all had an impact. For example, a 1916 polio outbreak in New York and adjacent states was blamed on Italian immigrants in a Brooklyn neighborhood called Pigtown, and even when children in wealthy homes in Pennsylvania and Connecticut started suffering from it, some blamed servants for bringing in the disease. But in the decades that followed, leading up to the stunning success of the polio vaccine in the late 1950s, bigotry informed attitudes toward the disease less and less—helped in part by Franklin Delano Roosevelt (who had isolated his family on a Canadian island to avoid the 1916 outbreak, only to catch it himself five years later). Meanwhile, the 1940s represented a turning point for Italian Americans, who now "began to identify and mobilize as whites en masse"; the slurs and stereotypes persisted, but by the 1950s, they were no longer immigrants but Americans of Italian descent.[16]

That also meant that racist mobs and supremacist terror organizations such as the Ku Klux Klan and the White League—which were responsible for the above-mentioned 1891 lynching of eleven Sicilians in New Orleans—turned their attention to newer immigrant categories as well as to their long-term primary target. Lynching in the Deep South is more often associated with Black victims, for good reasons. Whereas dozens of Italians were lynched in Louisiana and elsewhere in the Deep South in the 1880s through 1910s, the lynchings of Black men were in the many hundreds. Between 1882 and the Second World War, some 3,500 Black Americans were lynch murdered. Thirteen hundred "whites" were also murdered the same way during those decades, the majority being Mexican, Chinese, and Italian. Those three immigrant groups were the primary targets in the West (from Montana to California), but otherwise white mob hostility against Italians was most violent and fatal in the same southern states where Blacks were most frequently attacked (with the most dangerously racist states being Mississippi, Georgia, Texas, Louisiana, and Alabama—in that order). My point is not to compare lynchings by race, but to recognize a pair of related points: mob violence against Italian immigrants was part of a larger phenomenon of murderous racism by white Americans against non-whites; and Italians, especially those from Sicily and other regions of southern Italy, were often placed into the same racist category as Black Americans, or one adjacent to it, and discriminated against accordingly.[17]

The United States is a nation of immigrants with a deep and long history of violent discrimination against immigrants—a discrimination that is a logical product of the racism of early modern imperialism, subsequently adapted to U.S. expansionism. A nation that managed to incorporate slavery into its foundational philosophy of liberty, while also legitimating a series of genocidal campaigns against the nation's Indigenous population, was hardly likely to welcome new migrants. Newcomers were easily seen as "other" and thus inferior, especially if they had one or more of the following alien markers: speaking languages other than English, practicing a religion that was not Protestant Christianity, being dark-skinned, and forming their own communities (a necessary survival strategy that fomented suspicion). But by changing those markers, if possible—in other words, by assimilating—migrants could move out of discriminated categories.[18]

Italians arrived with little understanding of the American color line. But they learned that in a nation built upon processes of colonization, dispossession, and slavery, to be white meant protection from systematic humiliation and violence. Whiteness potentially offered preferential access to citizenship, property ownership, safer working conditions and fair pay, better housing and education, social mobility and even political power. According to writer James Baldwin, immigrants faced a "moral choice." The "price of the ticket" for full admission into American society was "to become 'white.'" And that required joining in "the debasement and defamation of Black people." Writers and scholars have long struggled with the question of whether Italian Americans did indeed face such a choice, and how to judge or avoid judging them for how they assimilated into white America; either way, there has been since the late nineteenth century "a very complicated history of collaboration, intimacy, hostility, and distancing between Italian Americans and people of color." That history extends into this century, offering crucial context to events such as the 2020 toppling of the Columbus statue outside Baltimore's Little Italy by Black Lives Matter protestors.[19]

AND THAT BRINGS US back to Columbus. In seeking ways to assimilate as Americans—while also defending their local communities from discrimination and maintaining cultural connections to their ancestral Italian homelands—Italian immigrants embraced Columbus as a convenient icon. His ubiquity as a symbol of American patriotism, combined with his Genoese origins, made him an ideal choice (many had even traveled through Genoa to reach the Americas). The heroic American Columbus was embraced, appropriated, and turned into an Italian American Columbus. Italian immigrants joined the Knights of Columbus. They campaigned for new statues that captured something of the American Columbus but also shifted him closer to their Italian American Columbus, a shift from an imperial hero to an ethnic one, from one almost Protestant by association to one defiantly Catholic. And they worked to establish Columbus Day as a local (then statewide, then national) holiday on which Italian Americans could perform their assimilation by publicly celebrating both their Italian origins and their fervent patriotism as Americans.

The Knights of Columbus, sometimes assumed to have been an Italian American organization, was created by Irish Americans, and it has always been a fraternal order of the Catholic Church. Irish Catholics had faced considerable discrimination, despite the fact that the logic of U.S. racism gave Irish immigrants the advantage of being English-speaking and often fairer skinned than Italians. Their Catholicism was treated as a stigma, but it was also their strength, as the Catholic Church was already rooted and well organized across the country. When Irish immigrants were denied access to banking and insurance, the church stepped in. That was the role played by a Catholic brotherhood in New Haven, Connecticut, which in 1882 renamed itself the Knights of Columbus—a name appropriated for the same reasons that it was embraced by Italian immigrants: Columbus was both an icon of American patriotism and a Catholic, with "Columbianism" defined as the very essence of Catholic American patriotism. Knights viewed themselves as "Catholic descendants of Columbus." The Columbus connection, and the gradual inclusion of Italians, meant the Knights played increasingly important roles in organizing Columbus Day parades and in other Columbus-related campaigns—especially those that defended the church against anti-Catholicism, such as the effort to canonize Columbus as St. Christopher.

Before the Knights, Italian American community organizations, offering banking and insurance and other services and aid otherwise denied to immigrants, tended to be localized, often based on regional identity back in Italy (late-nineteenth-century Chicago alone had more than seventy distinct Italian-origin-based communities). Because a minority of them drifted into racketeering, they were often smeared with the "Black Hand" label—which evolved in the 1920s into "Mafia." Racist conspiracy theorists found plenty of confirmation from the press and politicians that Italian migrants were part of an international criminal syndicate hell-bent on taking over America. Local and regional Italian American organizations still flourished—this was the era of "fraternal" organizations in the United States—but none matched the impact of the Knights of Columbus, which was fast becoming a nationwide organization. It was also Catholic, not Italian, tied not to organized crime but to a Catholic Church that faced prejudice but was nonetheless a conservative establishment.[20]

Because public monuments were such an overt manifestation of the American Columbus cult, it was natural that Italian Americans would adopt the practice, using local community organizations (sometimes with the involvement of the Knights) to raise funds to erect statues in prominent city squares and circles—or in their own Little Italy neighborhoods. In the century between the erecting of Philadelphia's statue in 1876 and Newark's in 1971, hundreds of public monuments funded by and for local Italian American communities were built across the United States.[21]

Columbus statues tended to seed local and then statewide Columbus Day celebrations. The earliest October 12 commemorations may have been in San Francisco, where in 1869 Italian immigrants first started an increasingly elaborate ritual reenactment of the First Landing. The 1882 monument to Columbus and Queen "Isabella" in the rotunda of the California State Capitol was the site of commemorations that soon spilled over into parades, with Columbus Day events in Sacramento, San Francisco, and elsewhere becoming part of a statewide holiday from 1909 on. Several years earlier, Colorado had become the first state to make Columbus Day official. By 1934, when Columbus Day became a national holiday, it was already marked as such in thirty-four U.S. states and in twenty-one other nations in the Americas. Columbus monuments had helped seed parades; Columbus Day events in turn helped perpetuate the building of monuments.[22]

But the relationship between these sibling Columbus avatars, the American and the Italian American, was an uneasy one. For some Americans, the rise of the Italian American icon tainted their patriotic hero. Even as the popularity of the American Columbus ascended to its 1890s peak, contrary voices looked to undermine it or replace it with a hero that was farther from Italian and Catholic, closer to white and Protestant. Not coincidentally, the beginning of a counter-historiography more critical of Columbus dates to the very decades of a rising tide of prejudice against increasing waves of Italian immigrants. Some historians, admittedly, were, like Henry Harrisse, motivated by more rigorous scholarly standards and methods. But others, like Justin Winsor, a member of the Boston Brahmins (the city's

Boxed In: The Columbus Monument in Philadelphia's Marconi Plaza: boxed up in the summer of 2020 to protect it from attack and in anticipation of a removal that was halted by a court order; and unboxed, with patriotic holiday decorations, in 2022. Originally erected in 1876 in Fairmount Park, six miles away, as part of the city's great Centennial Exhibition, the monument was moved a century later to what was then, but is no longer, a predominantly Italian American neighborhood. The park was named after Guglielmo Marconi upon the inventor's death in 1937; his statue, erected in 1975, stares at Columbus across the street that splits the park in two.

Anglo-American and Anglican élite), were at least partly moved by an antipathy toward Catholicism.[23]

Others saw ways to replace Columbus with a founding father who reflected their own racial, ethnic, or cultural identity. Elemental to Columbus's role as an American Adam was his primacy. As he never saw or set foot on any land that became the United States, his status as an American icon rested entirely on his position as "the first," reflected in the fetishizing of the First Voyage, its First Sighting, and its First Landing. But what if he had not in fact been the first?

While Chicago was awash in millions of visitors to 1893's World's Columbian Exposition, a re-created Norse longboat named *Viking* was sailed from Norway to North America, where it made various stops (later ending up in Chicago), promoting a counternarrative: America's discoverers and founders were white northern Europeans. The *Viking* is a revealing footnote in a disparate, multifaceted movement (very broadly defined) that goes back centuries and is very much alive today—ironically helping to keep alive the very figure that the movement seeks to displace. As one recent historian has detailed, "non-Columbian narratives of American discovery" have been driven by "a complex interplay" of hatred of immigrants and of Indigenous peoples, "of ethnic superiority, ethnic pride, as well as distrust of academic scholarship." Some of these narratives seek to sideline or "blot out the unwanted history" of genocide and slavery, some center imagined ancient migrants from Africa or China or even aliens from other planets. But most reflect some variant on the ideology of white supremacy, usually anti-Columbian and anti-Catholic, responding to white American anxiety over who founded—and thus who owns, runs, is allowed to live in and teach the history of—the nation.[24]

Racist antipathy toward the Indigenous peoples who were being dispossessed and slaughtered in the eighteenth and nineteenth centuries fed, and was fed by, elaborate claims that ancient sites—such as the human-made mounds in Ohio and Toltec and Aztec ruins in Mexico—were the work of earlier Celtic or Scandinavian migrants, ancient Greeks or Phoenicians, the Lost Tribes of Israel, or a great Lost (white) Race. The medieval legends of the Irish St. Brendan and the Welsh

Prince Madoc were revived as evidence that white Europeans not only discovered America before Columbus, but discovered what became the United States, and even influenced Indigenous cultures; the existence of Irish-looking and "Welsh Indians" was so widely believed that reports of their sighting were not uncommon.[25]

More popular and enduring was the claim that Viking adventurers had discovered America. Just as Portuguese and Galician claims on Columbus's birthplace, and Dominican and Spanish competition over his remains, have spurred each other on, so was there ethnic rivalry in the nineteenth-century United States over the Discovery. The Vinland sagas, offering apparent evidence of the pre-Columbian Viking "discovery," reached U.S. audiences in translation in the 1830s—just when Washington Irving was solidifying Columbus's status as Anglo-America's founding grandfather. For the next century or so, the more that Italian Americans appropriated Columbus and made him theirs, the more Scandinavians sought to displace his primacy. The movement's hotbed was the upper Midwest, where Americans of Scandinavian descent were concentrated. A small but steady stream of books—such as 1874's *America Not Discovered by Columbus* by a University of Wisconsin professor of Scandinavian languages—sought to undermine the Columbus cult. A parallel stream of archaeological "discoveries" also kept the movement going—from Minnesota's Kensington Stone, whose runic inscriptions told of Norse travelers reaching the Midwest (in fact, a Swedish farmer carved it there in 1898) to runes in Oklahoma (also forged) to a mysterious Viking tower in Newport, Rhode Island (a seventeenth-century windmill). As for Columbus, he knew only of "the lands which he pretended to discover" after visiting Iceland and learning of the older Norse voyages; Columbus's "discovery" was thus merely a chapter in the Viking discovery.[26]

At the same time, Anglo-Americans in late-nineteenth-century Boston, anxious over the influx of Irish and Italian Catholics, convinced themselves that Vikings had settled Massachusetts. They claimed to find the site of Leif Erikson's house on the Charles River, as well as evidence in Waltham of a great lost Norse city that had seeded white civilization. The Viking claims faded in the twentieth century, as it became clear that archaeological "evidence" reflected Indigenous and early English settlements, and as the Vikings as a cultural and racial ideal were appropriated by the Nazis. But then, in the 1960s, real archaeological evidence was uncovered of an eleventh-century Viking settlement on coastal Newfoundland, reviving the narrative that they (not Columbus, nor Indigenous peoples) had "discovered America." On October 9, 2020, President Trump called for that day to be named Leif Erikson Day; a year later, President Biden proclaimed it so. There remains no evidence that anyone by that name ever set foot in North America.[27]

But America's oldest inhabitants could not forever be ignored. The late-twentieth-century challenge to Columbus's primacy came from two sides, like a crab's pincers. Some wanted him replaced with a whiter Discoverer. Others wanted him replaced with the original discoverers, those who had crossed from northeast Asia into Alaska and spread across the hemisphere tens of thousands of years ago. In the late twentieth century, and especially in the current century, our grasp of that protracted process of migration and settlement has improved dramatically. By 2007, such knowledge had even reached books for young readers; "in recent years," asserted one, "as new evidence has come to light, our understanding of history has changed. We now know that Columbus was among the last explorers to reach the Americas, not the first."[28]

At the same time, the *indigenismo* movement in Mexico and some other Latin American countries was paralleled by a slow but steadily increasing recognition of Indigenous rights in the United States, contributing to the sidelining of Columbus's primacy and thus his efficacy as an icon. Indigenous objections to him echo the oft-quoted statement made by Russell Means of AIM (the American Indian Movement) in 1989 (protesting the "First Encounters" exhibit at the Florida Museum of History): "Columbus makes Hitler look like a juvenile delinquent." Asking Indigenous Americans to take a balanced view of Columbus Day, said Means, was like asking Jews to adopt a "balanced view of the Holocaust" on a day named after Hitler.[29]

Is that an extreme opinion? Obviously, it is, and deliberately so, designed to garner attention at a point when the United States seemed to be gearing up to a 1992 that promised to celebrate Columbus as unthinkingly (or thinkingly as a national-imperial hero) as in 1892. I suggest, however, that Means's provocation should not be buried within the Quincentennial context, but understood as a very clear expression of how different the Indigenous American Columbus is to the white American and Italian American Columbuses. One was set up as an Adam figure to represent a white America that excluded Indigenous and Black people as citizens. The other was set up as a symbol of migrants loyal to their adopted nation—not necessarily hostile, but at best indifferent, to Indigenous and Black Americans.

Looking at Columbus as a mirror long worked for many white and Italian Americans and for American men who saw themselves as embodying Columbus's legendary qualities—rugged individualism, adventurousness, the will to be first. To a lesser extent, it worked for other migrant groups, for whom Columbus could serve as a vague symbol of the ritual of discovery and hope for a new life. But it has never worked for many other Americans, especially those identifying with the two groups who did not voluntarily migrate in recent centuries to the continent: Indigenous peoples and African Americans.[30]

Hence the iconoclastic logic behind attacks on Columbus statues by Black Lives Matter protestors. Because of the success of late nineteenth-century and twentieth-century campaigns to erect Italian American Columbus monuments, it was those monuments—rather than the American Columbus statues, older, fewer in number, some hidden in parks or museums—that tended to be targeted. Many Italian Americans were appalled when *their* icon was spattered in blood or red paint, or battered and graffitied, or beheaded (as in Boston), or yanked down and dragged into the nearest river or bay. Local officials flip-flopped and dithered, unable to please all sides, often simply passing the buck to their successors. In some cases, as in St. Paul, Minnesota, battles in the 2020s over Columbus monuments were part of a war dating back to the Quincentennial.[31]

In 2018, Los Angeles city councilman Mitch O'Farrell, a member of the Wyandotte nation, managed to secure the legal removal of a statue that a local Italian American association had donated to Los Angeles in 1973. Although anti-Columbus protesters were there to cheer their

Face Down: Columbus pulled down in St. Paul, Minnesota, by Indigenous American protestors from the Chippewa, Ojibwe, and Dakota communities. Minnesota State Patrol officers, who had met with AIM leader Mike Forcia before the June 2020 protest, watched while the statue was toppled by ropes, spat on, danced around, and draped in the flag seen here—intervening only when a flatbed truck came to remove the fallen 1931 ten-foot statue. Democrat governor Tim Walz noted that the action was illegal (Forcia accepted 100 hours of community service in a plea agreement). Lieutenant Governor Peggy Flanagan (herself Ojibwe) declared "I will not shed a tear over the loss of a statue that honored someone who by his own admission sold nine- and ten-year-old girls into sex slavery." Three Republican congressmen condemned what they called a "lynching-like desecration."

support, this statue was not violently toppled but gently placed by city workers on a truck bed and taken into storage. The relatively quick and quiet process was called "furtive" and still prompted controversy, exacerbated by O'Farrell's characterization of Columbus as "someone who committed atrocities and helped initiate the greatest genocide ever recorded in human history." Italian Americans were still smarting from the council's replacement of Columbus Day with Indigenous Peoples' Day the year before (the state of California followed their lead soon after). Efforts by Italian Americans to secure a compromise—two separate days to represent each community—had failed, largely because of the rhetorical disconnect between the two sides. Councilmen representing Italian Americans saw their ancestors as innocents who migrated "to build something and not to destroy something." They were right; the decline in the Indigenous population that began in the 1490s had predated Italian migration by four centuries. In fact, the nation that had reluctantly accepted Italians, mistreating them for generations, had at the same time systematically destroyed Indigenous communities—even, in late-nineteenth-century California for example, enacted state policies of genocide. Anti-Columbus hostility, intended not for Italian Americans but colonialism and its five-century abuse of Indigenous peoples, thus seemed aimed at the wrong icon.[32]

These arguments were repeated in cities across the United States, with each side insisting that they were not opposed to the other side—olive-branch phrases like "all of our individual cultures matter" and "we seek not to create racial conflict but to end it" were regular refrains by Americans of Italian, Hispanic, Indigenous, and African ancestry. Common ground always seemed just out of reach, despite being visible. After all, the fifteenth-century Genoese Columbus was far removed from migrants leaving a unified Italy, and thus by acknowledging the separation of the historic Columbus from the Italian Columbus, might he be neutralized as a flashpoint for community conflict? As one Italian American told Councilman O'Farrell in the Los Angeles debates of the 2010s, "We never hurt you. We never wanted to hurt you."[33]

In the late twentieth and early twenty-first centuries, tensions also developed over Columbus statues stemming from demographic changes within city neighborhoods—changes related to the assimilation of Americans of Italian descent into white America. One example is Philadelphia, where Black, Asian, and Hispanic residents now outnumber whites, greatly outnumbering those of Italian descent. Another is Baltimore, a city with three Columbus monuments, although only one drew serious attention in 2020: as Fourth of July fireworks lit up the night sky, protestors used ropes to pull the statue over and then drag it to the harbor's edge, where they tipped its broken pieces into the murky waters. The rhetoric by the protestors deployed the usual mix of references to racism, slavery, genocide, colonialism, and white supremacy—while

Italian Americans were dismayed by the association of *their* Columbus with past events that they saw as unrelated to their community's recent immigration history. Local Knights of Columbus members retrieved the statue from the bay, but it was in twelve pieces and, at the time of writing, the plinth at the edge of Little Italy remained empty.³⁴

Another northeast city with a strong Italian heritage tied more to the past than present is Newark, New Jersey. During the 2020 protests, a Columbus statue in Newark was removed by the city "to avoid the potential danger of people taking it upon themselves to topple it" (explained the mayor; this was not the 1971 statue mentioned above). Three years later, the plaza in which Columbus had stood since 1927, Washington Park, was renamed Harriet Tubman Square, and an abstract sculptural representation of Tubman and the Underground Railroad was unveiled. In 1927, thousands had attended the unveiling of the *Cristoforo Colombo* statue (stone-carved dedication in Italian) and the reading of a congratulatory telegram from Mussolini. But by 2020, Newark had become 85 percent Black and Latino. "Italians needed Columbus a century ago, they don't need him now," opined the author of *Newark's Little Italy*. "For us to defend our historical identity by choosing someone from Genoa, historically makes no sense." Newark's mayor insisted that he had no objections to people celebrating Columbus; "we're just asking that you not make us celebrate him as well."³⁵

ONE OF THE RESULTS of Italian American assimilation has been the muting of Italian ethnic identity; or, as some scholars have put it, the shift from "traditional" ethnic identity (in which members of an ethnic group have no choice; their lives in the United States are largely determined by that identity) to "symbolic" ethnic identity (in which its members—in this case Americans of Italian descent—can choose to embrace their *italianità* or not, but it does not determine where they live and work, how much they prosper or whom they marry, any more than does English ancestry). In this sense, Italian Americans in this century are very different from Latino Americans, whose ethnic identities are highly disparate—ranging from Hispanic or Indigenous-Hispanic peoples whose ancestors settled in North America before the English did, to recent arrivals from the Iberian Peninsula, to migrants of every kind from every corner of Latin America. For every Latino with ancient American roots or in fully assimilated communities, there are migrants defying massive, militarized border control efforts. The challenges of finding a common heritage among such peoples are mind-boggling, but even if such an effort were seriously attempted, Columbus—now at best a bifurcated icon, at worst a multifaceted symbol of highly contested historical memories—would hardly help. In a United States whose multiculturalism is both growing and contested, Columbus's utility has expired—except to serve as a culture-war battleground.[36]

In the 2020s, political controversy delayed the construction of the $1 billion National Museum of the American Latino, assigned a prominent site beside other Smithsonian museums on Washington's National Mall. Latinos whose ancestors have been Americans for generations bristled at being associated with those labeled "illegal" and "alien," questioning the notion of "*the* American Latino," and objecting to any emphasis on the suffering and oppression experienced by immigrants. "Hispanics are not victims or traitors," declared two Republican congressmen—Latinos from Texas and Florida—in response to the museum controversy. "Instead, they are the backbone of our American society." A preview of the unbuilt museum's content prompted a number of objections along such lines; for example, three conservative commentators wrote in *The Hill* that "the Latino exhibit simply erases the existence of the Hispanic who loves, contributes to, benefits from and exemplifies the promise of American liberty."[37]

Columbus and the long history of Columbiana in the United States is predictable fertile ground for such conflicting attitudes. There is no single or dominant view of Columbus among Latino or Hispanic Americans, nor should we expect one. For some, he symbolizes the oppression of the poor, of the uprooted, of those who have suffered racial prejudice; for others, he symbolizes that "promise of American liberty," and attacking him smacks of "the classic oppressor-oppressed agenda of textbook Marxism." Every such articulation of a viewpoint serves to harden the contrary opinion. And that will continue for as long as Columbus is seen as a symbol or icon rather than a historical human being.[38]

It is striking how much conservative Latino perspectives on their identity as both Hispanic and patriotically American—claiming those two categories as synonymous—echo that of assimilating Italian Americans generations ago. Looking back to 1892 from today, the competition between New York Columbuses looks like a battle between differing visions of the icon—the Spanish Columbus, the American Adam, and the Italian migrant. Although the last of these arguably won in a New York City bustling in 1892 with recently arrived Italians, the long-term result was more of a standoff. All those Columbuses lived on, as the Columbiana conflict gradually widened to include white supremacists of various kinds, Indigenous-rights organizations and perspectives, and the BLM movement. Amid so many competing causes, the reductionist rationale behind the Italian Columbus—he was from Italy and crossed the Atlantic Ocean looking for something more—seems appealingly simple, yet also outdated.

Epilogue

"HE NEVER CAME BACK," THE GENOESE CHEESE SELLER TOLD ME. Taking a break from reading fifteenth-century Genoese documents, I wandered into the deli on a whim, with Columbus's connection to the business—his father had sold cheese for a little while, his sister married a cheese maker—fresh in my mind. Asked what I was doing in Genoa, I explained whom I was studying, and remarked that I was surprised by the ambivalence toward Cristoforo Colombo that I found in his hometown. I struggled to articulate that surprise in my broken Italian, but the cheese seller got the gist of it, responding with an insight as simple and sharp as her knives: he left, became rich and famous, and never returned. Then she made an untranslatable dismissive gesture, and I bought a small round of *formaggetta di stella*.

From the moment one arrives in Genoa, Columbus is present. The airport is named after him. His statue faces the train station across a traffic circle packed with bus stops, all very appropriate to his reputation as one of history's great voyagers. And yet he is also invisible. The airport's name is not leveraged into a theme or brand of any kind; like the United States' sixty cities, towns, and counties, and its hundreds of streets, named Columbus or Columbia, Aeroporto Cristoforo Colombo is just a name. Similarly, train and bus and taxi travelers ignore Genoa's pigeon-stoop of a Columbus statue, its adjacent parklet now a dog toilet and a receptacle for cigarette butts; even the nearby newsstands no longer sell postcards—and thus no longer carry Columbus postcards. There is a Via Colombo, but it might as well be "Dove Street" or "Pigeon

Road" (as it may be to Genoese who paid no attention in history class). Columbus references are lightly woven into the fabric of the city, but far from worn on its sleeve.[1]

Are the Genoese simply a step ahead of the rest of the world? Genoa embraced the Quincentennial with enthusiasm, but in the end, it was about almost everything *but* Columbus—a pretext for urban development and for various political and cultural projects. Whereas the 1992 controversies in the United States made its Columbuses a persistent battleground, in Genoa an embarrassed indifference permits a local Columbus to cling barely to life. By contrast, in Spain, the nation that succeeded more than any other in turning the Quincentennial into a commemoration of itself, Columbus is more present than ever.[2]

After all, Columbus opened the door to Spanish, not Genoese, imperial glory. The flip side to that Genoese cheese seller's flippant dismissal of her city's famous son is the fact that his descendants are Spaniards. The living don Cristóbal Colón—the Twentieth Admiral of the Ocean Sea—recently quoted Francisco López de Gómara, aforementioned sixteenth-century Spanish hagiographer of the first don Cristóbal Colón and the conquistadors who came after him. "He said that the greatest feat after the birth and death of *Jesucristo*, our Lord, is the discovery of the Indies," the Twentieth Admiral told a reporter for Colombia's *El Tiempo*. "And so, it really is," he continued. "The event that has influenced the history of humanity the most has been the discovery of America. They say that reaching the moon was a feat; and yes, it was, but what changed in people's lives? Nothing."[3]

That view certainly resonated in the 1990s, when on both sides of the Atlantic there was a Columbiana explosion in the creative arts—from exhibits and books to films and video games. But thereafter the drop-off was sharp—sharper than after 1892—and it has been sustained since then. In the fiction and film of the twenty-first century, Columbus is not completely absent, but he has ceased to be the central character. An incidental figure, he tends to be antiheroic or simply marginal. For example, amid the numerous historical characters in the *Night at the Museum* movies, Columbus is a minor one; in the first movie of 2006, he is a determinedly helpful figure, but speaks only Italian, and of so little significance that Larry (the main character played by Ben Stiller) cannot remember his name. He comes off worse in the BBC television series *Doctor Who* (active on and off since 1963), in which he featured in a 2013 episode as violent and dangerous, an instigator of genocide. At a time when colonialism itself is usually portrayed as far from laudable, the role of Columbus is uncertain at best; in the *Avatar* movies, for example, in which colonialism is genocidal and environmentally destructive, there is no role for a discoverer, let alone a heroic one.[4]

As for fiction, the prolific science-fiction writer Stephen Baxter included Columbus in 2007's *Navigator*, a title that suggests a central role to a character who is merely incidental. Arriving four-fifths into the book, Baxter's "Colon" is a "lusty," "shallow" buffoon, an instrument used by others rather than an architect of change. Remaining largely offstage, he is talked about by characters more important to Baxter's plot and, by extension, to history. Another way of sidelining Columbus is illustrated by Berry's aforementioned novel, *The Columbus Affair*, in which he is so heavily fictionalized that he can exist as both violently duplicitous and a Jewish hero.[5]

When the teaching of history shifted in the late twentieth century to emphasize larger, impersonal forces rather than heroic men, the experience of groups rather than individuals, Columbus was similarly avoided or sidelined. The post-1992 association of Columbus with genocide was usually ignored, especially in textbooks and—understandably—in children's literature. Having been idolized for two centuries in histories for children, in the Quincentennial's wake Columbus was given weak praise at best, for being bold and courageous, while "the encounter" was vaguely referenced as unfortunately violent. Museum exhibits and related publications have increasingly followed literature for the young in adopting the tactic of avoiding Columbus altogether, with Columbus nudged aside to devote space to topics far removed from his lifetime, such as Indigenous history before 1492 or Mexican multiculturalism today.[6]

In the twenty-first century, Columbiana is about so much that has nothing to do with Columbus that the references are as nebulous as Genoa's airport name. Even small changes can be indicative of larger cultural swings. For example, the term "Columbusing" was used throughout the nineteenth century and into the twentieth to refer to discovering something American. When Charles Dickens and his wife, Kate (née Catherine Hogarth), returned to London from their 1842 U.S. tour, taken without their children, their poet friend Thomas Hood joked that "Mrs. Dickens must enjoy being at home, & discovering her children after her Columbusing & only discovering America!" Columbusing could also be an intellectual, rather than physical, journey; in an 1899 Philadelphia lecture, the literary critic Bliss Perry derided readers excited by their discovery of new poets as "people who have just gone Columbusing over Whitman and Browning and imagine their new world out-marvels the old, until they sail back again."[7]

Perry's tone notwithstanding, the term's usage was neutral, not negative, slowly becoming rather arcane and then fading away in the early twentieth century. However, when it returned in this century, its usage was decidedly negative, reflecting the shift in the larger perception of Columbus. Thus, a 2012 *Washington Post* piece on the white "gentrification" of historically Black neighborhoods in Washington, D.C., was titled "Columbusing black Washington," with the term a reference to racially fraught invasion and appropriation. By 2014, it was common enough to be parodied in places such as the *College Humor* channel on YouTube, defined as "discovering something for white people" (so, "Columbus Columbused America"). NPR (the United States' National Public Radio) took a more earnest approach: "Columbusing can feel icky." Using the term to describe cultural appropriations ranging from henna tattoos to the hypocrisy of anti-immigration activists "relishing fajitas while believing the line cook should get deported," the article warned us "to be careful not to Columbus other culture's traditions." On the social media sites of the 2020s, Americans claiming to "invent" or "discover" foods or dishes that have for centuries been part of Mexican or other Latin American cuisines were accused of "Columbusing."[8]

As Columbus fades from being an icon to being a vague reference with minimal ties to historical reality, understandings of that reality will continue to warp. I have overheard a Spanish speaker remark, standing in front of the Columbus statue in Genoa, that "I thought he wasn't even from Genoa." And I have heard an English speaker in the U.S. Capitol rotunda point at Vanderlyn's painting of Columbus's First Landing and ask, "that's Florida, right?" A veteran park ranger at Plymouth Rock told another writer that the most common question asked by visitors was why "1620" was etched on the rock—and not "1492." Another frequent query was, "Is this where the three ships landed?" Tourists were not thinking of vessels sailing with the *Mayflower*: "They mean the *Niña*, the *Pinta*, and the *Santa María*. People think Columbus dropped off the Pilgrims and sailed home."⁹

Commonplace ignorance, perhaps, but also a reflection of Columbiana's misunderstandings passed on for generations by word of mouth, print, image, film, or website. As this book has sought to show, Columbiana is essentially the sum of the misconceptions and myths—the mythistory, folk history, faithistory, and pseudohistory—that in some cases originated centuries ago, but mostly developed in the past century and a half. Here, summarized from the book's chapters, is a selection of them.

There was never any mystery regarding the identity of Columbus's parents and his birthplace (he was born and grew up Genoese); the "mystery" is an imagined pretext for alternative agenda-driven or partisan re-creations of his life. He was not formally educated, and he was not literate until his twenties. Queen Isabel did not pawn her jewels to pay for his First Voyage. There was no romance between Isabel and Columbus. He was not mocked by a committee of Spanish Inquisition officials and university professors whose geographic knowledge was inferior to his, nor was his understanding of the world's sphericity in any way unusual. He had no fully formed vision of his voyages, prior to the First Voyage, and thus had no specialized or secret knowledge acquired from his late father-in-law (whose legacy of charts and maps is fictional) or from Paolo Toscanelli (with whom Columbus never corresponded) or from the Unknown Pilot (who did not exist). The *Diario* of his First Voyage is not a surviving "diary" or "ship's log" that gives us direct access to Columbus's thoughts and deeds but rather a complex and contrived account whose primary narrative voice and filter is that of fray Bartolomé de Las Casas. Columbus neither saw nor set foot anywhere that became part of the continental United States. He was a Catholic Christian, neither covertly Jewish nor a lifelong pious millenarian; he grew gradually motivated by Christian religious ambition after his First Voyage. He was no saintly or model family man (and he did not marry the mother of his second son), but neither was he a womanizer, rapist, or devotee of bestiality.

If I am right to challenge all (or even some) of these aspects of Columbiana's mythistory, there remains a question: Why did these myths evolve and survive for so long? Here again is a selection of answers.

The transformation of the Genoese Cristoforo Colombo into the Spanish don Cristóbal Colón, via the Portuguese Christovam Colom, was not odd or unusual in an age before nation-states established the controls of modern citizenship; but from a modern perspective (nineteenth to twenty-first centuries), it can easily be wrongly characterized as strange, suspicious, or mysterious. The long history of publicly known lawsuits within the Colón dynasty and between it and the Spanish Crown fed the notion that the details of Columbus's life were so debatable as to justify litigation—a misconception generated by those litigants. The early U.S. appropriation and re-creation of Columbus as an Adam, as a heroic and patriotic icon of national identity, further opened the lid of Pandora's box, making possible both the Italian American Columbus and the later antiheroic genocidal, slave-trading Columbus. The fetishizing of primacy, of Columbus as "the First," encouraged a parallel battle over primacy and thus origin mythology in the Americas; in the United States, that became part of a battle over the identity and ownership of the nation itself. Columbus was an instrument of empire, not its creator but a tool at its early modern dawn; he was thus easily misappropriated in service of modern ideologies. At the same time, the discipline of history emerged as a study of Great Men, providing fertile ground for competing visions of Columbus as a saint—a literal saint as well as variations on a secular one. The rise—at the end of the twentieth and in the early twenty-first centuries—of postmodernism in the scholarly world, of the spread of "outsider" pseudoscholarship (from pseudohistory and pseudoarchaeology to conspiracy theory movements), and of the virality offered by the internet, exploded any area of study that already featured mythistory into an ever-widening chaos field of invented mysteries and outlandish solutions; Columbiana has proved to be a prime candidate for such treatment.

To elaborate a little on this final point: the culture of conspiracy and fakery was not created by the internet; it is rooted in human behavior, with the internet simply creating a digital network for its maintenance. That is borne out by the history of Columbiana, which is heavily imbued with conspiracy theories and faked mysteries going back as far as Columbus's own lifetime. With every leap in communication culture and technology—the rapid spread of print culture in the sixteenth to eighteenth centuries, the spread of broadsheets and newspapers, the nineteenth-century emergence of academic disciplines out of amateur scholarship and their parallel rivalry into the twenty-first century, and the internet's invention—Columbiana has become increasingly fertile ground for such imaginative fakery, pseudoscience, and pseudohistory.[10]

We have seen that there are specific contexts that help explain Columbiana beliefs that run contrary to historical evidence. Examples include the international Catholic response to nineteenth-century anticlericalism as context for the conviction that Columbus married both his sons' mothers; micropatriotic sentiment and regionalist rivalry as context to the nexus of Iberian claims on Columbus's birthplace and origins; Iberian claims and anti-Semitism's complex history as parallel contexts for the assertion that Columbus was secretly Jewish; Black Lives Matter and other social movements of the early twenty-first century as context for the notion of Columbus as instigator of genocide and bestiality; the long, fraught history of immigration across the Atlantic as context for Columbus as an identity battleground in the United States.

But, as we've also seen, there is another layer of explanation to all these beliefs. Recall the notion of the mental rough corner that ended this book's Introduction. Psychologists theorize that pseudoscientific and pseudohistorical beliefs gain traction "when they hit a sweet spot of strangeness"—when they are neither too bizarre (prompting disbelief), nor too banal (prompting apathy), but instead are just counterintuitive enough to be "attention-grabbing, memorable, and likely to be passed on to others," while also reinforcing someone's perception of the world by giving it "an intriguing twist." Columbiana fits such analysis well, and it is likely to do so for many generations to come. Pseudo-historical beliefs often "coexist alongside incompatible prior beliefs" or evidence-based opinions. That is why a Corsican may both *believe* Columbus was born on that island and at the same time *know* that he was not; or why you may read this book and accept all its arguments, yet still paradoxically retain the belief that Columbus was a Gallego or was Jewish, that he abused manatees or had a love affair with Queen Isabel, that he is a heroic American founding father or that he should be held responsible for genocide and slavery in the Americas. Holding such beliefs "does not necessitate an implicit commitment to them in all contexts." Columbiana is not unitary; it contains innumerable perspectives, many of them paradoxical.[11]

Indeed, Columbiana isn't ultimately about Columbus. It is about all its contributors, including us, for we have kept alive his nine lives. Those lives are likely to be given periodic jolts of electric attention by historians and journalists, politicians and community activists, for generations to come—if not centuries. If the notion of public arguments over Columbus in 2192 seems outlandish, consider the story you have just finished reading.

Acknowledgments

AN ENORMOUS AMOUNT OF WORK WENT INTO THE MAKING OF THIS book—that is, work by archivists and librarians, tourism officials and publishing professionals, colleagues and students. That includes those, too numerous to list yet invaluable to the whole project, at: the Archivo General de Indias in Seville; Oxford's Bodleian Library; the Library of Congress in Washington, D.C.; the archive of the Curator of the Architect of the United States Capitol; the Archivo di Stato and numerous other research locations in Genoa; various historical and research sites in Galicia, in other parts of Spain, and in the Dominican Republic; Tulane University's Latin American Library; the Helmerich Center for American Research at the Gilcrease Museum and University of Tulsa; the John Carter Brown Library; and Special Collections at Penn State.

For numerous suggestions, comments, leads, and other forms of inspiration and assistance, I thank Sergio Aguirre, Robert Aronson, Mary Beard, Arne Bialuschewski, Jennifer Blancato, Richard Conway, Tom Cummins, Clara Drummond, Chet Van Duzer, Greg Eghigian, Felipe Fernández-Armesto, J. Michael Francis, Larry Gorenflo, Jana Gowan, Peter Hartman, Christopher Heaney, Joe Herb, Christine Hernández, Anthony Ilacqua, Mark Koschny, Michael Kulikowski, Francesco Lacopo, Lucía Lago, Kris Lane, Domingo Ledesma, Mark Lentz, Esteban Moctezuma, Bianca Nottoli, Kim Nusco, David Orr, Manuel Ostos, Al Rocco, Gonzalo Rubio, Frauke Sachse, Amanda Scott, Heather Scott, Charles Seguin, Hope Silverman, Amara Solari, Ian Spradlin, Camilla Townsend, Louis Warren, and Andrew Wood;

with a special thank you to those who read all or almost all of the manuscript with keen insight; and likewise special thanks to my Penn State research assistants Micaela Wiehe, Travis Meyer, and Ashley Smouse, as well as their peers in my "Empire and Society in Latin America" graduate seminar.

I am grateful to the funding support of the National Endowment for the Humanities, the United States Capitol Historical Society, and the Pennsylvania State University's College of Liberal Arts. The path that led to this being a Norton book was lit by Louis Warren, the agents at Writers House, and Dan Gerstle—Norton's editor extraordinaire, whose insight and guidance made the book what it is. I cannot thank him too much.

I am especially grateful to Lucía Lago for planting the seed for this book, for her support and contributions. I also embrace with thanks the Restalls and Solaris of my family support network, including my parents Robin and Judy (who lived to enjoy the story of my discovery of the Colón Gallego, passing respectively in 2023 and 2024), my magical sister Emma, my four spectacular daughters, my wonderful in-laws, and above all my spouse and partner, Amara Solari, who is unfailingly and astonishingly tolerant and supportive every time I disappear down a rabbit hole—including the many of Columbiana. Amara and Lucía planted that seed together, and Amara nurtured it in myriad ways from start to finish. Mil gracias, mil grazas, mille grazie.

Notes

EPIGRAPH

Inspector Torigai in Matsumoto, *Tokyo Express*, 119; Lamartine, *Christophe Colomb*, 53 ("Man, like the Ocean, has both a tendency to movement and an innate weight of immobility").

INTRODUCTION

1. Provost, *Columbus*, xvii, describes "Columbiana" as "various matters like Columbus's ships and his place of burial that lend themselves to separate study divorced from various other matters"; I use it to refer to the entire literary and visual industry surrounding Columbus, especially the promotion of particular versions of him.
2. As of 2024, the portrait was on view in Gallery 625 of the Met, whose website conceded only that its identification as Columbus "is debatable," while making no mention of the clergyman, and suggesting that Columbus's son Fernando Colón might have commissioned the painting in Rome in 1516–17 (metmuseum.org). Baker-Bates, *Sebastiano del Piombo*; Dugard, *Last Voyage*, 265–66. See also the witty post by Shoshone on sartle.com (from whom "carrying the burden" is paraphrased).
3. The portrait in the Uffizi was painted in 1556 (a half century after Columbus's death) by Cristofano dell'Altissimo. A copy of a seventeenth-century portrait was made for New York's senate chamber in Albany in the 1780s. Around 1880, U.S. ambassador Lucius Fairchild commissioned a copy of the Yanez portrait of Columbus then in Madrid's Biblioteca Nacional, which ended up in the Virginia governor's mansion and then its Historical Society. Numerous other examples are detailed in Butler, *Portraits*, 3–6; Thacher, *Christopher Columbus*, III, 3–83. Also see Curtis, *His Portraits*; Bushman, *America Discovers*, xiii, 130; Summerhill and Williams, *Sinking Columbus*, 65; LoC/HH, Container 4; various in Bedini, *Encyclopedia*, 315–37. Genoa's Museo del Mare holds what it claims, echoing the Met, is "the image that over time has established itself as the 'face' of the Navigator," while admitting that the portrait, attributed to Ghirlandaio and discovered in a street market in Florence in the nineteenth century, is only imagined to be Columbus—a hypothesis without evidence. The portrait is nonetheless used in the brochure for Genoa's Casa di Colombo—itself apocryphal (*Casa di Colombo*; my field notes, September 2023). On Columbus images in the U.S. Capitol, see Life Eight.

4. Quote by Horwitz, *Voyage Long and Strange*, 47. Paul Tetreault, the director of Ford's Theatre (where Lincoln was assassinated in 1865), claimed in 2012 that fifteen thousand books have been written on that U.S. president (replicas of half of them are towered in a sculpture in the lobby of the theatre's educational center), making him second only to Jesus Christ, book-wise (npr.org/2012/02/ 20/147062501). On toponyms, etc., see Shin et al., "Columbus Monuments," and Life Nine.
5. Biography attributed to Fernando Colón, the *Historie della Vita*, is RC: XIII (also Colombo, *Historie*). On fictography: Boyd, *The Romantic*, 459. "Cranks": Fernández-Armesto, *Columbus*, vii. For Phillips and Phillips, the problem is "scholars have not done a good job" of disabusing "the broader public" of the "simplistic notion of Columbus the unblemished hero" (*Worlds*, 7). I have capitalized First Voyage (etc.) for clarification, not to perpetuate a tradition of grandiosity.
6. Other examples of books committed to an agenda-free examination of evidence include Henige, *In Search of Columbus*, and Davidson, *Columbus*.
7. Harrisse, *Christophe Colomb*, I, x ("L'obscure origine de Christophe Colomb... d'une entreprise dont on commence seulement à calculer toute la portée"). All translations are mine unless otherwise stated. Harrisse readily decried the "uncritical" nature of earlier Columbus biographies yet saw Washington Irving's—its recasting of history into historical fiction so transparent today—as "a history written with judgment and impartiality" ("une histoire écrite avec jugement et impartialité") (*Christophe Colomb*, I, 136). Irving: see Life Eight.
8. Horwitz, *Voyage Long and Strange*, 47; Fernández-Armesto, *Columbus on Himself*, 17; Phillips in Bedini, *Encyclopedia*, 494–97.
9. Thacher, *Christopher Columbus*, III, 84–488, presents forty-two documents written and signed by Columbus; also see Wilson-Lee, *Shipwrecked Books*. Huyghe, *Columbus Was Last*, 213 ("so little"). On the nature of Columbus's so-called shipboard logs or diaries, see Life Two.
10. Berry, *Columbus Affair*, jacket flap and 425 (the "what is clear" quote is in Berry's "Writer's Note"); also see Sobral, "Columbus and Jamaican Jews."
11. Muldoon and Fernández-Armesto, *Medieval Frontiers*, xxviii ("ignorance").
12. Bartosik-Vélez, *The Legacy*, 10 ("symbol"), 6 ("inherently" in next paragraph), 14; Zahniser, "L.A. City Council."
13. Schreiber, *Theory of Everything Else*, xiii.

LIFE ONE: THE GENOESE

1. Non-Genoese claims discussed and detailed in Lives Six and Seven. The assertion that Domenico saw Cristoforo as looking like him is pure speculation by me, based on the well-evidenced typical reaction of fathers to newborns; there are no reliable descriptions of Columbus's parents' appearances or of him as a child.
2. CGCC, 6–69 presents in facsimile, with partial transcriptions and translations into English and German, some ninety references in sixteenth-century publications in six languages to Columbus as Genoese; 79–91 adds diplomatic correspondence of 1498–1502; 99–178 adds over thirty notarial documents attesting not only to the fact that the Genoese wool merchant Cristofero Colombo was the same man who would later become the Admiral don Cristóbal Colón, but also detailing some of his ongoing con-

nections to Genoese merchant houses; 181–284 adds dozens more documents, including some penned by Columbus. Notarial documents from Genoa and Savona, along with those written by Columbus in Seville and elsewhere, almost two hundred in total, are also presented with English translations in RC: IV. Also see LoC/HH, "Fasti" No. 1, IV; No. 1 B-C. Farina and Tolf, *Columbus Documents*, is a useful guide to the documents in the Savona and Genoa State Archives. Note that because spelling was not yet standardized in the fifteenth and sixteenth centuries, Columbus's original Christian name was styled variously as Cristofero, Christofforo, and—the most common, which I use—Cristoforo.

3. Dotson and Agosto, *Columbus and His Family*, 6–7, 448; Taviani and Nader in Bedini, *Encyclopedia*, 129–30, 161–67; Farina and Tolf, *Columbus Documents*; Airaldi, *Colombo*.

4. Taviani, *Great Adventure*, 3, continues "... and gives back only the service, tenacity, commitment, and genius of Genoese sons." Taviani (1912–2001) was an economics professor at the University of Genoa, a politician (a deputy, minister, senator, and senator-for-life, from 1946 until his death), and a passionate promoter of the Columbus legacy (writing numerous hagiographic works and spearheading Italy's efforts to celebrate the 1992 Quincentennial); see Summerhill and Williams, *Sinking Columbus*, 11–17, 152–56. On the city and coast in the fifteenth century, see Ferro, *Liguria and Genoa*.

5. RC: IV, 98–106, 275–82 (quoted phrase on 100, 277). At least seventy-seven notarial documents recording Domenico's business affairs have survived, in part because of his eldest son's later renown (RC: IV, 30–159, 212–334). The gate tower where Domenico was assigned still stands; it is unknown where the Colombo family lived at the time, and no adjacent fifteenth-century houses still exist; a fragment of a house near the gate tower is billed today as the Casa di Colombo—accepted by locals as a harmless fiction. Aged forty-one, Columbus claimed—in a paraphrase by Las Casas—that "I have traveled for 23 years upon the seas, without leaving it for any time long enough to count" (Yo he andado veynte y tres años en la mar, sin salir d'ella tiempo que se aya de contar; RC: VI, 378; also in Navarrete, *Viages de Colón*, 101; CVD: I, 149–296). Aged fifty, he claimed that he sailed "from a very early age" (de muy pequeña hedad), claiming that "for 40 years now" (ya pasan de XL años) he had been practicing "the art" (arte) of sailing. But this was in a letter to the monarchs in which he expounded at length on his own, great, God-given "intelligence" (ynteligençia) and immense "talent" (abondoso), learning, skill, and ability for navigation and all related sciences and arts (RC: III, 66–67). Columbus tended to exaggerate the years he had spent doing anything; perhaps he first sailed in 1469, but the nonstop "sailing the seas" surely began a few years later. His vague recollections and exaggerations must be offset by specific evidence. Likewise, the absurd claim that Columbus attended the University of Pavia is not be taken literally—although many have (e.g., Keen, *Life of the Admiral*, 35–36, 40–42; Flint, *Imaginative Landscape*, 42).

6. CGCC, xv–xx; RC: IV, 30–140, 212–316. Most of these records were in Latin; e.g., in August 1472, "Domenico Colombo, woolworker, resident of Savona, and his son Cristoforo" (Dominicus Columbus lanerius habitator Saone et Christoforus euis filius) agreed to pay "one hundred and forty in Genoese coinage" (centum quadraginta monete Janue) that "being the sale price of seven bundles of Sorlini and Biolante wool..." (sunt occaxione precii vendicionis cantariorum VII lane de Sorlinis et Biolante...) (facsimile and transcription in CGCC, 148–49). On "wines sold and delivered to the same Cristofero and the said Domenico" (vinorum eidem Christofforo et dicto Dominico

venditorum et consignatorum) in 1470, see RC: IV, 88–90, 266–68; CGCC, 132–33, 170; Harrisse, *Christophe Colomb*, I, 166–216.
7. Their sister, Bianchinetta, married a cheese merchant, Giacomo Bavarello. Her father and grandfather had at times been cheese mongers. Bianchinetta and her husband ran a cheese shop in Genoa. Although Bianchinetta outlived Cristoforo by eleven years, neither he nor his brothers ever mentioned her in correspondence or in their wills; she was apparently irrelevant to their business affairs. RC: IV, 86–152, 264–327 (on Bianchinetta Colombo and Bavarello, 150–52, 184–87, 325–27, 358–61, 436); Schoenrich, *The Legacy*, I, 202–4; Harrisse, *Christophe Colomb*, I, 199–200; II, 222–26. The Genoese named in Columbus's May 19, 1506, will, were all old creditors of his (AGI, *Patronato* 1, legajo 5/12, no. 6; CGCC, 249–53, facsimile 252–53 inter).
8. RC: IV, 30–159, 212–334; Dotson and Agosto, *Columbus and His Family*, 7–11.
9. Catalans played an early role in the process, and Florence's participation was significant too, albeit less than Genoa's (Fernández-Armesto, *Before Columbus*, 3–5, 151–222; Games, "Atlantic History").
10. Columbus becomes a more prominent part of this story in Life Two. Fernández-Armesto, *Before Columbus*, 127–28, 152–53; Dotson and Agosto, *Columbus and His Family*, 11–17; Rogers, "Vivaldi Expedition"; Lester, *Fourth Part*, 95.
11. Dotson and Agosto, *Columbus and His Family*, 17–22.
12. CGCC, xvi, 136–37, 172–73; RC: IV, 135–40, 312–16; Harrisse, *Christophe Colomb*, I, 254–67. In some sources, Centurione's first name is Lodisio (actually another family member), and Dinegro is styled di Negro.
13. AGI, *Patronato* 1, legajo 5/12, no. 6; CGCC, 249–53 ("A los herederos de Luis Centurion Escoto mercader Ginovés . . . a los herederos de Paulo de Negro, Ginovés").
14. RC: IV, 335. In a 1500–1501 property-dispute case in the Savona archives, Cristoforo and his brothers Bartolomeo and Giacomo (Diego) appear as *absentium* ("absent" or "in absentia") defendants (RC: IV, 344). San Giorgio letter in RC: IV, 172, 346 (facsimile on 347 mislabeled as this document; it is actually of a different letter Columbus sent to Genoa in March that year); his correspondence with the bank in RC: IV, 171–74, 176–79, 345–49, 351–54. Columbus wrote his letters to Genoa in a Portuguese-laced Castilian, being illiterate in his native Ligurian dialect without reason to learn what we think of as Italian.
15. And astonishingly, considering Fernando Colón's vast library was largely destroyed after his death (Wilson-Lee, *Shipwrecked Books*, 326–27; its remnant is BCC in my Bibliography; also see Pettegree and Der Weduwen, *Library*, 90–99). We return in Life Four to Fernando, his books, and the complex authorship of the biography attributed to him (Colón, *Historie della Vita*, in RC: XIII; Colombo, *Historie*).
16. This insight was made by Wilson-Lee, *Shipwrecked Books*, 305–7, 338, 373; also see Sabellicus, *Secunda pars enneadum*, Dec. X, Lib. 8; the copy owned by Fernando Colón is in BCC, Colombina 2-7-11.
17. Colón, *Historie della Vita* (RC: XIII, 240–42, quotes on 241; "Dio, che per alta maggior cosa l'aveva salvato; tanta cortesia e sí buona accoglienza che mise casa in quella città e tolse moglie"); Wilson-Lee, *Shipwrecked Books*, 307. Madariaga, *Christopher Columbus*, 69, is an example of how Columbiana embraced this moment as one when "a great soul" was "reborn."
18. Morison, *Admiral of the Ocean Sea*, 24, and *Southern Voyages*, 13; Berggren, *Columbus*,

55–56; Boyle, *Setting Sun*, 47. Examples of "careful scholars": Davidson, *Columbus Then and Now*, 30–34; Wilson-Lee, *Shipwrecked Books*, 305–7. Catz, mostly reliable, fails to resist the tale and begins *Columbus* with it (1–3). Phillips and Phillips, *Worlds*, 95, is a good example of careful scholars repeating the tale with insufficient skepticism.

19. Note that the date of that invented battle is always given as 1476. The claim that Bartolomeo was a professional cartographer in Lisbon, and that he had set up shop before his brother's arrival, seems to have originated with Antonio Gallo, Genoa's official historian during the time of the Columbus brothers; the relevant account was edited and not published until 1733–51 (Catz, *Columbus*, 4, 18). However, there is no evidence that Columbus arrived first.

20. Fernández-Armesto, *Columbus*, 17; Sale, *The Conquest of Paradise*, 54. Circumstantial evidence suggests a late-1478 or 1479 wedding. The loss of records means we cannot be completely sure of Filipa's father and his father. "Perestrelo" only appears in writing as part of Filipa's name from 1535 on (LoC/HH, "Fasti" No. 2, 2A, 3). As Davidson observed, in no surviving document does Columbus name his wife or her parents, nor does their son Diego in his will or any other document name Perestrelo as his grandfather (he calls his mother *doña Felipa Muñiz*); the unequivocal assertion of Filipa's parenthood may be "skating on very thin ice indeed" (*Columbus*, 46) but does not imply uncertainty regarding Columbus's own origins.

21. Phillips, *Testimonies*, 5; Fernández-Armesto, *Columbus*, 17.

22. Catz, *Columbus*, 25–34; Madariaga appears to have been correct in crediting ("perhaps maliciously," in Catz's opinion; 30) Perestrelo's governorship "less to his sea-adventures than to the love-adventures of his gifted sisters" (*Columbus*, 86).

23. Taviani, *Great Adventure*, 28; Fernández-Armesto, *Conquest of the Impossible*, xii; *Columbus*, 4; "Cristóbal Colón" (seeing Columbus inspired by chivalric romances popular at the time, such as Joan Martorell's *Tirant lo blanc*, with its romanticization of island kingdoms in the Atlantic).

24. The Fernando Colón–attributed *Historie della Vita* renders Bartolomeu Perestrelo's name as "Pietro Mogniz Perestrelo" (RC: XIII, 242); Davidson, *Columbus*, 44–46; Morison, *Admiral of the Ocean Sea*, 16; Phillips and Phillips, *Worlds*, 99 ("valuable").

25. CoH, 21–22, 25–26; Catz, *Columbus*, 4–6, 8–9. Was Columbus, therefore, a slave trader? We return to this thorny topic in Life Two.

26. In Taviani's words (*Great Adventure*, 28); Catz, *Columbus*, 5–7; Fernández-Armesto, *Columbus*, 32.

27. Taviani, *Great Adventure*, 56; Catz, *Columbus*, 8–11, accepts all three Toscanelli letters as authentic.

28. Taviani, *Great Adventure*, 56; Thacher, *Christopher Columbus*, I, 301–24; Vignaud, *Toscanelli and Columbus*, quote on 249. Davidson's analysis of the Toscanelli letters is outstanding (*Columbus*, 50–59, 74–80). Examples of scholars accepting the first letter but not the other two are Fernández-Armesto, *Columbus*, 30, and Phillips and Phillips, *Worlds*, 108. Sale, *The Conquest of Paradise*, 357, is cautiously skeptical, as is Wilson-Lee, *Shipwrecked Books*, 23. Wey Gómez, *Tropics of Empire*, 357–58, 384–88, and Boyle, *Setting Sun*, 60, are not. Both Las Casas and the Colón-attributed *Vita* offer a clue as to how Columbus or his brother might have acquired a long-lost document, one that they later rewrote to look like a letter in which Toscanelli gave credence to Columbus's pre-1492 geographic assertions: both sources claim the first Toscanelli latter came from a mem-

ber of the Berardi family; the Berardis were Florentines connected to the Vespuccis and Colombus brothers through slave trading, other business dealings, and the Genoese and Florentine diaspora in Lisbon and Seville; Vespucci, Berardi, and Bartolomeo Colombo likely overlapped in Paris in 1479–80. This is admittedly speculative, but it helps us to envision a scenario of forgery in the 1493–1506 years, when Columbus was struggling to prove that his voyages fulfilled his contracted claim to have sailed to Asia.

29. RC: VII, 43, 273 ("descubriria grandes tierras...e gentes infinitas"); Fernández-Armesto, *Columbus on Himself*, 19, 23; Catz, *Columbus*, 11–16; Davidson, *Columbus*, 60.
30. Catz, *Columbus*, 12–15; Taviani, *Great Adventure*, 59.
31. Catz, *Columbus*, 15–16 (translation hers); LoC/HH, "Fasti" No. 4, A.
32. Fernández-Armesto, *Columbus*, 45; Restall, *Seven Myths*, 6–18.
33. Herrera, *Historia General de los Hechos*, Dec. I, Libro I, Cap. 7, 13 (1601), 11 (1726) ("por mas comoda pronunciacion").
34. Muldoon and Fernández-Armesto, *Medieval Frontiers*, xxviii. The 1946 work is Toro, *Historia Colonial de la América Española*, I, 215 (translations mine).
35. Chiari, *Christopher Columbus*, 5; also quoted in Russell, *Flat Earth*, 3.
36. Boorstin, *The Discoverers*, 100, 107; also quoted in Russell, *Flat Earth*, 4.
37. Scott, *1492*; this point also made in Restall, *Seven Myths*, 5–6 (Obregón painting on 5; original in the Museo Nacional de Arte, Mexico City).
38. Muldoon and Fernández-Armesto, *Medieval Frontiers*, xxviii; Morison, *Admiral of the Ocean Sea*, 89. Washington Irving is often credited with spreading the Flat Earth myth through his depiction—in his influential 1828 biography—of Columbus's imagined run-in with "monastic bigotry." In fact, Irving's derision of the "council of clerical sages" centers on a cluster of alleged "deficiencies" and "prejudices," some being the Flat Earth belief but most being the belief that no lands could exist on the earth's other side for they would fall off, and sailing back from there would be an impossible journey up "a kind of mountain" of chaotic waters (*Life and Voyages* [1828], I, 73–81).
39. United Nations A/37/PV.83 (November 29, 1982), 1371; Russell, *Flat Earth*, 53–54; Huyghe, *Columbus Was Last*, 2. Also see Eco, *Serendipities*, 4–7, on the late-twentieth-century persistence of the nonsense that Columbus "wanted to prove" to the flat-earther "Salamanca sages" that the earth was round.
40. Morison, *Admiral of the Ocean Sea*, 88; Razaf, "Christopher Columbus"; Restall, *Seven Myths*, 1, 6. In 1937, while Waller's hit was still playing on the radio, filmgoers heard these opening lines, written by Ira Gershwin, to "They All Laughed," a hit song from the Rogers/Astaire movie, *Shall We Dance*: "They all laughed at Christopher Columbus, when he said the world was round." Russell, *Flat Earth* (quote on 3), cites myth-busting studies going back to the 1920s, along with 1980s textbooks that correctly ducked the myth, but he argues that all were overwhelmed by popular belief and numerous repetitions of the error.
41. John of Holywood's early-thirteenth-century *The Sphere* even illustrated how all sailors knew through observation the world's sphericity. Russell, *Flat Earth*, 15–16; Lester, *Fourth Part*, 27–30, 84.
42. CoH, 161–64, 202–6; Fernández-Armesto, *Columbus*, 35–41; Crosby, *Measure of Reality*; Muldoon and Fernández-Armesto, *Medieval Frontiers*.
43. Fernández-Armesto, *Columbus*, 54; Russell, *Flat Earth*, 8; Ballesteros Beretta, *Cristóbal Colón*, I, 446–55. At the time of the committee's meetings, Hernando de Talavera

was not yet archbishop of Granada, which was still the capital city of a Muslim kingdom, although he is almost always given the title in modern depictions of the commission. Another Columbiana myth that is sometimes attached to "Talavera Commission" mythistory (e.g., Boyle, *Setting Sun*, 111) is that of Columbus's Egg: Giorgio Vasari had told of architect Filippo Brunelleschi cleverly making an egg stand on its end by slightly crushing it; the tale was stolen by Girolamo Benzoni and in his 1565 *Historia del Mondo Nuovo* applied to Columbus; it then took on a life of its own, repeated in drawings (most famously by Hogarth in 1752) and accounts for centuries, even reflected in modern Columbus monuments such as Seville's *Huevo de Colón* and the *Monument al Descobriment d'Amèrica* in Ibiza.

44. Russell, *Flat Earth*, 9; Fernández-Armesto, *Columbus*, 54 (Maldonado's testimony was given in the lawsuits between the crown and the Colón family discussed in Life Three).
45. Russell, *Flat Earth*, 8–10; Fernández-Armesto, *Columbus*, 53–54; Wey Gómez, *Tropics of Empire*, 109–58.
46. Fernández-Armesto, *Before Columbus*, 195–213; *Columbus*, 48–49.
47. Fernández-Armesto, *Before Columbus*, 203–22; Valen, *Heart of the World*, 27; Ribot García, *El tratado de Tordesillas*, 2:928–34.
48. Drayson, *Moor's Last Stand*.
49. Drayson, *Moor's Last Stand*, 20–21, 24–38, 47; Ladero Quesada, *Granada*, 117–39.
50. RC: II, 4; Drayson, *Moor's Last Stand*, 53–112, 126–36; Ladero Quesada, *Granada*, 138–50; Coleman, *Creating Christian Granada*, 1–7, 15–19, 34–43. Boabdil was permitted to remain in his former kingdom, but in 1493, he fled to North Africa, where he soon died; like the remains of Columbus (see Life Three), the location of his burial is still disputed (is it in Fez or in Tlemcen?). Within ten years of Granada's capture, the Spanish monarchs reneged on their promise to permit its Muslims to practice their religion.
51. Council's letter to the king in Coleman, *Creating Christian Granada*, 29. The Granada conquest anticipated that of Mexico, which similarly *began*—rather than ended—in 1521; see Restall, *Seven Myths*, 64–76, for that argument applied to Spanish America. Fernández-Armesto has long noted the Granada-Canaries-Americas connections (e.g., *Before Columbus*, 208, 212; *Canary Islands After the Conquest*); see also Macías Hernández, "Expansión Ultramarina."
52. Quotes by Fernández-Armesto, *Columbus*, 46. Some have argued that Isabel was persuaded of Columbus's proposal as early as 1489, but she made him wait three years at court until the timing was right for her, thereby dissuading him from seeking support elsewhere (e.g., Taviani, *Great Adventure*, 73–83).
53. Fernández-Armesto, *Columbus*, 2, 14–15; Davidson, *Columbus*, 12–13; Ladero Quesada, "I Genovesi a Siviglia."
54. Bernáldez, *Memorias*, 269 (1962 ed; also see HCAR); Fernández-Armesto, *Columbus*, 5.
55. Madariaga, *Christopher Columbus*, 19; quoted often, e.g., Horwitz, *Voyage Long and Strange*, 48. On Fernando's inventions, see Wilson-Lee, *Shipwrecked Books*, 81, 104, 113, 136, 298–310.

LIFE TWO: THE ADMIRAL

1. CVD: II, 32 (doc. XV, 30 March 1493); RC: VI, 307–8 ("que dende en adelante yo me llamase Don y fuesses Almirante Mayor de la mar Océana [*sic*] y Vissorey e Governador

perpetuo de todas las islas y tierra firma que yo descubriese y ganasse ... y asi sucediese mi hijo mayor"); RC: VII, 70, 300.

2. CGCC, 212–13 (the image at this chapter's start is taken from the facsimile that is on 212–13 inter). Believing that the bank had ignored his letter, which offered to tithe 10 percent of his estate's income to the bank, he withdrew the offer in 1504; the 1502 letter had in fact never reached Genoa; CGCC, 216–17; RC: IV, 172–74, 346–49. Acronymic signatures were a medieval Christian tradition and not uncommon, the best known being the S.C.C.R.M. (Holy, Imperial, Catholic, Royal Majesty) of King Carlos, usually styled S.C.R.M. by his son Felipe II. Columbus's S./S.A.S./XMY, being unique to him, was thus not the most common kind, but it can—and frequently has—been easily read, most obviously as *Sub Scripsit Armiratus Xristoferens Maior Indicarum* (Signed by Christopher, Admiral of the Indies). It could also be read as "Servant, I am, of the Highest Savior, Christ, Son of Mary," and it would be very like Columbus to have imagined several such readings. There are two dozen versions of the signature written at the end of documents by Columbus, none before 1498; on some he adds his name as *XpōFERENS* (Christopher) (e.g. Colón, *Libro Copiador*, 76), on others he signs off as *el Almirante* or as *el Virrey*. That sign-off, like the acronymic, was not cryptic or intended to disguise a mystery of religious or national origins, but was instead meant to convey openly recognized high rank and status conferred by the Christian God and His monarchs. Fernández-Armesto, *Columbus*, 118–19; Davidson, *Columbus*, 278–79; Milhou, *Colón y su Mentalidad Mesianica*, 79–80; Flint, *Imaginative Landscape*, 184–85; Catz, *Columbus*, 84–87.

3. The poet—Freneau, *The Poems*, 41—is discussed further in Life Eight.

4. On the endless debate over the First Landing, see Rocca, *Historical Geography*, 291–304; Davidson, *Columbus*, 223–24; Fernández-Armesto, *Columbus*, 81.

5. In 1493–96, while Columbus was on Cuba, ships reached Hispaniola from Spain (including one with his brother Bartolomeo), while Spaniards started to abandon the new colony and take vessels back to Spain; Phillips, *Testimonies*, 10. Columbus's original account of the First Voyage, called here the *Diario*, has not survived, nor do we know the length or content of the document he gave the monarchs in Barcelona in April 1493. We know that copies were made (perhaps altered or edited; none have survived), one of which the monarchs gave back to Columbus, and one was acquired by Las Casas, from which he made the heavily redacted version that is the *Diario*—which thus cannot be taken as unfiltered access to Columbus's thoughts and deeds on the First Voyage.

6. Fernández-Armesto, *Columbus*, 98 ("lionized" is his term, and his too are the Martyr reference, and the St. Thomas one by Jaume Ferrer). Martyr: Anghiera, *Extraict*; RC: V.

7. Wilson-Lee, *Shipwrecked Books*, 57 ("Mosquitoes" quote).

8. Irving, *Life and Voyages* [1828], III, 202; [1951], 569.

9. Fernández-Armesto, *Columbus*, 3 ("purpose"); Schoenrich, *The Legacy*, I, 35–51, 57; Gil, *Columbiana*, 555–57.

10. As Mentz, *Ocean*, 32, observes, Alfred Crosby's "Columbian Exchange" concept was an ecological one, often broadly misapplied, along with the "overfamiliar date of 1492" (Mentz's phrase; Crosby, *Columbian Exchange*). The 1493 letter to Santángel (introduced in Life One; I have drawn from it the image labeled by me "Naked and Afraid?") was published in at least seventeen editions in Latin, Spanish, Italian, and German by 1500; a 1494 Rome edition is in HCAR (printed and bound with an account of

"King Fernando's Conquest of Granada"); see *Et de Insulis in mari*, appended to *In laudem Serenissimi Ferdinandi*; Robbins, *Columbus Letter*; and Willingham, *Mythical Indies*; CoH, 103–10. Also see Life Eight for the depiction of cowering "Indians" in John Vanderlyn's *Landing of Columbus*—in the U.S. Capitol Rotunda, but the basis for numerous representations; see Schlereth, "Columbia, Columbus, and Columbianism," 946–50—showing how the Western perception of Indigenous peoples as naked, fearful savages remained unchanged or centuries.

11. Oviedo, *Historia General*, 1851: I, 13.
12. Morison, *Admiral of the Ocean Sea*, 62–63; Ballesteros Beretta, *Cristóbal Colón*, I, 358; Las Casas, *Historia de las Indias*, Lib. I, Cap. XIV (1951: I, 72); Catz, *Columbus*, 75–80. In the Colón-attributed *Vita*, e.g., the pilot is the Portuguese navigator Vicente Dias de Tavira; in Francisco López de Gómara's 1552 *Historia de las Indias*, the tale is presented as detailed fact, with Columbus receiving the dying Unknown Pilot in his house on Madeira (there is no evidence that Columbus ever lived on Madeira). See also Manzano Manzano, *Colón y su secreto*; Gil, *Mitos*.
13. An even larger perspective views European expansion beginning ca.1000, prompted by the end of invasions *into* Europe and by the onset of the Medieval Warm Period (950–1300); Viking settlements in Greenland, leading to the first European settlements in the Americas (in Newfoundland), date from ca.1000. The stalling and contraction of European expansion in the fourteenth century—marked by climate cooling, famines, and plague (especially the Black Death, which killed a third of Europeans)—thus becomes a temporary reversal of a millennium-long phenomenon that resumes in the mid-fifteenth century. Muldoon and Fernández-Armesto, *Medieval Frontiers*, xiv–xxvi; Fernández-Armesto, *Before Columbus*, 11–202.
14. Fernández-Armesto, "Los libros"; *Before Columbus*, 155, 160, 173, 221; *Columbus*, 4, 19, 21, 79; Irizzary, *Christopher Columbus*, 39–41.
15. Catz, *Columbus*, 65–74; Fernández-Armesto, *Before Columbus*, 157, 162, 195–200; Lester, *Fourth Part*, 88–89, 180–235; Regal, *America's Origin Story*, 78–80. On why "primitive celestial navigation," not the use of technology such as the astrolabe, was more important before, during, and after Columbus's day, see Fernández-Armesto and Lucena Giraldo, *How the Spanish Empire*, 67–69.
16. Previous two paragraphs: Pedro de Velasco also shared details with Columbus of his many voyages and those of other Portuguese navigators—according to Las Casas (RC: VII, 39, 269; also, 1951 edition of Las Casas, *Historia*, I, 68–69). Catz, *Columbus*, 66–68; Fernández-Armesto, *Before Columbus*, 153–70, 192–202; *Columbus*, 26–27, 46–47; Harrisse, *Christophe Colomb*, I, 307–29; Morison, *Portuguese Voyages*, 31–32; Lester, *Fourth Part*, 226–35. On claims that the Portuguese were the true discoverers, see, e.g., Cortesão, "Mystery of Columbus," 326–30. According to Las Casas, while Columbus was in the Canaries on the very brink of crossing the Atlantic for the first time, prominent locals swore to him that "every year they saw land to the west of the Canaries" (cada año vian tierra al Vueste de las Canarias), with people on Maderia and the Azores swearing the same (RC: VI, 309).
17. AGS, *Real Patronato*, legajo 2, f. 9; CGCC, 80–81 ("Los de Bristol ha siete año que cada año an armado dos, tres, cuatro caravelas para ir a buscar . . . yo he visto la carta que ha fecho el inventador que es otro genoves como Colon").
18. Cronau, *América*, 204–10; Fernández-Armesto, *Columbus*, 19–20, 27, 73; Lester,

Fourth Part, 208–13, 239–40, 282–87; Boyle, Setting Sun, 16–20, 39–42, 64–68, 113–15, 178–81, 280–81.

19. The role played by cod fleets in the history of European expansion into the Atlantic tends to be eclipsed by Columbiana; one is reminded of the nickname for J. H. Parry's course on seaborne empires, taught for decades in the twentieth century at Harvard: fish and ships (Fernández-Armesto and Lucena Giraldo, *How the Spanish Empire*, 57).

20. Fernández-Armesto, *Columbus*, 21, 27; on sailing into the Atlantic west vs. south, see Wey Gómez, *Tropics of Empire*.

21. Stevens-Arroyo, "Inter-Atlantic Paradigm"; Phillips, "European Expansion Before Columbus"; Muldoon and Fernández-Armesto, *Medieval Frontiers*, 327–71; Fernández-Armesto, *Before Columbus*, 165–67. Portuguese–Castilian rivalry as a backdrop to the First and Second Voyages is well evidenced in such sources as the *Diario* (RC: VI, 131, 134–35).

22. Taviani, *Great Adventure*, 41 ("brilliant"); Fernández-Armesto, "Maritime History," 30 ("code"); Phillips, "Maritime Exploration," 56 ("steps of others"); Phillips and Phillips, *Worlds*, 70–75.

23. Fernández-Armesto and Lucena Giraldo, *How the Spanish Empire*, 60–69 ("The history of the world, as conventionally written, has too much hot air and not enough wind," 61); Winchester, *Atlantic*, 115–20.

24. A third gyre at around latitude 60 degrees north helped medieval Norse navigators reach Newfoundland and aided Cabot in the 1490s but was not helpful to Columbus or early modern Spaniards. Fernández-Armesto, "Maritime History," 30 ("reach out"); Muldoon and Fernández-Armesto, *Medieval Frontiers*, xxv–xxvi; Catz, *Columbus*, 71–72; Phillips and Phillips, *Worlds*, 81–82; Mentz, *Ocean*, 60–61; Butel, *The Atlantic*, 48, 64–91.

25. CGCC, 116–17; CoH, 89; Fernández-Armesto, *Columbus*, 1–2, 9; Taviani in Bedini, *Encyclopedia*, 130–37.

26. RC: VIII (testimony of 1512–15 and 1535–36); Larducci in RC: VI, 15, 30; *Diario* entries in RC: VI, 350, 398, 410, 415; Varela, *Diario*, 17–21, 29–38; CoH, 40–41, 45–47, 61; Lemos in Bedini, *Encyclopedia*, 545–49; Fernández-Armesto, *Columbus*, 62, 94; Pagden in RC: VII, 5, 7; Phillips and Phillips, *Worlds*, 138–43, 153. Columbiana, especially by non-Spaniards, has tended to follow the Lascasian tradition—ignoring, slighting, or criticizing the Pinzón brothers, especially Martín Alonso (Dugard, *Last Voyage*, 79–86, is a recent example; an exception is Rocca, *Historical Geography*, who remains neutral).

27. RC: VIII; Ballesteros Beretta, *Cristóbal Colón*, I, 547–53 ("no justifican el que se diga que a Pinzón se debió la iniciativa del descubrimiento"); Manzano Manzano and Manzano Fernández-Heredia, *Los Pinzones*; Pennock, *On Savage Shores*, 50; Ramos, "Los Contactos Trasatlanticos Decisivos." Boyle, *Setting Sun*, imagines the transatlantic voyages as a competitive yet collaborative enterprise of "joint plans" (68) and "almost identical" (67) ones by Columbus, Vespucci, and Cabot (who indeed "knew each other," 3).

28. CGCC, 182–83, letter to Geronimo Annari in Savona ("Navicando adoncha verso la Spagnola io fui primo a discoprire terra... il Signore Armirante... li pose nome el cavo de San Michele Saonese per mi respeto, e cossi notto nel suo libro"). Columbus merely noted that he named the point "Cabo de Sant Miguel" (RDSP I, i, 196). Cuneo also claimed that he was the first to see another island, encountered soon afterward, and that Columbus therefore named it "la Bella Saonese." On the Porras brothers: Dugard, *Last Voyage*, 115–16.

29. The historian is William D. Phillips, Jr. (quotes in "Maritime Exploration," 47–48; also see 56–58); Carla Rahn Phillips makes a similar (but not identical) argument ("Exploring," 64–68).
30. Kadir, *Columbus*, 71–73.
31. Fernández-Armesto and Lucena Giraldo, *How the Spanish Empire*, 60 ("bickering"); Pennock, *On Savage Shores*, 33–73; Crosby, *Columbian Exchange*.
32. The first usage in print of *indios* as a label for Indigenous Americans comes seven lines into the letter Columbus wrote (published soon after) to royal minister Luís de Santángel while aboard the *Niña* on his 1493 return voyage, right after he states that he had sailed to *las indias* (the Indies); Willingham, *Mythical Indies*, 146–47 inter (facsimiles), 153 (transcription), 191 (translation); Rocca, *Historical Geography*, 99–118. On "India" and "the Indies" (*las Indias*), see Wey Gómez, *Tropics of Empire*.
33. AGI, *Contratación*, legajo 5575 on the settlers who died at that camp, the "town" of La Navidad. On Taíno political culture, see Deagan, *En Bas Saline*; Mira Caballos, *El Indio Antillano*.
34. RC: VI, 319–417; 1493 letter in Willingham, *Mythical Indies*, 146–47 inter (facsimiles), 153–61 (transcription), 191–96 (translation). See careful analysis in Zamora, *Reading Columbus*, 9–20. The 1493 letter, the sole text written (albeit in edited form) by Columbus to be published in his lifetime, "became something of an early bestseller" (Lester, *Fourth Part*, 272); Las Casas line from his *Brevísima Relación* in previous paragraph ("humility") is often quoted (e.g., *Fourth Part*, 267).
35. On the significance of gold, and how the seemingly obsessive Spanish quest for it is often misunderstood, see Restall, *Seven Myths*, 22–23. For a hilarious spoof on the Spanish imperial gold motive, see the October 22, 2023, *Saturday Night Live* sketch at youtube.com/watch?v=G31E4ML9ViY.
36. RC: VI, 357–96 (dozens of examples through the *Diario* but concentrated in these pages); RC: VII, 7 (pope quoted by Pagden); 1493 letter in Willingham, *Mythical Indies*, 146–47 inter (facsimiles), 156–57 (transcription), 193 (translation); CoH, 57–89; RC: X, 30–37, 93–100 (three papal bulls of 1493); Payne, *Spanish Letter*. The Santángel letter and its variations are part of a nexus of texts from 1493 through the early sixteenth century, published in multiple European languages, showing how heavily early accounts of Columbus and his voyages were imbued with *misinformation* (as detailed by Harrisse in LoC/HH, "Fasti" No. 6 to No. 9); also see RC: X. See also Muldoon, *Spanish World Order*, 22–23, 38–65, 111–18.
37. Pennock, *On Savage Shores*, 48; Stone, *Captives of Conquest*, 2–53.
38. Winchester, *Atlantic*, 88 ("tyrant"; I recognize that I am taking Winchester's rhetoric literally). Phillips and Phillips are thus right in asserting that "Columbus invented nothing when he called for settlement" (*Worlds*, 171). Or, as Bartosik-Vélez put it, Columbus's claims in his letters to the monarchs that he and his voyages were "fundamental to Spain's drive to universal Christian empire" were "politically astute" expressions of loyalty (*The Legacy*, 16). See also Altman, "Revolt of Enriquillo," 589–93.
39. On 1494 executions: Las Casas in RC: VII, 116, 348. On the slaves-for-cows scheme, Antonio de Torres was sent to Spain with the January 1494 letter, returning that autumn with an August 16, 1494, reply informing Columbus that "this project is to be set aside" (August letter in AGI, and in Taviani, *Voyages of Columbus*, I, 159; CoH, 111–27). On Taíno households: a *real cédula* of 1512, e.g., clearly recognized the distinction between

esclavos and traditional dependents (*naborías*), but its weak insistence that Hispaniola colonists also recognize it reflected how much it was ignored (quoted in Francisco Moscoso, "Parentesco y clase en los cacicazgos Taínos: El caso de los naborías," unpublished MS); also see Stone, *Captives of Conquest*, 13–28; Moscoso, *Caciques, aldeas y población taína*; Sued Badillo, *Caribe taíno*; Keegan, *Taíno Indian Myth*; Oliver, *Caciques and Cemí Idols*; Kadir, *Columbus*, 113 (on *rescate*); Rocca, *Historical Geography*, 320–29; Mira Caballos, *El Indio Antillano*; and Deagan, *En Bas Saline*.

40. RC: X, 47, 54–58, 109, 113–17. The 1504 license reproduced in Fernández, *Capitulaciones Colombinas*, 377; also, Kadir, *Columbus*, 118 ("salvo los que por nuestro mandado son pronunciados por esclavos que son los ... que se dicen canibales").

41. RC: VI, 353, 404 ("*Caniba*" and "*Caribe*"; but dozens of cannibal references throughout the *Diario*); also see the Columbus/Torres 1494 letter cited in note 39. On the early European invention of Indigenous cannibalism: Isaac, *Aztec Cannibalism*, 31–43.

42. Las Casas, *Historia*, quoted by Padgen in RC: VII, 3; 1493 letter in Willingham, *Mythical Indies*, 146–47 inter (facsimiles), 160 (transcription), 195 (translation); Payne, *Spanish Letter*; Stone, *Captives of Conquest*, 5, 29, 37.

43. Oviedo, *Historia General*, Lib. VIII, Caps. I–II; Fernández, *Capitulaciones Colombinas*, 369–73 ("pueda tomar en qualesquier partes que descubrieren yndios e yndias para lenguas de aquellas tierras con tanto que no sean para escablos [sic] ny para les haser mal ny dapno [sic] e que los tomen los mas a su voluntad que ser pueda"); Kadir, *Columbus*, 115–17; Pennock, *On Savage Shores*, 74–115. Oviedo (Lib. I, Cap. VII) advanced an even larger loophole to justify the systematic abuse of Indigenous peoples in the Americas: as the apostles had centuries earlier preached "the evangelical truth" to "these Indians," he asserted, "they cannot pretend ignorance" and are thus fair game.

44. For example, Royal, *1492 and All That*, 38–42, for a three-part such list.

45. RC: VII, 116, 348 (Las Casas: "en estas Indias ... el comienço del derramiento de sangre, que después tan copioso fue en esta ysla ... e después désta en todas las otras infinitas partes dellas"); Royal, *1492 and All That*, 40 (less a defense of Columbus than an effort to be balanced); Altman, *Life and Society*, 12; Deagan, *En Bas Saline*; Varela, *La caída*, 146–48 (the steward or *despensero* was Pedro Gallego; in addition to him and the First Admiral's majordomo, his sister-in-law's husband Miguel Muliart was also tortured to death, under his orders, for translating into Spanish a letter by a French friar criticizing the regime of the Columbus brothers).

46. The 1903 historian is Thacher, *Christopher Columbus*, II, 292–96; Barry, *The Life*, 264 ("capons"; the 1870 English version of Roselly de Lorgues, *Christophe Colomb*).

47. Fernández-Armesto, *Columbus*, 112; Royal, *1492 and All That*, 41.

48. On the early Caribbean: Moya Pons, *Después de Colón*; Altman, *Life and Society*; Stone, *Captives of Conquest*. The literature on genocide is vast, but see Kinstler, "Defining an Erasure," for a succinct summary of the (currently contested) distinction between "genocide" as requiring the *dolus specialis* or specific intent to annihilate, as exemplified by the Holocaust, and a broader definition that includes a wide variety of mass killings and other actions designed to eliminate "the essential foundations of the life" of a group (in the words of Raphael Lemkin, who coined the term in 1944). Spanish colonialism was thus not a genocide on the basis of the narrower definition, as the goal was to subjugate and exploit Indigenous peoples, not displace or eliminate them as in

Anglo-America. But it was genocidal on the basis of the broader definition. For this argument with respect to the Spaniards in Mexico, see Restall, *When Montezuma Met Cortés*, 328–30, 347–48.

49. Adams, *Columbus*, 6.
50. "Event" is Taviani, *Voyages of Columbus*, I, 31 (Columbian apologist), copying Francisco López de Gómara (apologist for Spain's early-sixteenth-century empire) (Restall, *Seven Myths*, 2).
51. RC: VI, 318 ("gran multitude de aves de la parte del norte al sudueste, por lo qual era de creer q. se yvan a dormir a tierra"); Caserio: CGCC, 192–93 (folio lines 12–14: "se Enrico Settimo Ri d'Inghilterra avesse attesto a Christofforo Colombo Genovese la mettà del Mondo sarebbe al presente non de Spagnoli ma de Inglesi"); Caserio makes the same argument regarding King João II of Portugal failing to lend an ear (*dato orecchie*) to Magellan. Irving, *Life and Voyages* [1828], I, 142–48. The "Spanish and Catholic" phrase is Alexander Humboldt's, quoted by Taviani, *Voyages of Columbus*, I, 25–26. That Florida was colonized first by Spaniards and still ended up English-speaking further exposes the argument's absurdity.
52. Lopez, *The Rediscovery*, 5, 10; Pagden in RC: VII, 19; Van Duzer ("Columbus," 4) is an example of a scholar bucking that tradition.
53. Reséndez, *Other Slavery*, 13–45; Van Deusen, *Global Indios*; Pennock, *On Savage Shores*, 46–47, 49 (Cuneo and Ojeda); Stone, *Captives of Conquest*, 40–45. D'Aguiar quoted in Mentz, *Ocean*, 38.

LIFE THREE: THE REMAINS

1. His difficulties in traveling, due to poor health, beginning with his return to Seville in November 1504, are revealed in letters he wrote to his son Diego (AGI, *Patronato*, legajo 295; CoH, 239–50). He was home in Seville in 1505, then followed the court to Segovia in May, to Salamanca in October, and in March 1506 to Valladolid—from where, dying and despairing and full of self-pity, he wrote to his friend Diego de Deza that his efforts to reach the king's ear were like "whipping the wind." Gil, *Columbiana*, 555–57; Davidson, *Columbus*, 464; Varela, *Textos y documentos*, 312, 320, 531; Stavans, *Imagining Columbus*, 46 ("neglect"). With respect to gout or *la gota*: the disease was well known in Columbus's day, as were the symptoms of which he complained, such as chronic pain and insomnia; gout likely killed him through renal failure or congestive heart failure. See Espinoza and González, "La enfermedad del Almirante Mayor." Arthritis is commonly cited as Columbus's predominant ailment (e.g., Stavans, *Imagining Columbus*, 45); note that gout is a type of inflammatory arthritis, marked by intense joint pain. The evidence that he died in the building that is today labeled Casa de Colón is circumstantial, at best. Columbus's crusade obsession is discussed in Life Four.
2. The story that Columbus was interred with those chains, by his wishes, has been repeatedly exposed as "pure legend" (as Harrisse, *Christophe Colomb*, II, 146, put it: *une pure légende*). Diego Colón's will confirmed that his father's body was moved to Seville in or before mid-March 1509; Thacher, *Christopher Columbus*, III, 511–21.
3. Harrisse, "Les restes mortels," 17; Wilson-Lee, *Shipwrecked Books*, 1, 313, 325, 374 (also, Fernando's death, AGI, *Patronato*, legajo 10, n. 2, r. 3); Muratore, *The Remains*, 14.

Fernando's plans to escort his father's remains to Santo Domingo were well advanced; his license to bring four enslaved Africans with him is in AGI, *Indiferente General* 423, legajo 19, fs. 4v-5r.

4. CVD: II, 422–29 (# CLXXVII, 1795/1796); Nader in Bedini, *Encyclopedia*, 78–79; Real Academia de la Historia, *Los restos de Colón*, 24–25, has the bones shipped across the Atlantic in 1536, then waiting four or five years in Santo Domingo until the new tomb was completed.
5. Thacher, *Christopher Columbus*, III, 526; quoted with errors, misdated 1555, in Asensio Toledo, *Cristóbal Colón*, II, 829.
6. Thacher, *Christopher Columbus*, III, 527 ("segun la tradicion de los ancianos..."). Colón palace: built in 1509–10, thrived in the sixteenth century, ceased to be a Colón residence in the seventeenth, became an overgrown ruin in the eighteenth, largely dismantled in the nineteenth, restored in the 1950s, and again in the 2020s as the Museo Alcázar de Colón. Schoenrich, *The Legacy*, I, 169–73.
7. Thacher, *Christopher Columbus*, III, 534–37 ("La tradition communiquée par les anciens du pays...").
8. CVD: II, 422–29 (# CLXXVII, 1795/1796) ("las cenizas de aquel héroe"); Asensio Toledo, *Los restos de Cristóval Colon*, 17–21; Thacher, *Christopher Columbus*, III, 538–45; Harrisse, "Les restes mortels," 15–16.
9. Cocchia, "Acta del descubrimiento"; Nader in Bedini, *Encyclopedia*, 78–79 (in which the canon's name appears as Bellini; I have followed Cocchia, "Acta del descubrimiento"); Cronau, *América*, 391–401. The drawing shown at the start of this Life was made in 1877 and sent to Madrid for the benefit of the Royal Academy investigation (reproduced in Real Academia de la Historia, *Los restos de Colón*, 203; see Cronau, *América*, 400, for an 1892 redrawing with added label).
10. Real Academia de la Historia, *Los restos de Colón*, 71–72, 207–11; Asensio Toledo, *Los restos de Cristóval Colon*, 23–24, 32–39, 44–49.
11. Lamb in Bedini, *Encyclopedia*, 419; Dozier, "The Controversy on Whereabouts"; Thacher, *Christopher Columbus*, III, 546–66.
12. Real Academia de la Historia, *Los restos de Colón*.
13. Asensio Toledo, *Los restos de Cristóval Colon*, 7–8 (translation mine). Asensio's massive 1888 Columbus biography incorporated his position and his 1881 pamphlet on the remains (*Cristóbal Colón*, II, 618–26, 823–61). A few Spaniards ducked the patriotic position: Rodolfo Cronau, e.g., traveled to Santo Domingo and gained access to the tomb there, publishing in 1892 his conclusion that "the respectable remains of the great discoverer" had never left Santo Domingo's cathedral (*América*, 301–401). Likewise, Thacher, *Christopher Columbus*, III, 546–612, made a thorough survey of the evidence and concluded that the remains "remain to this day" (1903) in Santo Domingo.
14. Thacher, *Christopher Columbus*, III, 567–71.
15. Dvarchive.com, Clip 921–2001. The urn with the glass vial spent the rest of the century in Genoa's city hall, and early in this century was moved to the Columbus room in the basement of the dockside Museo di Mare—where it is labeled "Ampolla contenente le ceneri di Cristoforo Colombo" (Vial containing the ashes of Christopher Columbus), with one short paragraph recounting the story told in this chapter, ending with the 1877 events explaining the urn. When I visited Genoa in 2023, nobody working either

in city hall or in the tourism offices around the city or in the Casa di Colombo admitted to having heard of the urn, let alone its location.
16. Horwitz, *Voyage Long and Strange*, 79–85, 88–94; Summerhill and Williams, *Sinking Columbus*, 171; my field notes, March 2024.
17. In addition to the sources cited above, also see Lugo, *Los Restos de Colón*, and Muratore, *The Remains* (both arguing the Dominican case); Gil, *Columbiana*, 61–62, 568–93, summarizes much of the dispute. There will always be claims that Columbus's "true" burial place is a mystery that the claimant has now solved (e.g., there periodically resurfaces the "solution" that his bones never left Santa María de las Cuevas in Seville). As Davidson (*Columbus*, 465) noted, Spanish authors tend to favor Seville as the final resting place, authors of other nationalities tend to favor Santo Domingo.
18. RC: VIII.
19. Masters, *We, the King* ("the many" quote is his, 261); Fernández-Armesto and Lucena Giraldo, *How the Spanish Empire*.
20. Schoenrich, *The Legacy*, I, 35 ("evasions").
21. Schoenrich, *The Legacy*, I, 64–87, 91–104. Don Diego's sense of entitlement and self-confidence is evident in his correspondence at the time; see, e.g., his 1512 letter to Cardinal Cisneros in HCAR, CC #126.
22. RC: VIII, 49–211, 275–426; Phillips, *Testimonies*, 13; Schoenrich, *The Legacy*, I, 92–104.
23. Further emboldened by the king's favor, Diego wrote a letter to Charles from Santo Domingo in 1520 that boasted of his alliance with fray Bartolomé de Las Casas; he promised that in a decade he and the friar would convert every "Indian" in the stretch of Central America then known to Spaniards, without raping or enslaving or killing a single one, this kind treatment guaranteeing a steady flow of tribute to the crown; they also requested that five hundred "Black slaves" be divided among the new Indigenous "pueblos," which they would establish at the rate of three a year (original manuscript in HCAR, CC #127; Colón, *Memorial*). A fairly flattering mini biography of Diego is Hoffman in Bedini, *Encyclopedia*, 133–35.
24. RC: VIII, 211–71, 427–83; Phillips, *Testimonies*, 14; Schoenrich, *The Legacy*, I, 104–9, 113–41; Oviedo, *Historia General*, IV, VI, I (1851 edition, 113) (the fact that as don Diego crossed the Atlantic, he was "already very ill and thin" [ya muy enfermo é flaco] suggests disease rather than stress alone).
25. Schoenrich, *The Legacy*, I, 134–41.
26. Schoenrich, *The Legacy*, I, 141–54. A sixteenth-century Spanish ducat contained 3.5 grams of gold or 35 grams of silver, equivalent to 375 maravedís (copper coins); the 17,000-ducat annuity was thus more than 6 million maravedís, compared with the approximate cost of Columbus's First Voyage at 2 million maravedís. A modern equivalent cannot easily be calculated, but based on the price of gold in 2023, 17,000 ducats is equal to just under $1.2 million; but based on Schoenrich's 1949 calculations, adjusted to the 2020s, 17,000 ducats (6 million maravedís) was more like $5 million. That estimate is supported by the buying power in sixteenth-century Spain of 17,000 ducats: the average sale price of an enslaved person in Seville was 20 ducats in the first decade of the century, rising to a hundred by century's end (due primarily to inflation; Pike, "Sevillian Society in the Sixteenth Century").
27. Otto Schoenrich (*The Legacy*, I, 26), who also asserted it "was at the time considered the

richest estate under litigation in the courts of Spain" (I, 158, citing Harrisse, *Christophe Colomb*, II, 269, but see 265–70). One assumes that by "celebrated," Schoenrich meant "renowned" rather than "acclaimed." A Baltimore-born lawyer, Schoenrich (1876–1977) served as a judge and governmental official in U.S.-occupied territories such as Cuba and Puerto Rico, before settling down as a law-firm partner in New York in 1916.

28. Wilson-Lee, *Shipwrecked Books*, 258 ("cruelty").
29. My suggestion that Diego and Felipa left no heirs because they were first cousins is speculation; although they were both in their twenties and apparently in good health, they lived together in the Columbus Palace in Santo Domingo for four years before Felipa died (in Valladolid in November of 1577; Diego died two months later in Madrid). Numerous reasons could explain the lack of a child in that time. Note that Diego's opposition to their marriage was dropped when Felipa and María's lawyers sued for the entire Columbus legacy on the grounds that Diego's father had—prior to Diego's birth—been sleeping with both his wife and his wife's cousin, making Diego "adulterated, incestuous" (espurio, incestuoso) and thus legally "illegitimate" (no era descendiente legitimo) (Schoenrich, *The Legacy*, I, 311–27, lawsuit quotes on 317).
30. CGCC, 275–77; CoH, 250–52; Schoenrich, *The Legacy*, I, 177–91 ("obscure" quote is Schoenrich's, on 188); II, 17–91.
31. Schoenrich, *The Legacy*, II, 20–31. Doña Francisca lived to 1616, battling four of her close family members—her aunt, her daughter, her granddaughter, and eventually her great-grandson—as well as cousins Álvaro and Cristóbal. After Álvaro died in 1581, his daughter, both his sons, and a nephew filed claims. After Cristóbal died in 1583, his sister, and then her husband, joined in, as did two more cousins, both children of Fourth Admiral don Luís: María Colón y Mosquera, whose status as a nun undermined her claim; and his illegitimate son Cristóbal, whom the other claimants dismissed as "the bastard" and "nothing but a bastard" (*The Legacy*, II, 31–64).
32. The logic of the 1605 judgment was that the awards of 1580 and 1586 to don Cristóbal de Cardona y Colón were correct, but that the legacy should *not* have then passed to his sister doña María, instead going to his cousin Jorge Alberto de Portugal y Córdova, upon whose death in 1589 it then went to don Nuño, thereby Sixth Admiral (Schoenrich, *The Legacy*, II, 64–92). The three-century litigation story is full of accusations by one Colón party that another had, in the course of having documents copied by notaries, made small but critical changes; see examples of some of these altered documents (often called forgeries, although I think that is a misleading label for these documents) in the JCB as Codex-SP 1 (an early sixteenth-century copy of the *Capitulaciones* claiming a 1494 date); Codex-SP 124 (a 1757 copy—with a little rewording—of the 18 April 1493 *Privilegio* documents given to Columbus in Barcelona by Queen Isabel); MsAmer 1590 S Oversize (a handsome 1590 copy—cataloged as a forgery—of that 1493 *Privilegio* document); and Codex-SP 125 (an 1841 copy of a 1590–1620 copy of the *Privilegio* also cataloged as forged). Another cache of documents, including originals and copies from the litigation story, ended up preserved but understudied in a private archive in Pennsylvania (CCBM). In 1608, lesser payments were also made to other claimants. "The bastard" Cristóbal ended up with 400 ducats a year, while his mother Luisa de Carbajal received a one-time payment of 6,000 ducats. A 2,000-ducat award went to Italian claimant Baldassare Colombo of Cuccaro, a town some fifty miles from Genoa, claiming descent from an uncle of Columbus. His case was stronger than that of the other

Italian claimant, Bernardo Colombo, whose claim to be descended from Columbus's brother Bartolomeo was so spurious that Baldassare was able to have him arrested and prosecuted for fraud; in 1584, the Council of the Indies permanently expelled Bernardo from the proceedings. Baldassare provided testimony from more than forty witnesses in the Genoa region affirming his identity as a Colombo. He may have been distantly related to Columbus, but his hope of being made Admiral of the Ocean Sea and of the Indies was unrealistic, to say the least. It was also revealed that he was funded by Spanish claimants (as was Bernardo), enemies of doña Francisca. Presumably the court hoped that the one-time payment would send Baldassare back to Cuccaro for good—which it did (Schoenrich, *The Legacy*, II, 37–38, 57–60, 70–71, 78–81, 90–91; LoC/HH, "Fasti" No. 1, C, II, 1).

33. Schoenrich, *The Legacy*, I, 158–74.
34. The saying may have roots going back to Columbus's son, Fernando, who had commented that litigation with the crown by him, his brother, and his father (el pleyto de su padre) was so chronic as to be "immortal" (que aquel es ynmortal; Schoenrich, *The Legacy*, I, 195–96); Fernando's letter in AGS, Estado, *legajo* 21, f. 22; Wilson-Lee, *Shipwrecked Books*, 284, 372, led me to it.
35. Schoenrich, *The Legacy*, II, 86, 95.
36. Previous two paragraphs: Schoenrich, *The Legacy*, I, 95, 150–53.
37. Previous three paragraphs: Schoenrich, *The Legacy*, II, 129–89 (Jovellanos quotes on 147).
38. Schoenrich, *The Legacy*, II, 154–57.
39. Schoenrich, *The Legacy*, I, 154; on the Exposition, Rydell, *All the World's a Fair*, 38–71. A fraction of its value centuries earlier, the annuity was still equivalent to several hundred thousand dollars today.
40. Schoenrich, *The Legacy*, II, 163–69; Peris, "Yo soy el Cristóbal Colón número 20."
41. Peris, "Yo soy el Cristóbal Colón número 20."
42. NBC News, nbcnews.com/news/amp/wbna12871458. Harrisse made a similar observation a century and a half ago regarding the state of knowledge, evidence, and confusion back then ("Les restes mortelles," 19–23)—a reminder of how partisan agendas and deliberate misunderstandings of history only create and deepen mysteries rather than correctly identify and solve them.
43. Bradley, "Archaeology breakthrough"; October 12, 2024, report on bbc.com/news/articles.

LIFE FOUR: THE SAINT

1. Leo XIII, *Quarto Abeunte Saeculo* ("eius opera / centena mortalium millia / ex fero culte ad mansuetudinem ... nihil fere aggressus nisi religione duce, pietate comite"); Mattei, "Columbus noster est!"
2. The painting (oil on canvas, 48 × 52 centimeters) was acquired by the Museo Lázaro Galdiano, in Madrid, prior to 1913. As is customary in the world of Columbiana art, the museum (flg.es/museo/la-coleccion; also see mismuseos.net/en/community/metamuseo/resource/la-virgen-de-cristobal-colon) claims that the image of Columbus is "the oldest surviving" one, while conceding that its dating means that the Columbus portrait within the painting is nonetheless a copy of a supposed lost original. Fernández-Armesto

(*Columbus on Himself*, frontis) notes that it "reputedly" portrays Columbus. Its title is traditional, not marked on the painting, which is unsigned, painter not known, and purportedly dates to 1540, although it has also been vaguely dated 1501–50. It certainly could not predate 1540, as in that year the cathedral depicted here was completed, but the *escudo* of Charles V held aloft by angels suggests it was painted before 1556, when Charles passed the Spanish throne to his son Phillip; thus, in the caption I have dated it 1540s. The photograph of the Plaza Colón was taken by me in 2024.

3. Carpentier, *Harp and the Shadow*, 151 (italics in original).
4. Carpentier, *Harp and the Shadow*, 157–58. Doria lived 1466–1560.
5. My citations to pages of the *Vita* are to the translation *and* transcription in RC: XIII, but I have also consulted the 1728 Venetian edition (see Colombo, *Historie*). The title glosses as "History of the Life and the Deeds of the Admiral don Christopher Columbus."
6. RC: VII, 7, 24, 254 ("que quiere dezir traedor o llevador de Christo"); The *Historia* was first published, in Spanish, in 1875–76; it has yet to be published in full in English (RC: VI; VII; and XI present and translate all the Columbus portions). On "Lascasian scholarship," see Orique and Roldán-Figueroa, *Bartolomé de las Casas*, especially 1–25.
7. Caraci in RC: XIII, 1 ("the basis"); Keen, *Life of the Admiral*, v (other quotes; "rich and faithful" is Keen himself). Ramón Iglesia's prologue to his 1947 edition in Spanish offered a pithy summary of the authorship debate to that point, but it was far from enough to offset the impression that the paperback provided Fernando's own words (Colón, *Vida del Almirante*).
8. Thacher, *Christopher Columbus*, I, 396; Harrisse, *Christophe Colomb*, I, 108 ("des anachronismes, des contradictions et des erreurs dont les *Historie* fourmillent"); RC: XIII, 30, 234.
9. RC: XIII, 30, 31, 34, 35, 66, 234, 235, 238–39, 268 (as with all *Vita* quotes, translations are mine but benefit from those in RC: XIII).
10. Dedication by Giuseppe Moleto, a professor at Padua: RC: XIII, 28, 231 ("delle più rare cose di tutta Europa"); Wilson-Lee, *Shipwrecked Books*, 259–62; Fernández-Armesto, *Columbus*, 23 ("wastrel").
11. The second of one hundred eight short chapters is a ranting denunciation of a book by Genoese bishop Agostino Giustiniani, published less than two years before the death of Fernando Colón—whose hectic schedule during those months (including making plans to sail for a third time across the Atlantic) and fatal illness preclude the possibility that he wrote the *Vita* then.
12. RC: VI; Caraci in RC: XIII, 16–20 (Ilaria Luzzana Caraci's name is wrongly styled Ilaria Caraci Luzzana on the title page of RC: XIII). We know such a lost Colón manuscript existed because both Las Casas and Gonzalo Fernández de Oviedo refer to it. Also see Symcox in RC: XI, 4–7; Taviani in Bedini, *Encyclopedia*, 136; Varela, *Textos y documentos*; Irizarry, *Christopher Columbus*, 29–30. For early insight into the care with which the historian must use Las Casas's telling of the Columbus story ("plutôt polémiste qu'historien," 133), see Harrisse, *Christophe Colomb*, I, 108–34.
13. Caraci in RC: XIII, 21–24; also see Caraci, *Colombo vero e falso*; and Lardicci in RC: VI, 14–30.
14. My lines on "storytellers" are inspired by Gopnik, "The Great Interruption," 66; also see Sehgal, "Tell No Tales."

15. Drayson, *Moor's Last Stand*, 119–20 (I here follow Drayson's translation of King Fernando's letter).
16. My translation from a 1494 Rome edition of *Et de Insulis in mari* (in HCAR), but I have also used the facsimiles in Willingham, *Mythical Indies*, 146–47 inter, 153–61 (variorum transcription), 191–96 (translation); also see CoH, 101–10; Payne, *Spanish Letter*; and Robbins, *Columbus Letter*.
17. CoH, 231 (202–35 for full 1503 letter from which "gold" quote comes; frequently quoted, e.g., Crosby, *Measure of Reality*, 71); Russell, *Flat Earth*, 6 ("combination").
18. Fernández-Armesto, *Columbus*, x (quoted), 33 (paraphrased). Writers "misled" are too numerous to list, but Delaney, *Quest for Jerusalem*, is one example.
19. RC: VI, 314 ("Así que muy neçessario... de Egipto contra Moysén"); Fernández-Armesto, *Columbus*, 79, 85–86.
20. RC: VI, 390; RC: XI, 51, 185; Fernández-Armesto, *Columbus*, 88, 90–91, 118, 126–27.
21. RC: II, 161, 378 ("O ombre de poca fee, levantate, que Yo soy, non ayas [sic] miedo"); Fernández-Armesto, *Columbus*, 147, 167–70.
22. CoH, 202–6. I paraphrase Fernández-Armesto ("to take Islam in the rear"; *Columbus*, 26); also see Lester, *Fourth Part*, 280.
23. Restall et al., *Friar and the Maya*, 282–85; RC: III, 23–26. The affectation of the Franciscan habit lasted on and off until Columbus was buried in it. On Columbus and the Franciscans, see Milhou, *Colón y su Mentalidad Mesianica*; Watts, "Prophecy and Discovery"; Delaney, *Quest for Jerusalem*. Watts and Delaney backdate Columbus's millenarianism too far, making it his entire life's focus; see Flint, *Imaginative Landscape*, 191–202. I concur with Bartosik-Vélez's view of Columbus's millenarian rhetoric as strategic evolution, going from 1493's emphasis on *Reconquista* and empire to a prophetic rhetoric in 1498 that in 1500 became apocalyptic ("Three Rhetorical Strategies").
24. Doña Juana de la Torre (or de Torres y Ávila) was a senior handmaiden to Queen Isabel (the record of a 1501 gift from Isabel to her of fifteen religious statues is in NGAL, Rare Books MS 11), having served as *ama* to the household of royal heir Prince Juan, in whose court Columbus's sons were pages starting in 1493 or early 1494 (Juan died in October of 1497, aged nineteen); her brother was the Antonio de Torres on Columbus's Second Voyage. Columbus's letter to her in RC: II, 161–69, 377–87 ("mill combates me ha dado... con crueldad me tiene echado al fundo... el spiritu de ynteligençia y esfuerço grande... me hizo dello mensagero y amostro aqual parte"); and CoH, 188–97. Also see Fernández-Armesto, *Columbus*, 34–35, 39, 150, 154–55; *Ferdinand and Isabella*, 58–59; Thacher, *Christopher Columbus*, II, 423–38. I have translated *me tiene echado al fundo* as "throwing me to the bottom," while Fernández-Armesto leans further into the metaphor, with "holding me underwater" and "cast me down to the depths" (CoH, 198); Nader opts for "has cast me into darkness" (imagining this as Columbus's self-pity over deteriorating eyesight; RC: II, 161, 211).
25. RC: II; RC: III, 4–5.
26. RC: III, 3–51, 66–73 ("despues paro lo que Ihesu Christo nuestro redentor diso, e de antes avia dicho por boca de sus santos profetas... avra de seneçer el mundo").
27. CoH, 250–52; Fernández-Armesto, *Columbus*, 3, 170, 182. Wilson-Lee, *Shipwrecked Books*, 72–73 (also see 59–76), cautions against dismissing *The Book of Prophecies* "as an expression of Columbus's narcissistic insanity"; millenarian thinking was common at the time, and Gorricio and Fernando Colón were not alone in seeing the Bible

as a source of prophecies that could make sense of current events (narcissistic, then, but not insane).

28. Oviedo, *Historia general*, Lib. II, Cap. VII; 1851, 29 ("la verdad evangélica"); Lib. III, Cap. IX; 1851, 80–81 ("platado y exerçitado... por medio é industria del almirante don Chripstóbal [sic] Colom... tantos tesoros de oro, é plata, é perlas, é otras muchas riquezas é mercaderias"). Oviedo's history of Columbus is Lib. II, Caps. II–IX (1851 edition, 13–81).

29. Griffin in RC: VII, 16–19; Restall, *When Montezuma Met Cortés*, 82–84, 240–41.

30. Watt, *Dante, Columbus*, 113 ("supra-imperial power"). On early reactions in Italy to Columbus's voyages, see RC: X; RC: XII.

31. Watt, *Dante, Columbus*, 130–65, 169 (Garibaldi quote). The original caption to the Stradano figure I have captioned "Holy Hero" reads "Christopher Columbus of Liguria: with the terrors of the Ocean overcome, he passed to the Spanish king almost a whole other World, which he himself found" (Christophorus Columbus Ligur terroribus Oceani superaris alterius pene Orbes regiones a se inventas Hispanis regibus addixit); also see Markey, "Stradano's Allegorical Invention of the Americas."

32. CVD: II, 423 (# CLXXVII, 1795/1796) ("primer descubridor de aquel nuevo mundo... la dilatación de la verdadera Religion y Sagrado Evangelio").

33. Quotes from Campe (1746–1818) are my translations from an 1817 edition (Corradi, *Descubrimiento*, I, 21; I, 276–77), checked against 1820s editions of *Die Entdeckung*, and the 1799 English edition (*Discovery of America*, e.g., 225, 251–53); I loosely translate the final phrases ("los que conozcan el precio de unas virtudes como las que caracterizaron").

34. Roselly de Lorgues, *Christophe Colomb*; Winsor, *Christopher Columbus*, 53–55; Adams, *Columbus*, 12; *Christian World*, vol. 27 (1877), 35–37.

35. Roselly de Lorgues, *Christophe Colomb*, I, i–iii ("L'histoire de Christophe Colomb est la glorification du génie catholique"); an early U.S. edition is Barry, *The Life*; an 1878 Spanish edition was a lavishly oversized and illustrated three volumes (Roselly, *Monumento*). Provost, *Columbus*, 170, characterizes this as "the biography that led to the nineteenth century movement to canonize" Columbus, and I've found no evidence to the contrary. Claim that Pius IX commissioned Roselly in Toro, *Historia Colonial de la América Española*, I, 445, repeated in Carpentier's novel, *Harp and the Shadow*, 8. Letter by the archbishop of Bordeaux, Cardinal Donnet, published August 19, 1876, in *The Tablet*, quoted in Campbell, "Nineteenth century Movement to Canonize."

36. Roselly de Lorgues, *Christophe Colomb*, I, 57 ("grande sagacité d'observation"); De la Torre y del Cerro, *Beatriz Enríquez*, 27.

37. Roselly de Lorgues, *Christophe Colomb*, I, 44–45 (I have very freely glossed "était devant l'Église l'épouse de" and "nous le prouvons à l'instant même"); other Columbus apologists of the age embraced as fact the fiction of the Columbus-Enríquez marriage (e.g., Casanova de Pioggiola, *La vériteé sur l'origine*, Ch. XIX [1889 edition, 145–51]).

38. Roselly de Lorgues, *Christophe Colomb*, I, 172–76 (quotes on 173–75: "appartenait à la noble maison des Arana... vers la fin de novembre 1486"). Perhaps Columbus felt guilt, not over the absence of marriage, but over taking her brother and cousin, both named Diego, to Hispaniola on his First Voyage, and then leaving the cousin there—where he died along with all the men left in that first, failed settlement (RC: VI, 102, 108; RC: VII, 182, 416; CoH, 89). The Columbus-Enríquez marriage myth had roots

in a rumor in a 1606 suit filed by Cristóbal de Cardona y Colón, a great-grandson of Columbus, part of Life Three's great web of lawsuits; and in the imperial history that Spanish Crown historian Antonio de Herrera started publishing in 1601, which stated that Columbus "married doña Filipa Muñiz de Perestrelo, and had don Diego Colon by her, and then by doña Beatriz Enríquez, native of Cordoba, don Fernando, a gentleman great in virtue and letters." The reference to a second son, not a second marriage, is obvious. Still, its slight ambiguity, read as an implication of a second marriage, sat there in print for others to quote—including Roselly, centuries later. Herrera, *Historia General de los Hechos*, Dec. I, Libro I, Cap. 7, 14 (1601) ("casò con doña Filipa Muñiz de Perestrelo, y huuo en ella a D. Diego Colon, y despues en doña Beatriz Enriquez, natural de Cordoua, a D. Hernando cauallero de grã virtud y letras"); De la Torre y del Cerro, *Beatriz Enríquez*, 24–27.

39. Roselly de Lorgues, *Christophe Colomb*, II, 395–563 (471: "tous les saints ne sont pas arrivés au ciel par la même voie").
40. Asensio, *Cristóbal Colón*, I, lxxxii; De la Torre y del Cerro, *Beatriz Enríquez*, 27; Winsor, *Christopher Columbus*, 54; *Christian World*, vol. 27 (1877), 36; Murray, *Lives of the Catholic Heroes*, 17–178 (quotes on v, 175, 176; italics in original).
41. Roselly de Lorgues, *Satan contre Christophe Colomb*; *Histoire posthume*; De la Torre y del Cerro, *Beatriz Enríquez*, 27; Brocken, *Des Vicissitudes Posthumes*; Lamartine, *Christophe Colomb*; Casanova de Pioggiola, *La véritée sur l'origine* (1889 edition), xxxvii (I have loosely translated "les chansons déshonnêtes le révoltaient"). Bloy's book published simultaneously in 1884 in French and English: *Le Révélateur du Globe* and *The Revealer of the Globe*. On Bloy: Introvigne, "Modern Catholic Millennialism," 555–56.
42. Asensio, *Cristóbal Colón*, I, lxix–lxxi; Barry, *Life of Christopher Columbus* (the translation of Roselly de Lorgues's *Christophe Colomb*); Elliott, "Catholic Publication Society" ("ardent" quote about society founder Isaac Hecker [1819–88]); Winsor, *Christopher Columbus*, 53–54, 505.
43. Sanguineti (sometimes spelled Sanguinetti), *La canonizzazione*; *Vita di Cristoforo Colombo* (which included a truncated version of his Enríquez position, as well as another short book, bound with his *Vita*, on the birth debate, titled *Della Patria di Cristoforo Colombo*). Provost, *Columbus*, 170, called *La canonizzazione* a "classic refutation of the arguments for canonizing" Columbus.
44. Toro, *Historia Colonial de la América Española*, I, 445–47; Leo XIII, *Quarto Abeunte Saeculo*; Campbell, "Nineteenth century Movement to Canonize."
45. CGCC, xxi; Taviani, *Great Adventure*, 69.
46. Odoardi, "Il processo di beatificazione"; Roncari, *Nine Arguments* (illustration on 28, quotes on 193).
47. "The Canonization of Columbus," *Sacramento Daily Record-Union*.

LIFE FIVE: THE LOVER

1. On one ban campaign, see Maheshwari, "In Big Sky Country."
2. Taviani, *Great Adventure*, 31; Barrientos, "Colón," 52.
3. Phillips, *Testimonies*, 5.
4. RC: XIII, 242; Colón, *Vita*, Cap. V; Harrisse, *Christophe Colomb*, I, 267. Analysis of

the passage is complicated by the fact that Fernando likely wrote little of the first part of the *Vita*. Fernández-Armesto (personal communication) suggests *amicitia* (*amicizia* in modern Italian) implies more than friendship, which may be the case here but the implication is not evidence, and I am mindful of how biographers have spun threads from the *Vita* into great yarns (e.g., Morison fully embraces the imaginary romance as "one of those sidewalk-to-window affairs"; *Admiral of the Ocean Sea*, 38); Catz, *Columbus*, 24, is a rarer example of a suitably skeptical scholar.

5. Taviani, *Great Adventure*, 33; Taviani in Bedini, *Encyclopedia*, 24–25; Harrisse, *Christophe Colomb*, I, 269–96.

6. Harrisse, *Christophe Colomb*, I, 299–300 ("nous voulons croire; qu'àpres la mort de sa femme légitime"). The reference in Columbus's 1497 *majorat* document to Diego, and *otros vuestros hijos* was standard legal language that covered any future children, not evidence of mysterious additional ones. Thacher, *Christopher Columbus*, I, 405–7 (my "regretted"); Fernández-Armesto, *Columbus*, 17; Phillips and Phillips, *Worlds*, 114, and Bergreen, *Columbus*, 69, are two further examples among many modern biographers giving Columbus much benefit of the doubt. The circumstantial evidence had Davidson, *Columbus*, 48, "inclined to believe" that Felipa was alive when Columbus left Portugal.

7. Harrisse, *Christophe Colomb*, I, 355. Taviani in Bedini, *Encyclopedia*, 24, states that Columbus met Enríquez de Arana and his son Diego (a cousin of Beatriz's) in the store of Genoese brothers named Esbarroya, and Diego introduced Columbus to Beatriz; pure speculation, but a reasonable guess. Five years later, Diego de Arana (Beatriz's cousin) would sail on Columbus's First Voyage, with the other Diego de Arana (Beatriz's brother) as marshal (responsible for order) on the *Santa María* (CoH, 89; Bedini, *Encyclopedia*, 25).

8. The codicil was dictated and notarized on May 19, 1506: "E le mando que haya encomendada a Beatriz Enriquez, madre de D. Fernando, mi hijo, que la provea que pueda vivir honestamente, como persona á quien yo soy en tanto cargo. Y esto se haga por mi descargo de la conciencia, porque esto pesa mucha para mi ánima. La razon dello no es lícito de la escribir aquí." Thacher, *Christopher Columbus*, I, 421–22; Harrisse, *Christophe Colomb*, I, 355; II, 157–58; CVD: II, 354–61 (# CLVII, May 4, 1506); also see CoH, 249–53, for the 1505 codicil. I have followed Columbus himself in styling his son's name as Fernando (the assertion that he was named after the king of Aragon is my speculation, but it rings true to Columbus's ambitions); styling it as Hernando (as does Wilson-Lee) is equally accurate, as sixteenth-century pronunciation of the initial F/H made them the same name.

9. Beatriz died around 1521; in 1523, documents show her heirs successfully suing to have the pension payments restarted and paid to them. De la Torre y del Cerro, *Beatriz Enríquez*, 64–66; Fernández-Armesto, *Columbus*, 52–53; Taviani, *Great Adventure*, 69–70.

10. Thacher, *Christopher Columbus*, I, 422 ("mystery"). In the *Vita*, Fernando's mother is virtually absent (as Wilson-Lee, *Shipwrecked Books*, 28, observes). He likely had no memory of his infant years spent with her, just as his stepbrother Diego would have had no memory of his Portuguese mother, and little in the way of childhood memories either of stepmother Beatriz or of the aunt who took care of him between his mother's death and Columbus's placing him with his stepmother in Cordoba.

11. De la Torre y del Cerro, *Beatriz Enríquez*, 28 ("una amante"), 63 ("poco fiel a la memoria de su primer amante").
12. Fernández-Armesto, *Columbus*, 52.
13. RC: VI, 416 ("Dize mas q. tambie. le dava gran pena dos hijos q. tenia en cordova al estudio q. los dexava guerfanos de padre y madre en tierra estraña").
14. For example, Roncari, *Nine Arguments*, 28.
15. The jewel-pawning myth originates, like many Columbiana legends, in Colón, *Vita*: in a line (60, 262) asserting that the queen offered such collateral (for which there is no evidence, nor is it at all plausible), but that a loan from Luís de Santángel spared her having "to commit the jewels" (d'impegnar le gioie). Despite the fact that even the apocryphal story's source has the pawning not happening, it was seized upon and embellished by Spanish chroniclers such as Herrera, then repeated with imaginary details in numerous accounts; Ballesteros Beretta, *Cristóbal Colón*, I, 520–22; a seventeenth-century English example is Molloy, *De Jure Maritimo*, prefatory 28 [unnumbered]. United Nations A/37/PV.83 (November 29, 1982), 1376. The 1892 Columbian Series (Wanamaker, "Postage Stamps"; Justice, "Stamps"; Durnin, *Columbus*) paralleled the World's Columbian Exposition in Chicago, beginning a long tradition of postal stamp issues in conjunction with World's Fair or Exposition events in U.S. cities—more than twenty-five such philatelic tie-ins during the twentieth century (Stage, "Many Fairs").
16. Barrientos, "Colón" (Juan José Barrientos, b. 1944, is the Mexican critic); Heck, "Operatic Christopher Columbus"; Carpentier, *Harp and the Shadow*, 65–71; Peris, "Yo soy el Cristóbal Colón número 20." One such example historian is Morison, *Admiral of the Ocean Sea*, 96–107 (also see Stavans, *Imagining Columbus*, 36).
17. Lawton Campbell (1896–1980) was an Alabama-born New York City advertising executive who also wrote plays, of which three were staged (including *Immoral Isabella?*—although it was not published); Stavans, *Imagining Columbus*, 73–74 (on Rushdie).
18. Irizarry, *Christopher Columbus's Love Letter* (also see 2011 edition, *La Carta de Amor de Cristóbal Colón a la Reina Isabel*, 114–20). Irizarry's book inspired Ecuadorian songwriter Luís Padilla Guevara to develop a 2015 musical theater production about the Columbus-Isabel romance, based also on his and Gustavo Pacheco's 1985 hit song "La Pinta, La Niña, y la Santa María" (eluniverso.com/vida-estilo/2015/09/21/nota/5139062/musical-30-anos-cancion-pinta-nina-santa-maria/).
19. AGI, *Autógrafos*, legajo 43; Fernández-Armesto, *Columbus on Himself*, 198, and CoH, 198–200; Varela, *Textos y documentos*, 303. Irizarry had earlier published another Columbus book: *The DNA of His Writings*, arguing that he was a Catalan of Jewish origins, found Spanish readership. Believers in the Columbus-Isabel romance myth have read much into his 1500 letter to Juana de la Torre, in which he claims that God gave the queen "the spirit of understanding" his plan, which he then executed "in her royal name" (RC: II, 161–69, 377–87; CoH, 188–97); in fact, this was Columbus strategically pandering to Isabel via doña Juana, and there is no evidence of Isabel favoring Columbus and his voyages more than did King Fernando—who, on the contrary, was notably supportive of Columbus and his son Diego after Isabel's death (also noted by Fernández-Armesto, *Columbus on Himself*, 188).
20. Various online stories relate to the campaign: e.g., eldebate.com/historia/20220922/isabel-catolica-biografia-marcada-fakes_61455.html; and fsspx.news/en/news-events/news/spain-cause-beatification-isabella-catholic-relaunched-79516.

21. Peris, "Yo soy el Cristóbal Colón número 20."
22. Menéndez, "Isabel la Católica se enamoró de Cristóbal Colón." An example of the Bobadilla fable repeated as fact: Bergreen, *Columbus*, 130. Bobadilla was aunt to the Francisco de Bobadilla who arrested and replaced Columbus on Hispaniola in 1500.
23. RC: XII, 50, 176: letter of October 15, 1495, to Gerolamo Annari ("... de triunfi et tiri de bombarde et lanzafochi ... et questo fu facto per cagione de la signora del dicto loco, de la quale fu alias il nostro signor armirante tincto d'amore").
24. Fernández-Armesto, *Columbus*, 53.
25. Taviani, *Great Adventure*, 70; *Voyages of Columbus*, I, 16–17.
26. RC: XII, 52, 177–78; Annari letter ("Li altri Camballi insieme cum dicti schiavi mandassimo poi in Spagna. Essendo io ne la barca, presi una Camballa belissima, la quale il signor armirante mi donò ... ultimate, fussimo de acordio in tal forma, che vi so dire che ne; facto parea amaestrata a la scola de bagasse").
27. See, e.g., Bergreen, *Columbus*, 143, and online reactions to his verdict that "So began the European rape of the New World."
28. Restall, *When Montezuma Met Cortés*, 283–93, 305–11.
29. Zamora, *Reading Columbus*, 170 ("metaphor"), 177 ("head," from the *Diario*); Fernández-Armesto, *Columbus*, 52 ("common standards").
30. RC: XIII; Wilson-Lee, *Shipwrecked Books*, 28, 35, 113.
31. Wilson-Lee, *Shipwrecked Books*, 132, 324 ("pitiful inheritance" is his phrase).
32. Wilson-Lee, *Shipwrecked Books*, 126–28, 137, 146 ("wrong damsel" quote is his), 178.
33. Previous two paragraphs: Schoenrich, *The Legacy*, I, 253–85.
34. In 1571, María de Orozco was finally free to marry Villareal; they married in Venice, where they had settled and had two sons; six months later, she died (a few months before Luís). On behalf of those sons, Villareal sued Luís's daughter and nephew (who were married to each other), winning their mother's property in Santo Domingo. Meanwhile, Luís's third wife, doña Ana de Castro, also predeceased her duplicitous husband—just weeks after he was transported to Africa. It was said that she died of humiliation. Schoenrich, *The Legacy*, I, 284–93, 305–6.
35. Six months later, Luisa filed for an annulment on the grounds that the marriage was forced upon her; but just as the case seemed to be going her way, she dropped it. Four decades later, she was still married to the same man—and still participating in the Colón lawsuits on behalf of herself and her son Cristóbal. Schoenrich, *The Legacy*, I, 293–204; II, 87–88.
36. Spanish saying (*Si quieres hacerte immortal, hazte pleito sobre el estado de Veragua*) quoted in Life Three. Schoenrich, *The Legacy*, I, 304 on Petronila.
37. Schoenrich, *The Legacy*, I, 253–307; Thacher, *Christopher Columbus*, I, 346; Fernández-Armesto, *Columbus*, 52. A dispassionate, mini biography of don Luís is Hoffman in Bedini, *Encyclopedia*, 137–38.
38. Fernández-Armesto, *Columbus*, 34, 41–42. Columbus's copy of Plutarch's book (aka *Parallel Lives* or Plutarch's *Lives*) became part of his son Fernando's library, surely influencing parts of the Colón-attributed *Vita*; Wilson-Lee, *Shipwrecked Books*, 300. The postils marking homosexual references or insinuations are in RDSP, I, iii, 6, 13, 18, 28, 47, 57, 66, 75, 88, 92, 112, 243, 329. Flint, *Imaginative Landscape*, 75, detected in the postils distaste for "homosexual practice."

39. Fernández-Armesto, *Columbus*, 58–59; Ballesteros Beretta, *Cristóbal Colón*, I, 455–58; Wey Gómez, *Tropics of Empire*, 231–32.
40. Fernández-Armesto, *Columbus*, 59. For two 1504 letters from Columbus to son Diego mentioning close ties to Deza, see CoH, 239, 244; and 248 for an excerpt, preserved by Las Casas, of a letter from Columbus to Deza (in which the former's implied criticism of King Fernando reflects his trust in the latter). Deza's name is more properly (but seldom in modern works) styled Diego de Deza y Tavera (1444–1523). Deza and Columbus are companions in death: their tombs are today not far apart inside Seville's cathedral.
41. Mantle, "Gay Columbus Played Too Dully," F1; youtube.com/watch?v=gDAIFFIotno (the *Match Game* ran on and off, with slightly different names on three successive networks, between 1962 and 1981).
42. Maurer, "Christopher Columbus."
43. McKenzie, *Manatees*, 4–6.
44. RC: VI, 400 ("dixo que vido tres serenas que salieron bien alto de la mar pero no eran tan hermosas como las pintan que en alguna manera tenian forma de hombre en la cara"); Flint, *Imaginative Landscape*, 83–85, 136–41; Mentz, *Ocean*, 93–102.
45. Philiponus, *Nova typis transacta navigatio*, its JCB catalog entry; Wilson-Lee, *Shipwrecked Books*, 84–87. Martyr: Anghiera, *Extraict*; RC: V.
46. Mann, *1491*, 351–53; Stavans, *Imagining Columbus*, 37. Díaz de Isla's book (*Treatise on Venomous Sicknesses*) was first published in 1539, but the 1493 allegation circulated earlier.
47. Bradley, "Archaeology breakthrough."
48. Kelly, "Conspiracy Culture"; Hsu, "Conspiracy Theories"; Scheirer, *A History of Fake Things*.

LIFE SIX: THE LOCAL

1. Casanova de Pioggiola, *La vériteé sur l'origine* (1889 edition), xxv–xxvi (on the plaque's creation); Taviani, *Christopher Columbus*, 16; an example, among numerous websites, is france.fr/fr/corsica/article/maison-natale-christophe-colomb-calvi, warning visitors "not to tell the people of Calvi that Columbus was born in Genoa: they'll laugh in your face!" (Ne dites pas aux Calvais que Christophe Colomb est né à Gènes, ils vous riraient au nez!)
2. Taviani, *Christopher Columbus*, 15.
3. In 1884, Harrisse, *Christophe Colomb*, I, 217, listed Pradello, Cuccaro, Cogoleto, Savone, Nervi, Albissola, Bogliasco, Cosseria, Finale, and Oneglia for the Italian claims (plus Genoa). In 1993, Catz, *Columbus*, 97n2, listed 16 (with Chinese making 17) non-Italian regions or nations, with an Italian list of ten cities (Genoa included). By my count in 2024, there are now twelve Italian and twenty-one non-Italian claims—so far. Most of the one hundred fifty books mentioned above are listed in Ballesteros Beretta, *Cristóbal Colón*, I, 174–76.
4. Casanova de Pioggiola, *La vériteé sur l'origine*; Harrisse, *Christophe Colomb et la Corse*, 1–2; Ballesteros Beretta, *Cristóbal Colón*, I, 13–32.
5. Casanova de Pioggiola, *La vériteé sur l'origine* (1880 edition), vi, 17, 139 ("encore mieux les textes écrits... elle est la plus forte"); Harrisse, *Christophe Colomb et la Corse*, 2;

Desimoni, *Di Alcuni Recenti Giudizi*; Ballesteros Beretta, *Cristóbal Colón*, I, 130–34; Cesarini, *Cristóbal Colón Nació en Corcega*. One Corsican solution to the Calvi claim issues is to argue that Columbus was from the other end of Corsica; the author of *Christophe Colomb Corse du Niolu* (2017), Victor Geronimi, is also from Niolu and claims descent from Columbus.

6. Harrisse, *Christophe Colomb*, I, 217.
7. Quotes from Sable, "Columbus, Marrano Discoverer from Mallorca."
8. RC: XIII, 30–31, 234–35. In 1514 and 1521, Fernando visited Genoa, where he bought books and met with local scholars and relatives (although the *Vita* only mentions a meeting with two very old Colombos in Cugureo) (Taviani in Bedini, *Encyclopedia*, 136; Wilson-Lee, *Shipwrecked Books*, 176).
9. Herrera, *Historia General de los Hechos*, Dec. I, Libro I, Cap. 7, 13 (1601) and 11 (1726) ("i para maior inteligencia de ella conviene saber.... Unos quieren que fuse de... adonde se litiga, se determinará"); Ballesteros Beretta, *Cristóbal Colón*, I, 135–41.
10. Ballesteros Beretta, *Cristóbal Colón*, I, 134–35.
11. CGCC, viii–ix; Taviani, *Christopher Columbus*, 16; Franzoni, *La Vera Patria*; LoC/HH, "Fasti" No. 1, 1–5.
12. CGCC, viii–x; Irving, *Life and Voyages* [1828], I, 3–6. "Notoriety": Italian senator Eugenio Broccardi (1867–1959), introducing the English and German edition of the Genoese government's publication of hundreds of documents detailing Columbus's Genoese origins; CGCC, v; Airaldi, *Colombo*; Ballesteros Beretta, *Cristóbal Colón*, I, 135 ("en extremo ridículas"). Harrisse lamented that by the 1880s, "most historians" were asserting Columbus was of noble birth, despite all evidence to the contrary (*Christophe Colomb*, I, 137–65).
13. On early modern piracy, see Lane, *Pillaging the Empire*.
14. LoC/HH, "Fasti" No. 1, I, 1–6; Cateras, *Greek Prince*, 27–34; Canoutas, *Christopher Columbus*, 80–90; Davidson, *Columbus*, 5, 12–14, 487.
15. CGCC, ix; Davidson, *Columbus*, 14; Cateras, *Greek Prince*, 15–17 (who also asserted that "archaeologists and learned researchers found that one-third of the language of the Mayans was pure Greek of the Homeric period").
16. RC: XIII, 238; Cateras, *Greek Prince*, 44–54, 58; Canoutas, another proponent of the "Greek prince" claim, renders his "real name" as Dishypatos (*Christopher Columbus*, 61–78).
17. Sannes, *Christopher Columbus*; Associated Press, "Writer Suggests Columbus Was Son of Norway"; Huyghe, *Columbus Was Last*, 213–14; Mellgren, "Was Columbus Really a Norwegian Son?"; Forskning.no (Norway's Research Council online newspaper).
18. Davidson, *Columbus*, 487; Ballesteros Beretta, *Cristóbal Colón*, I, 122; Huyghe, *Columbus Was Last*, 213.
19. Ballesteros Beretta, *Cristóbal Colón*, I, 122.
20. "Doubts" quote is by Huyghe, *Columbus Was Last*, 214.
21. "Italian" quote is by Huyghe, *Columbus Was Last*, 213; Ballesteros Beretta, *Cristóbal Colón*, I, 107; Cateras, *Greek Prince* (another example of the erroneous birth certificate and language arguments).
22. Ballesteros Beretta, *Cristóbal Colón*, I, 123, on Catalan claimant Luís Ulloa ("Cuando a Ulloa le molesta un documento lo dictamina falso y de esta manera desaparece el obstáculo").

23. Active 1956–67, Michael Flanders (1922–75) and Donald Swann (1923–94) were English musicians who wrote more than one hundred comic songs together, giving some two thousand theatrical performances around the English-speaking world.
24. Quotes from Catz, *Columbus*, 101.
25. Molloy, *De Jure Maritimo*, prefatory 27 [unnumbered]. Davidson, *Columbus*, 487, cites "Malloy" [sic] with no citation; Taviani, *Christopher Columbus*, 15, omits the author's name, citing an unnamed later edition; Ballesteros Beretta, *Cristóbal Colón*, I, 134, also cites the later edition but by "Charles Mollow" [sic].
26. Catz, *Columbus*, 101. Catarina (1638–1705), daughter of King João of Portugal, thereby became Queen Catherine of England, Scotland, and Ireland.
27. Burton, *Wanderings*, I, 59.
28. "One document": Las Casas, *Historia*, I, cap. cxxx (RC: VII, 182, 416). I follow Catz's use (*Columbus*, 111) of Pereira, "Cristóvão Colombo."
29. Catz, *Columbus*, 104–5.
30. Catz, *Columbus*, 105–6.
31. The novel (first published in Funchal, 1909) and play (first performed there, 1913), written by J. Reis Gomes, centered on an imaginary character named Guiomar, whose husband was brother to Columbus's wife (Catz, *Columbus*, 106–7).
32. Catz, *Columbus*, 108.
33. Catz, *Columbus*, 106, 108.
34. Quoted in Catz, *Columbus*, 111.
35. Catz, *Columbus*, 112; Statista.com; dozens of tourist sites detail Columbus-related places, including those mentioned in the preceding paragraphs: totraveltoo.com, getyourguide.com, visitmadeira.com, madeira-web.com, madeiraislandnews.com, etc. The embrace of Columbiana in Madeira yielded minor American tourism dividends; in 2022, German tourists, who numbered fewer than British visitors, outnumbered Americans by 10 to 1.
36. Casanova de Pioggiola, *La véritée sur l'origine* (1889 edition), vii–viii ("de sincere éloges pour le soin minutieux avec lequel il cherche des arguments en faveur de sa these. . . . La question est résolue et le doute n'est plus possible: le Révélateur du Nouveau-Monde est né à Calvi, dans l'Ile de Corse"); Peris, "Yo soy el Cristóbal Colón número 20."
37. Quoted in Catz, *Columbus*, 112.

LIFE SEVEN: THE IBERIAN

1. Spanish newspaper and online news reports on the story were widespread in autumn 2022, starting with local press (e.g., *Galicia Confidencial*, 9/11/22; *La Voz de Galicia*, 11/13/22) (see "Exhuman 7 ósos" and González, "El sepulcro") and spreading to stories in such outlets as *El Independiente* (11/21/22), *Diario de Sevilla* (11/21/22), *El Debate* (11/22/22), and *El País* (11/28/22) (quote above used in multiple pieces: "una pieza clave para poder confirmar las investigaciones que apuntan at origen gallego de Cristóbal Colón"). Quote in caption to the figures opening this chapter: Moreno Ortíz, "En busca del origen gallego" (restos óseos de Cristóbal Colón u su hijo).
2. González, "El sepulcro"; Moreno Ortíz, "En busca del origen gallego"; "Exhuman 7 ósos"; see also turismopoio.com/que-ver/cultura-y-patrimonio/casa-museo-de-colon/.
3. Killgrove, "DNA Testing of Skeleton"; publications by Fernando Branco, a civil engi-

neer (and self-identified "amateur historian") at the University of Lisbon, active in the campaign; a seven-minute video packed with Columbiana tropes at bbc.com/reel/video/p08jgfdg/the-truth-about-christopher-columbus. Ataide is briefly mentioned as a dead end in Barreto, *Portuguese Columbus*, 367 (see Mascarenhas Barreto in the bibliography). Dugard, *Last Voyage*, 264, is an example of an author optimistic (as of 2007) about Lorente's work, despite "inconclusive" results.

4. Heaney, *Empires of the Dead*, 229–45 (Smithsonian quote on 239). NAGPRA was passed in 1990.
5. "Dejando un legado histórico y un enigma sobre su verdadero origen" (from *El País*, cited above). Rogers, "Vivaldi Expedition," 31–32, noted how the study of Atlantic discoveries had become nationalized by the time he was writing (1955).
6. Canel (1857–1932), *La Cuna de Colón*, 34 ("ha encontrado en Galicia datos y pruebas feacientes para probar que era Colón Gallego"). García de la Riega's (1844–1914) initial 1898 paper was published months after his death as a small book titled *Colón español: su origen y patria*. Solari: 1899–1980.
7. Canel, *La Cuna de Colón*, 48 ("todos dudaban y muchos sonreían"); Carbia, *La Patria de Cristóbal Colón*, 11 (*adulterados*); Mosqueira Manso, *La cuna gallega*, 13–14 (*retocado; adulterado*; etc.).
8. Ballesteros Beretta, *Cristóbal Colón*, I, 103–12; Taviani, *Christopher Columbus*, 17.
9. Ballesteros Beretta, *Cristóbal Colón*, I, 108–9; Horta y Pardo, *La Verdadera Cuna* (7, *vindicación*, and 53, *estúpida leyenda*); Carbia, *La Patria de Cristóbal Colón*; Bustos y Bustos, *La Patria de Cristóbal Colón*; Mosqueira Manso, *La cuna gallega*.
10. Carbia, *La Patria de Cristóbal Colón*, 49–52 (a skeptical appraisal); Mosqueira Manso, *La cuna gallega*, 38–74 (an unambiguous defense); Ballesteros Beretta, *Cristóbal Colón*, I, 106–7 (debunking).
11. Hedgecoe, "'A cock-and-bull story.'" Sites include tintineando.com/casa-museo-cristobal-colon and turismoriasbaixas.com/en/-/inspirate-revive-las-hazanas-de-pedro-madruga.
12. Cortesão, "Mystery of Columbus" (quotes on 322).
13. Catz, *Columbus*, 84–89; Ribeiro, *A Nacionalidade portuguesa de Cristovam Colombo*, was a bilingual, expanded 1927 version of his 1915 paper, "The Mysterious Character of Colombo and the Problem of his Nationality" (O Carácter Misterioso de Colombo e o problema da sua Nacionalidade).
14. Taviani, *Christopher Columbus*, 15–16.
15. These works include (but are by no means limited to) Ribeiro, *A Nacionalidade portuguesa de Cristovam Colombo*; Pestana, *D. Cristóbal Colón*; Gaspar da Naia, *Cristóbal Colón* and *D. João II e Cristóbal Colón* (Naia was a captain in the Portuguese merchant marine); Rumeu de Armas, *El "Portugues" Cristóbal Colón en Castilla*. See also the summaries in Catz, *Columbus*, 88–97 (90–92, on the books published 1930–42 by G. L. Santos Ferreira, Antonio Ferreira de Serpa, and Saul Ferreira) and in Ballesteros Beretta, *Cristóbal Colón*, I, 127–30.
16. Catz, *Columbus*, 93–97; Mascarenhas Barreto (his long name thus reduced on his publications), *"Colombo" Português*; *O Português Cristóvão Colombo*; and *The Portuguese Columbus*. On other Zarco advocates: Regal, *America's Origin Story*, 50–51. I thank historian Trevor Hall for sharing with me his plans for a book "confirming the theory that Columbus was a Portuguese spy working for King Joao II of Portugal" (email of 7/10/24).

17. Kamen, *Imagining Spain*, 27–28.
18. Ballesteros Beretta, *Cristóbal Colón*, I, 112–27; Ulloa, *Cristòfor Colom Fou Català* (published first in French as *Christophe Colom Catalan*). P. Catalá i Roca found four Colom brothers in local archives, asserting in a 1978 book titled *Quatre germans Colom, el 1492* that they were Columbus's brothers.
19. Costa-Amic, *Colom* (quote on 77: "la verdad es que los reyes aceptaron privadamente eso hecho como una realidad, y por ello decidieron su ayuda a los proyectos de Colom"); *La tesis mallorquina* is by J. Suau Alabern (1967), not in my Bibliography, but see Gil, *Columbiana*, 46; Ballesteros Beretta, *Cristóbal Colón*, I, 124–27. Verd: Hedgecoe, "'A cock-and-bull story'" (quote of same phrase); and a 2024 interview with a Palma newspaper at ultimahora.es/noticias/cultura/2024/01/21/2089497/cristobal-colon-primavera-sabremos-mallorquin.html. The repeated publication of slightly different versions of the same book is a feature of Columbiana birthplace literature, of which Verd is a good example: 1984's *Recopilación del enigma de Don Cristóbal Colón* and 1986's *Cristóbal Colón y la revelación del enigma*, both printed in Palma de Mallorca, were versions of the same book, with English versions published with similar names. Pohl, *New Columbus*, promoted the Majorcan claim outside the island; also see Sable, "Columbus, Marrano Discoverer from Mallorca," 8–9. True to Columbiana's tradition of regional rivalry, Majorca claims soon prompted a journalist on Ibiza to begin that island's own competing set of claims (Fernández-Armesto, *Columbus*, 196n6).
20. Summerhill and Williams, *Sinking Columbus*, 16–17 (the historian was Luís Arranz Márquez); Kubal, *Cultural Movements*, 84–89.
21. Kamen, *Imagining Spain*, 6, 122–25; Adorno, "Washington Irving's Romantic Hispanism." Examples include Irving's *Life and Voyages* (1828, in Spanish in 1834), his *Conquest of Granada* (1829), and Prescott's *Ferdinand and Isabella* (1838). The Dominican Republic adopted Día de Colón in 1912, e.g., and Mexico the Día de la Raza the following year. Proposals to rename the October 12 holiday in Spain began at least from 1977; in 1981, it was officially renamed Fiesta Nacional de España y Día de la Hispanidad; and in 1987 reduced to Fiesta Nacional.
22. Gil, *Columbiana*, 33–46 (quote on 33: "marcó de manera indeleble el curso de los estudios americanistas"); Ballesteros Beretta, *Cristóbal Colón*; Madariaga, *Christopher Columbus*.
23. Madariaga, *Christopher Columbus*, 53–57; also see Fernández-Armesto, *Columbus*, ix.
24. Fernández-Armesto, *Columbus*, ix ("defiance"). The Colón in García de la Riega, *Colón español*, was a crypto-Jewish Gallego, as he was in Carbia, *La Patria de Cristóbal Colón* (1923) and Andre, *Columbus* (1928). The confusing of Piacenza (Italy; but called Plasencia in Castilian) with Plasencia (Extremadura, Spain) underpins one minor claim; on Vicente Paredes and others, see Madariaga, *Christopher Columbus*, 434n23; Ballesteros Beretta, *Cristóbal Colón*, I, 98–103. See also a trio of works published in the 1920s: Calzada, *La Patria de Colón* (1920); Otero Sánchez, *España, Patria de Colón* (1922); and Beltrán, *Cristóbal Colón* (1925).
25. Madariaga, *Christopher Columbus*, 114, 56, 57, 20, 159, 91 (in order of the quotes).
26. Madariaga, *Christopher Columbus*, 329; Roth, "Who *Was* Columbus?"; Sarna, "Columbus & the Jews"; Lawson, "Cecil Roth," 48, 148–49; Wassermann, *Columbus*. Madariaga on Roth: *Christopher Columbus*, 434–35n23. Madariaga was also influenced by German Jewish novelist Jakob Wassermann (1873–1934), whose 1929 *Christoph Columbus*

biography came out in English a decade before Madariaga's and used similar—but less bluntly expressed—arguments regarding Columbus's characteristically Jewish traits. Madariaga borrowed Wassermann's book subtitle (*Don Quixote of the Seas*) for a book chapter, acknowledging the loan, but then dismissed his predecessor's work as "marred by a total incomprehension of Ferdinand and Isabel" (somewhat true), and sympathetic to "those who think that Colón was Jewish; but he did not go into the matter at all" (not really true) (*Christopher Columbus*, 447n9).

27. Ballesteros Beretta, *Cristóbal Colón*, I, 98–103. Madariaga's *Life* sold well despite Morison's insistence that "there is no more reason to doubt that Columbus was a Genoese-born Catholic Christian... proud of his native city, than to doubt that George Washington was a Virginia-born Anglican... proud of being an American" (*Admiral of the Ocean Sea*, 7).

28. Lawson, "Cecil Roth," 148–49.

29. Sarna, "Columbus & the Jews," 38–39. AOC, Cabinet 11, "Works of Art, Statues: Capitol & Grounds," Subfolder 1 ("Discovery Group: General") (hereafter AOC, 11, *Discovery*, Sub 1) (Baird especially liked Persico's *Discovery*, to which we return in Life Eight). García de la Riega's work was not published in English, but in 1933 Maurice David made available to U.S. readers its core ideas (*Who was "Columbus"?*).

30. Wiesenthal, *Sails of Hope*, was the 1973 U.S. edition of the original Amsterdam publication, *Zeilen der hoop: De geheime missie van Christoffel Columbus*; Summerhill & Williams, *Sinking Columbus*, 13.

31. Irizarry, *Christopher Columbus*, 7–9 ("silences"), 47–70; Fernández-Armesto, *Columbus on Himself*, 34–35 ("insecure"); Huyghe, *Columbus Was Last*, 214–15; Kadir, *Columbus*, 156–57. On the coded signature: Life Two; Fernández-Armesto, *Columbus*, 118–19; Davidson, *Columbus*, 278–79; Milhou, *Colón y su Mentalidad Mesianica*, 79–80; Flint, *Imaginative Landscape*, 184–85; Catz, *Columbus*, 84–87. On portraits: see the Introduction; a 1936 example, "It must be admitted quite frankly that *all existing portraits* of the discoverer give him a Jewish cast of countenance; though, again, this is not conclusive or valid evidence" (Duff, *The Truth*, 63).

32. Posse, *Dogs of Paradise*; Marlowe, *Memoirs*, 1–6; Kritzler, *Jewish Pirates*; Sobral, "Columbus and Jamaican Jews," 263; Berry, *Columbus Affair* (quote on 425, in "Writer's Note"). Berry mentions neither Madariaga, *Christopher Columbus*, nor Marlowe, *Memoirs*, but he cites Wiesenthal, *Sails of Hope*, as the key work that "convincingly... postulated the premise" of the Jewish Columbus. Also see Israeli novelist Youval Shimoni's 2021 *MiBa'ad LaKarka'it HaShkufa* (*Beyond the Transparent Bottom*), in which Columbus is sidelined by a focus on the history of anti-Semitism and on a fictional 1506 letter from him to the king requesting support for a crusading voyage to save Jerusalem from Muslim rule.

33. Amler, *Columbus's Jewish Roots*, 247–48; Irizarry, *Christopher Columbus*, 146 ("widely accepted"); Huyghe, *Columbus Was Last*, 215 ("strongly point"); Arana, *Latinoland*, 176; Sarna, "Columbus & the Jews," 41.

34. Summerhill & Williams, *Sinking Columbus*, 128 ("all other nations").

35. The previous paragraphs draw on Summerhill & Williams's excellent *Sinking Columbus*, 127–49 (128, "protagonist"). See Gil, *Columbiana*, 131ff. on the Italy/Spain rivalry.

36. Summerhill & Williams, *Sinking Columbus*, 147–49, 168–73.

37. Zavala, "Destrucción y daños"; Summerhill & Williams, *Sinking Columbus*, 177; Kubal, *Cultural Movements*, 90–97.
38. Candidatura d'Unitat Popular (CUP) calls were issued every couple of years, beginning in 2016; in 2020, Barcelona's mayor, Ada Colau, insisting the monument would remain untouched, instead supported the "slavery tour" of the city launched four years earlier, designed to educate locals and visitors about the *memoria incómoda* of Barcelona's historic involvement in the slave trade ("uncomfortable memory"; local history professor Oriol López's phrase); theguardian.com/travel/2016/apr/13/barcelona-slave-trade-history-new-walking-tour-catalonia-spain-ramblas; theguardian.com/world/2020/jun/19/columbus-statue-will-stay-but-slavery-tour-aims-to-address-barcelonas-past; theolivepress.es/spain-news/2020/06/15/podemos-politician-calls-for-columbus-statue-to-be-removed-from-spains-barcelona-amid-anti-racism-protests/.

LIFE EIGHT: THE ADAM

1. When Union Station opened in 1907, the Knights of Columbus had already lobbied for a Columbus monument to go in front of it, facing toward the Capitol. The *Columbus Fountain*, created by sculptor Lorado Taft, surrounded by Columbus Circle, was unveiled over a three-day celebration in 1912. On the Capitol's Columbuses, see Chambers, *Columbus in the Capitol*; Fryd, *Art and Empire*, 18–20, 54–56, 80–83, 89–105, 125–43; Bushman, *America Discovers*, 131–45; and other sources cited below. Greenough's *George Washington* was moved in 1964 inside the National Museum of American History. My analysis of these artworks was also made possible by fellowships from the United States Capitol Historical Society in 2017 and 2024–25, for which I am grateful. A separate study based on my research in the Capitol (tentatively titled "A Capitol Columbus") is under way.
2. Bushman, *America Discovers*, 132–36; Ponce de León, *Columbus Gallery*, 125–27.
3. AOC, II, 3 (*Landing*), Sub 1; Fryd, *Art and Empire*, 54–56; Schlereth, "Columbia, Columbus, and Columbianism," 946–50.
4. AOC, II, 3 (*Landing*), Sub 1 (1962 article in *Kingston* [NY] *Freeman*); Ponce de León, *Columbus Gallery*, 153; Bushman, *America Discovers*, 138–39.
5. The panels' artists were Antonio Capellano, a Florentine, and Enrico Causici, from Verona, supervised by Francesco Iardella of Carrara; as the icons were carved in 1824–29, the Columbus bust is the oldest representation of him in the Capitol. On them and Brumidi's *Frieze*, see Fryd, *Art and Empire*, 17–21, 142–54; Restall, *When Montezuma Met Cortés*, 24–27; Restall, "Montezuma Surrenders in the Capitol."
6. Bushman, *America Discovers*, 141–43. A. G. Heaton's oil painting *The Recall of Columbus* was installed in 1882, the Capitol's most recent Columbus depiction.
7. Bushman, *America Discovers*, 24–30; Bartosik-Vélez, *The Legacy*, 13, 48–59, 69. Americanus was Samuel Nevill (d. 1764), later speaker of the New Jersey Assembly, who published mostly in his own *New American Magazine*. Martyr: Anghiera, *Extract*; RC: V.
8. Raynal's 1770 *A Philosophical and Political History of the Settlements and Trade of the Europeans in the East and West Indies* was first published in English in 1776; Robertson, *History of America*, I, 59–175 (his "Book II"); Russell, *History of America*, I, 15–39. Also see Bushman, *America Discovers*, 30–40; Bartosik-Vélez, *The Legacy*, 66–105.

9. Russell has Taíno leader Hatuey, right before being burned alive in an act of "unjust vengeance," calling Spaniards "that accursed race" (*History of America*, I, 50); Robertson, *History of America*, I, 175; on Robertson's influence, see Bushman, *America Discovers*, 32–40; Provost, *Columbus*, 43, 195.
10. Freneau, *The Poems*, 57; and *A Poem*, 25; republished in 1788 as part of "The Pictures of Columbus: The Genoese" (partially quoted by Stavans, *Imagining Columbus*, ix); Barlow, *The Columbiad* (a reworking of his earlier, rather turgid epic poem, *The Vision of Columbus*); Bushman ("one scholar"), *America Discovers*, 60–80 (quotes on 62, 74, 80); Bartosik-Vélez, *The Legacy*, 60 ("paradigmatic"); Dennis, "Eighteenth-Century Discovery"; Blakemore, *Joel Barlow's Columbiad*.
11. Belknap, *A Discourse on the Discovery*, quoted phrases on (in order) 12, 20, 16, 27, 50, 47.
12. "Ode" in Belknap, *A Discourse*, 56–58; Belknap's depiction of Columbus as a misunderstood hero went into the entry included in his 1794 *American Biography* (also see Bushman, *America Discovers*, 88–91).
13. Dennis, "Eighteenth-Century Discovery," 216–22. In 1792, Londoners enjoyed performances in Covent Garden's Theatre Royal of *Columbus*, a play by Thomas Morton. Although the theme was familiar to English audiences, Morton felt the need to include on Columbus's First Voyage a fictional Englishman named Harry Herbert—a protagonist second only to Columbus himself, the two forming a heroic pair, in contrast to the greedy and treacherous Spaniards. Audiences presumably cheered and jeered as rebel leader Roldán was called everything from a "scoundrel" and "coward" to a "reptile" and "cannibal." Morton anticipated core elements of the lives that would later be crafted for Columbus in the United States, such as his separation from Spanish scoundrels and the rulers who wronged him. The play closed with a reminder that "To your support alone he trusts his cause, / And rests his fame, on Englishmens [*sic*] applause"; the sentiment would have worked as well across the ocean, with "Americans" substituted for "Englishmen" (quotes scattered throughout, 3–66).
14. Clarke, *Old and New Lights* (an 1892 book, the quoted phrase used countless times over the preceding century); on poetry and naming, Bushman, *America Discovers*, 20–21, 41–59; Schlereth, "Columbia, Columbus, and Columbianism," 937–42. Today's District of Columbia dates from 1871. George Washington left provisions for a university in the new capital, and when it was finally chartered in 1822 it was named Columbian College (University in 1873); the name was changed in 1904, due to confusion with New York's Columbia University, to George Washington University.
15. Numbers based on Shin, "Columbus Monuments," and on my searches through various Wikipedia sites, including nation-specific sites and "List of" sites relating to "Christopher Columbus" and "Cristóbal Colón." There are municipal districts called Colón in Mexico, Honduras, Panama, Colombia, Venezuela, and Argentina. I likely missed some toponyms in Latin America, but the pattern is so striking that it is unlikely a more comprehensive statistic would change the conclusion. "Colombia" as a rule-proving exception: Venezuelan independence leader Francisco de Miranda's specific and obsessive vision was that all Spanish America, from Mexico to Patagonia, be an independent empire named Colombia—a name inspired by travel in the United States and life in exile in England (including in Joel Barlow's London home, soon after *The Vision of Columbus* came out); although Miranda died a Spanish captive in 1816, successors such as Simón Bolívar adopted the name to designate the continent and eventually a part of

it (Bartosik-Vélez, *The Legacy*, 106–43; Racine, *Francisco de Miranda*; Cock Hincapié, *Historia del nombre*).

16. Mucher, *Before American History*, 363–72 ("occupation" quote on 370).
17. Irving, *Works*; *Life and Voyages* [1828], I, vi–vii; Adorno, "Washington Irving's Romantic Hispanism"; Regal, *America's Origin Story*, 40–43. Irving relished what he called the "zealous controversy" surrounding aspects of Columbus's life, exaggerating the mysteries and muddles before providing the solution, a Columbiana tradition that he may not have invented but which he brought to generations of readers and helped ensure its survival into this century.
18. Irving, *Life and Voyages*, I, vi; Bushman, *America Discovers*, 107–26.
19. As of 2023, Boyd had published four "whole life" novels—although their technique appears to a lesser extent in many of his other works—the most recent of which includes an interview in which he discusses "fictography" (*The Romantic*, 459).
20. Muñoz, *Historia del Nuevo-Mundo*, 42–342. Irving had access to this in Madrid. It contained many of the errors yet to be corrected by scholars (e.g., Columbus born in 1446), and the laudatory tenor of Columbus histories that was typical of the eighteenth and nineteenth centuries; yet it was remarkably restrained, considering the era and circumstances of its commission. Had Muñoz (1745–99) lived to write subsequent volumes, and thus extended his account beyond 1500 and through Columbus's death, he may have resorted to the kind of eulogizing expected at that time. Bushman, *America Discovers*, 112 ("pernicious"). Adorno, "Washington Irving's Romantic Hispanism," 82–86.
21. Irving, *Life and Voyages* [1828], I, 78; Bushman, *America Discovers*, 111; Russell, *Flat Earth*, 29–40, 49–62, 75–77. On the 1892 stamps: Durnin, *Columbus*; Justice, "Stamps"; Maineri and Hamill, *Checklist*; Wanamaker, "Postage Stamps."
22. The four books are in my Bibliography as *History of the Life* and in *Works*, with the above quotes from 1832's *Tales of the Alhambra*; also see Drayson, *Moor's Last Stand*, 147.
23. Prescott, *History of the Reign* [1838], III, 245 (this passage is not in the abridged 1962 edition also consulted).
24. Harrisse, *Christophe Colomb*, and LoC/HH; Winsor, *Christopher Columbus*.
25. Horwitz, *Voyage Long and Strange*, 48, 390 (quoting Stephen Jay Gould, writing about baseball; "The Creation Myths of Cooperstown," November 1989 issue of *Natural History*).
26. Campe, *Die Entdeckung* (and its English editions, e.g., *Discovery of America*); *Stories of Great Men*, 36; Bushman, *America Discovers*, 98–107, 145–52, 170–75. Campe's Columbus bears little resemblance to historical reality, fitting instead the model of a late-eighteenth-century Christian gentleman (e.g., *Discovery of America*, 252–53). Regal, *America's Origin Story*, 21–30. Bartosik-Vélez, *The Legacy*, 66–67. On flag and pledge: Summerhill and Williams, *Sinking Columbus*, 111; Kubal, *Cultural Movements*, 2; Schlereth, "Columbia, Columbus, and Columbianism," 955–60.
27. Bushman, *America Discovers*, 159–61.
28. Rydell, *All the World's a Fair*, 38–71; Summerhill and Williams, *Sinking Columbus*, 9 ("greatest"), 64–66; Bushman, *America Discovers*, 161–64.
29. "Modern civilization": As expressed in a review in the November 1892 issue of the *Magazine of American History*, quoted in Bushman, *America Discovers*, 166; Summerhill and Williams, *Sinking Columbus*, 66–67; Kubal, *Cultural Movements*, 11–29.

30. Summerhill and Williams, *Sinking Columbus*, is a brilliant study of the Quincentennial (styled "Quincentenary" by them) (34–126, U.S. case; 39, "infighting; 181, "bust"); see also Kubal, *Cultural Movements*, 11–29. On the stamps: Durnin, *Columbus*.
31. AOC, 11, *Discovery*, Sub 1; and Sub 2.
32. Representative James Besler of Alabama, quoted in Fryd, *Art and Empire*, 100.
33. AOC, 11, *Discovery*, Sub 1, var. docs.; Fryd, *Art and Empire*, 89–91. Persico lived 1791–1860, Buchanan, 1791–1868. On Buchanan's expansionist sentiments, see Greenberg, *Wicked War*, 97, 247, 260. Buchanan was subjected to homophobic slurs and gossip throughout his career in the capital (as was the man with whom he lived, Senator William Rufus King, called "Mrs. Buchanan" and "Mrs. B" by Washington gossips; Greenberg, *Lady First*, 91–92, 112); while I may be overreading expressions of affection commonly expressed in nineteenth-century letters, it seems to me that Buchanan could barely disguise the depth of his fondness for Persico (their "private friendship," the "many agreeable hours" spent together). Was this the motive behind his year-long campaign to persuade a majority of senators to give Persico the commission? The Persico-Buchanan friendship in Lancaster included future state politician Benjamin Champneys (1800–1871), Persico's letters to Champneys being likewise effusively affectionate (Hensel, "An Italian Artist").
34. Fryd, *Art and Empire*, 77–83, 90–91; Genetin-Pilawa, "Curious Removal," 3–4; Hensel, "An Italian Artist."
35. AOC, 11, *Discovery*, Sub 1, var. docs.; Henderson, *Art Treasures*, 305–9; Fryd, *Art and Empire*, 92–93; Cohen, "Preservation Dilemma," 132–37; Restall, "Trouble with 'America,'" 9–11.
36. Fryd, *Art and Empire*, 98; Greenberg, *Wicked War*.
37. Fryd, *Art and Empire*, 99.
38. Previous two paragraphs: Fryd, *Art and Empire*, 100–103. An Indigenous tribal chief had formally objected to *Discovery* and *Rescue* in 1855, but his words fell on deaf ears. The century's printed comments were overwhelmingly positive, with criticism focusing on the artworks' artistic failings (e.g., Leupp, *Walks About Washington*, 58; Ponce de León, *Columbus Gallery*, 125). Controversy over the Indigenous woman centered on her nakedness, not the negative depiction of Indigeneity: Persico was "stung" (his word) by an 1844 *Baltimore Sun* criticism of her nudity; and an 1850 congressional resolution to move *Discovery* to "a suitable place in one of the public squares" was a response to the nudity complaints; AOC, 11, *Discovery*, Sub 2 ("Requests"); Sub 8 ("Newspaper & Magazine Articles").
39. AOC, 11, *Discovery*, Sub 1; Fryd, *Art and Empire*, 104–5. On the genocide against Indigenous peoples in California, e.g., see Madley, *American Genocide*.
40. AOC, 11, *Discovery*, Sub 1; Genetin-Pilawa, "Curious Removal," 4–7; Fryd, *Art and Empire*, 104–5, 233n32–33; Chambers, *Columbus in the Capitol*, 4. "Almost a century away": e.g., Columbus statue in Chula Vista, California, not removed until 2020, its Discovery Park renamed Kumeyaay Park after a local Indigenous group.
41. AOC, 11, *Discovery*, Sub 1, var. docs. ("ravages" is Wolanin, December 23, 1985; "disgrace is a very strong word, but I think it can be applied to what has happened to these priceless parts of the US Capitol's art and history" is May 4, 1983, the curator to the architect); Genetin-Pilawa, "A Curious Removal."
42. AOC, 11, *Discovery*, Sub 1; Sub 2.

43. AOC, 11, *Discovery*, Sub 2 ("Requests"; e.g., *Atlanta Constitution*, October 11, 1977, p. 2–A); Sub 4 ("Description").
44. AOC, 11, *Discovery*, Sub 2 ("Requests").
45. AOC, 11, *Discovery*, Sub 1 (1978 letter by Paul Perrot, the Smithsonian's assistant secretary for museum programs).
46. AOC, 11, *Discovery*, Sub 1, var. docs. (Lewis comments in phone calls to Wolanin, October 9, 1991, noted by her; "relevance" letter to Wolanin from the museum, June 6, 2006).
47. Cohen, "Preservation Dilemma," 131–39; my field notes (AOC, various months, 2023–25).

LIFE NINE: THE ITALIAN

1. Four opening paragraphs: "The Voyager in Marble" (the Oct. 13, 1892, *Times* report); Dickey, *His Monument*, 243–47; Bushman, *America Discovers*, 181–83.
2. Seguin and Nardin, "Lynching of Italians"; Seguin and Rigby, "National Crimes"; Seguin, "Making a National Crime," 59–74. On most covered stories, see Dell et al., "American Stories," 9.
3. Dickey, *His Monument*, 249–50; Bushman, *America Discovers*, 181–82. This rivalry has surfaced many times; in Philadelphia during the 1926 Sesquicentennial's Columbus Day events, e.g., the mayor was obliged to shuttle between rival lunches in the same hotel, for "the Italian and Spanish groups, in a slightly different way, celebrate the same day" (as an assistant district attorney put it at the time; Conn, "Italy on Display," 174).
4. Bushman, *America Discovers*, 182–83; "New Shrine for Italians" in *New York Times* (28 November 1892). The king in question was Alfonso XIII, who died in 1885 aged twenty-seven, leaving his pregnant widow, Maria Christina, as regent until 1902; she did not attend the statue's unveiling. See also Paul, *Myths*, 60–68.
5. Quotes by John Viola ("belittling") in a 2017 *New York Times* op-ed contribution, also quoted in Ugorji, "A Controversial Monument," and by Angelo Vivolo ("vital role") in a 2017 interview in *The Guardian*, quoted in Cormaney, "Columbus Monument."
6. Previous two paragraphs: Borrman, "Naming, Blaming, and Claiming," 5–12. Also, Irving, *Life and Voyages* [1828], I, 68–277.
7. Borrman, "Naming, Blaming, and Claiming," 13–14; Banks, "Stolen Heads" and "Hot Heads"; Baker, "Columbus committee couldn't agree."
8. Hill's statement (August 12, 2020) at onondaganation.org/news/2020/onondaga-nation-statement-on-columbus-statue-in-downtown-syracuse/. Borrman, "Naming, Blaming, and Claiming," 1–4, 15 ("public space" quote is hers); Shin, "Columbus Monuments," tallied almost forty in 2018–21, noting that some six thousand monuments, toponyms, etc., still exist in the United States.
9. The KKK blocked Richmond's Columbus statue, eventually erected (1927), later toppled by protestors (2020). Borrman, "Naming, Blaming, and Claiming," 13–16.
10. Numbers drawn from online U.S. Census Bureau data, and from loc.gov/classroom-materials/immigration/italian/; also see Marinari, *Unwanted*; LaGumina, *Wop!* 181–246; Kubal, *Cultural Movements*, 104. The Immigration Acts restricted migration in 1921 to 3 percent of 1910 numbers already in the United States in 1910, then in 1924 to 2 percent of numbers in country in 1890; due to the massive pre-1920s influx of Italians, this dramatically reduced their permitted numbers relative to northern and western

Europeans (and to other regions of the world where the restrictions did not apply). My "greatest migration" claim may raise eyebrows, as history is packed with mass migration episodes: it is based on migrant numbers, on U.S. and Italian population sizes, and on chronological concentration. Poster image: "L'Italia Delle Migrazioni, 1861–2011" [MMG].

11. Cinotto, *Making Italian America*, 1 ("fascinated").
12. LaGumina, *Wop!* (Harrison quote on 107); Jacobson, *Whiteness*, 43; J. Guglielmo and T. Guglielmo in Guglielmo and Salerno, *Are Italians White?* 10–11, 29–43; Guglielmo, *White on Arrival*, 3–11. Scholarship on the "inbetweenness" of immigrants to the United States, especially Italians, is further referenced by Peter Vellon in Connell and Gardaphé, *Anti-Italianism*, 26.
13. "Spreading" quote is William Hornaday in his *Our Vanishing Wildlife*, quoted by Warren, *Hunter's Game*, 26; J. Guglielmo in Guglielmo and Salerno, *Are Italians White?* 10; Watson, *Sacco and Vanzetti* (one of numerous books on this famous case).
14. That came to mean, as pediatrician-scholar Perri Klass explained, "the domestic, hygienic, and maternal skills of the mother"; the solution was thus "to teach mothers how to do their job properly" (Klass, *Good Time to Be Born*, 75, 81–82, 102–3, 107–8).
15. Klass, *Good Time to Be Born*, 114; Jacobson, *Whiteness*, 78, 82–85.
16. Klass, *Good Time to Be Born*, 207–13; J. Guglielmo in Guglielmo and Salerno, *Are Italians White?* 13.
17. Lynching statistics are widely found, and also hotly debated, but see: NAACP website, at naacp.org/find-resources/history-explained/history-lynching-america; statista.com; Seguin and Rigby, "National Crimes"; Jacobson, *Whiteness*, 52, 56–62; Peter Vellon in Connell and Gardaphé, *Anti-Italianism*, 23–32. Data categories tend to use "white" as a catch-all non-Black demographic.
18. Prejudicial anxiety over immigration has tended to coincide with higher influx periods (Filkins, "Borderline Chaos," 32), with the economic and cultural benefits of such influxes widely understood yet too often prevented from influencing policy; as Hernandez (*Truth About Immigration*, 19) argues, anti-immigration laws such as 1882's Chinese Exclusion Act and 1924's National Origins Act, in addition to being "disgraceful on moral grounds," did lasting social and economic damage to the United States.
19. J. Guglielmo in Guglielmo and Salerno, *Are Italians White?* 2–4 (including Baldwin quotes, taken from *Price of the Ticket*, 660–67, and "On Being 'White' . . . and Other Lies, in *Essence* [April 1984], 90–92); Campbell and Opilo, "Christopher Columbus statue."
20. Above two paragraphs: Warren, *Hunter's Game*, 21–22, 33–43; Summerhill and Williams, *Sinking Columbus*, 14–15; Kauffman, *Patriotism and Fraternalism*, 2–3; Kubal, *Cultural Movements*, 31–38, 104–7.
21. Murrell, "Who Was the Real Christopher Columbus?"; Bushman, *America Discovers*, 7; Thacher, *Christopher Columbus*, III, 80–83. I include neither pre-1876 Columbus statues erected as heroic American Columbuses, as discussed in Life Eight, nor the handful of lesser pre-1876 examples (such as the small marble statue on Boston's Beacon Hill, given to the city by a Greek merchant in 1849); Groseclose, "Monuments and Memorials," in Bedini, *Encyclopedia*, 475–85.
22. Baltimore, first parading on Columbus Day 1890, claims the longest unbroken such annual event in the United States. In 1972, President Nixon moved Columbus Day from October 12 to the second Monday of the month. Initial years for Columbus Days

vary due to the ambiguity of declarations by presidents, governors, and mayors (since 1930, there have been at least fifty-three Columbus Day presidential proclamations; Kubal, *Cultural Movements*, 11). I have simplified the complex history of the Knights of Columbus (which is today a $120 billion international insurance company, charitable organization, and Catholic brotherhood) and of Italian American organizations such as the Columbus Citizens Foundation (New York based and a longtime organizer of that city's Columbus Day parades); kofc.org; Kubal, *Cultural Movements*, 31–56, 103–33; Kauffman, *Patriotism and Fraternalism*; Paulmier and Schauffler, *Columbus Day*, xiv; LindaAnn Loschiavo in Connell and Gardaphé, *Anti-Italianism*, 151–62; Summerhill and Williams, *Sinking Columbus*, 14–16, 111.

23. Harrisse, *Christophe Colomb*, and other works; Winsor, *Christopher Columbus*. On Catholicism as "an obstacle to assimilation," see LaGumina, *Wop!* 163–80.
24. Regal, *America's Origin Story*, 3 ("interplay"), 4–7, 279 (the *Viking*), 243–46, 284–86. Regal explains how television and the internet allowed a resurgence of older racist ideas under the guise of nonsensical, but popular, claims regarding ancient aliens. The literature on pseudoarchaeology—both its publications and those dismantling its false premises and logic—is too vast to cite here, but on the example of China see the work of Gavin Menzies, his apologists and detractors. An example of the books, mostly published close to the Quincentennial, that give credence to many of the claims that Asians, Africans, and Europeans "discovered" the Americas dozens of times before Columbus, is Huyghe, *Columbus Was Last*.
25. Regal, *America's Origin Story*, 53–101.
26. Anderson, *America Not Discovered by Columbus* ("pretended" on 91); Freedman, *Who Was First?*; Wyatt, *Who Discovered America?*; Horwitz, *Voyage Long and Strange*, 14–17; Huyghe, *Columbus Was Last*.
27. Regal, *America's Origin Story*, 103–66, 261. In some parts of the United States, a Leif Erikson Day has been celebrated since 1964 (Winchester, *Atlantic*, 89). The Vinland site in Newfoundland, L'Anse aux Meadows, was radiocarbon dated to around the year 1000 and named a UNESCO World Heritage site in 1978.
28. Freedman, *Who Was First?*, vi. Accessible introductions to the Early Americas fast-moving field include Dillehay, *Settlement of the Americas*; Mann, *1491*; and Raff, *Origins*.
29. Means statement, November 24, 1989, often cited; e.g., Royal, *1492 and All That*, 19. On twentieth-century "American Indian" mobilization against Columbus Day and the spread of Indigenous Peoples' Day, see Kubal, *Cultural Movements*, 57–75.
30. Bushman, *America Discovers*, 21, 190, argues for a similar way of seeing Columbus "as a mirror for Americans"; also see Regal, *America's Origin Story*; Paul, *Myths*, 68–75.
31. See fightbacknews.org/2020/6/11/native-americans-pull-down-christopher-columbus-statue-minnesota-capitol; mprnews.org/story/2020/06/10/minnesota-protesters-pull-down-columbus-statue-at-capitol; video footage can easily be found online.
32. Morales, "Christopher Columbus"; Zahniser, "L.A. City Council"; Madley, *An American Genocide*.
33. Zahniser, "L.A. City Council."
34. Campbell and Opilo, "Christopher Columbus statue near Little Italy." Shin, "Columbus Monuments," includes video footage of the toppling. Baltimore's (and the nation's) oldest Columbus monument is an obelisk, which the French consul placed on his private

estate in 1792, and which was moved in 1964 to Herring Park (Bushman, *America Discovers*, 92–95); although it was apparently attacked in 2017, its location among pine trees is obscure, its Columbus connection is only visible when one is very close, and its impact is further minimized by an adjacent monument to fallen police officers. The second Baltimore monument, a statue placed on Druid Hill in 1892, was inaccessible in 2020 (and still so in 2023) due to a massive public-works project in the area.

35. Bellamy, "In Newark, a Harriet Tubman monument"; Tobia, "Anti-Racism Protests" (quoting that author, Michael Immerso).
36. Summerhill and Williams, *Sinking Columbus*, 109–16.
37. Schuessler, "Smithsonian's Latino Museum"; americanlatinomuseum.org. "Preview": opened 2022 in the American History Museum. "Congressmen": Representatives Mario Diaz-Balart and Tony Gonzales. The August 2022 opinion piece in *The Hill* called for the new museum to be defunded.
38. Schuessler, "Smithsonian's Latino Museum." In the internet age, a digital diaspora of "Latin" communities has emerged, one that has both expanded and exploded the nexus of hemispheric identities that includes Chicanx, Latinx, Afro-Latin American, Indigenous American, and so on. Whereas the experiences of northward migrants from Latin America in the early twentieth century may have been somewhat similar to those of Italians in the 1861–1921 era, those experiences have since increasingly diverged (Dalton and Ramírez Plascencia, *Imagining Latinidad*).

EPILOGUE

1. My field notes (2023) (on Genoa); Shin, "Columbus Monuments" (on U.S. names).
2. Summerhill and Williams, *Sinking Columbus*, 11–12, 16–18, 150–59.
3. Peris, "Yo soy el Cristóbal Colón número 20."
4. The relevant scene in *Night at the Museum* is at 0:28:06 to 0:28:20 (Levy, *Night at the Museum*; numerous websites detail the franchise). Columbus is a flawed figure in *Assassin's Creed II: Discovery* (see assassinscreed.fandom.com/wiki/Christopher_Columbus #Legacy). *Doctor Who* ran 1963–89, 2005–present; Columbus in the audio-only "Trouble in Paradise."
5. Baxter, *Navigator* (quotes on 262); Berry, *Columbus Affair*.
6. Phillips and Phillips, *Worlds*, 6–7; Royal, *1492 and All That*, 120–29, 141–46. I surveyed dozens of examples of children's literature: In the years leading to 1992, the older trend persisted of presenting Columbus as visionary, uniquely gifted, even heroic, often leaning heavily on the First Voyage's "log" (Roop, *I, Columbus*; Adler, *A Picture Book*; and Scott, *The Log* [1989 reprint of a 1938 book]). But post-Quincentennial, a notable shift occurred; in some books, Columbus was now a flawed hero (Larkin, *Christopher Columbus*; Macdonald, *You Wouldn't Want*), while in most books he was buried beneath coverage of other explorers (Greenwood, *I Wonder Why*) or other discovery narratives—mostly about Indigenous Peoples and Vikings (Wyatt, *Who Discovered America?*), or speculation on discovery theories ranging from the Chinese to ancient aliens (Freedman, *Who Was First?*; Hart, *Who Really Discovered America?*). By the 2020s, an Indigenous Peoples' Day book could ignore Columbus completely, never explaining why the holiday occurs on October's first Monday (Sabelko, *Indigenous Peoples' Day*).

7. Letter from Hood to Dickens of October 12, 1842—written three hundred fifty years to the day after Columbus's First Landing—reproduced in Whitley, "Hood and Dickens," 397–98. Perry, "On Poetry," published in Van Dyke, *Counsel Upon the Reading of Books*, quote on 254.
8. In a 1926 usage, e.g., the term required scare quotes, as "the 'Columbusing' of" something—as in a snide reference to discovering something, but in this case without any reference to the Americas (Wimberly, "Spook English," 318). References from the 2010s include Yates, "Columbusing black Washington," and follow-up *Post* stories in the weeks that followed; "Columbusing" on youtube.com/watch?v=BWeFHddWL1Y; Salinas, "'Columbusing.'"
9. Horwitz, *Voyage Long and Strange*, 3 (the park ranger's name was Claire Olsen).
10. As recently argued: Scheirer, *A History of Fake Things*. On the link to Columbiana: Regal, *America's Origin Story*, especially 233–59.
11. Quotes mix phrases from Spencer Mermelstein, "The mind's meta-data" (PhD dissertation, University of California–Santa Barbara, 2021), 84; and St. George, "The Long Tale," 35; also see Regal, *America's Origin Story*, 233–59; Scheirer, *A History of Fake Things*.

Bibliography of Sources

ARCHIVES, MANUSCRIPT COLLECTIONS, AND THEIR ABBREVIATIONS

AGI	Archivo General de Indias, Seville, Spain
AGS	Archivo General de Simancas, Simancas, Spain
AOC	Archive of the Office of the Curator, the Architect of the Capitol, Washington, DC, USA
APRL	American Philatelic Research Library, Bellefonte, PA, USA
BCC	Biblioteca Capitular y Colombina, Seville, Spain
BL	British Library, London, UK
BM	British Museum, London, UK
Bodl	Bodleian Libraries, University of Oxford, Oxford, UK
CCBM	Columbus Chapel and Boal Museum, Boalsburg, PA, USA
CGCC	City of Genoa, Commissione Colombiana (see City of Genoa under Publications)
CoH	Fernández-Armesto, *Columbus on Himself* (see under Publications)
CVD	*Colección de los viajes y descubrimientos* (see Navarrete under Publications)
HCAR	Helmerich Center for American Research, Gilcrease Museum, Tulsa, OK, USA
JCB	John Carter Brown Library, Providence, RI, USA
LoC	Library of Congress, Rare Books and Manuscript Collections, Washington, DC, USA (HH: Henri Harrisse Collection; JBT: John Boyd Thacher Collection)
MMG	Museo di Mare, Genoa, Italy
NGAL	National Gallery of Art Library, Washington, DC, USA
PSUL	Pennsylvania State University Libraries, PA, USA
RC	*Repertorium Columbianum* (see by editors under Publications: II, Nader; III, Rusconi; IV, Dotson and Agosto; V, Eatough; VI, Lardicci; VII, Griffin; VIII, Phillips; IX, Carrillo; X, Symcox; XI, Symcox and Carrillo; XII, Symcox and Formisano; XIII, Caraci)
RDSP	*Raccolta di documenti e studi pubblicati*, cited by part (I–VI), volume (i–xiv), and item or page number as indicated (see Lollis under Publications)
TU/HT	Howard-Tilton Memorial Library Rare Books and Manuscript Collections, Tulane University, New Orleans, LA, USA
TULAL	Latin American Library, Tulane University, New Orleans, LA, USA

PUBLICATIONS

Place of publication is included for pre-1925 books only, their library locations indicated by abbreviations in brackets.

Adams, Charles Francis. *Columbus and the Spanish Discovery of America: A Study in Historical Perspective.* Pamphlet. Cambridge, MA: John Wilson and Son, 1892 [JCB].
Adler, David A., and John and Alexandra Wallner. *A Picture Book of Christopher Columbus.* Holiday House, 1991.
Adorno, Rolena. "Washington Irving's Romantic Hispanism and Its Columbian Legacies." In Richard L. Kagan, ed., *Spain in America: The Origins of Hispanism in the United States*, 49–105. University of Illinois Press, 2002.
Airaldi, Gabriella. *Colombo: Da Genova al Nuovo Mondo.* Salerno Editrice, 2012.
Altman, Ida. *Life and Society in the Early Spanish Caribbean: The Greater Antilles, 1493–1550.* Louisiana State University Press, 2021.
Altman, Ida. "The Revolt of Enriquillo and the Historiography of Early Spanish America." *The Americas* 63: 4 (April 2007), 587–614.
Amler, Jane Francis. *Christopher Columbus's Jewish Roots.* Jason Aronson, 1991.
Anderson, Rasmus B. *America Not Discovered by Columbus: An Historical Sketch of the Discovery of America by the Norsemen.* Chicago: S. C. Griggs & Co. (London: Trübner & Co.), 1874 [LoC].
Andre, Marius. *Columbus.* Eloïse Parkhurst Huguenin, trans. Knopf, 1928.
Anghiera, Pietro Martire d' [Peter Martyr]. *Extraict ou recueil des isles nouuelleme[n]t trouuees en la grand mer oceane ou temps du roy Despaigne Ferna[n]d & Elizabeth sa femme / faict premierement en latin par Pierre Martyr de Millan, & depuis translate en languaige francoys.* Paris: Simon de Colines au soleil dor, 1532 [HCAR].
Arana, Marie. *Latinoland: A Portrait of America's Largest and Least Understood Minority.* Simon & Schuster, 2024.
Asensio Toledo, José María. *Cristóbal Colón: su vida, sus viajes—sus descubrimientos.* 2 vols. Barcelona: Espasa y Compañia, 1888 [but n.d.] [JCB].
Asensio Toledo, José María. *Los Restos de Cristóval Colon estan en La Habana.* Valencia: Domenech, 1881 [JCB].
Associated Press. "Writer Suggests Columbus Was Son of Norway." *Roanoke Times* (March 6, 1991), A-6.
Baker, Chris. "Columbus committee couldn't agree on fate of Syracuse statue, left choice to Walsh." Syracuse.com (October 13, 2020), syracuse.com/news/2020/10/columbus-committee-couldnt-agree-on-fate-of-syracuse-statue-left-choice-to-walsh.html.
Baker-Bates, Piers. *Sebastiano del Piombo and the World of Spanish Rome.* Routledge, 2016.
Baldwin, James. *The Price of the Ticket: Collected Nonfiction, 1948–1985.* St. Martin's Press, 1985.
Ballesteros Beretta, Antonio. *Cristóbal Colón y el descubrimiento de América.* 2 vols. Salvat Editores, 1945.
Banks, Adelle M. "Hot heads returned to monument." *Orlando Sentinel* (October 12, 1992), A-1, A-4.

Banks, Adelle M. "Stolen Heads Found in Snow Warehouse." *Orlando Sentinel* (October 10, 1989).
Barlow, Joel. *The Columbiad: A Poem*. Philadelphia: Fry & Kammerer, 1807 [LoC].
Barlow, Joel. *The Vision of Columbus: A poem in nine books*. Hartford: Hudson and Goodwin, 1787 [PSUL and LoC].
Barrientos, Juan José. "Colón, personaje novelesco." *Cuadernos Hispanoamericanos* 437 (November 1986), 45–62.
Barry, J. J. *The Life of Christopher Columbus, from authentic Spanish and Italian documents*. [Loose translation of Roselly de Lorgues, *Christophe Colomb*.] Boston: Catholic Publications Society, 1870 [LoC; TU/HT].
Bartosik-Vélez, Elise. *The Legacy of Christopher Columbus in the Americas*. Vanderbilt University Press, 2014.
Bartosik-Vélez, Elise. "The Three Rhetorical Strategies of Christopher Columbus." *Colonial Latin American Review* 11: 1 (2002), 33–46.
Baxter, Stephen. *Navigator*. Gollancz, 2007.
Bedini, Silvio A. *The Christopher Columbus Encyclopedia*, 2 vols. Simon & Schuster, 1992.
Belknap, Jeremy. *A Discourse Intended to Commemorate the Discovery of America by Christopher Columbus*. Boston: Belknap and Hall, 1792 [JCB; Bodl].
Bellamy, Claretta. "In Newark, a Harriet Tubman monument replaces Christopher Columbus." NBCnews.com (March 10, 2023), nbcnews.com/news/nbcblk/newark-harriet-tubman-monument-replaces-christopher-columbus-rcna74167.
Beltrán y Rozpide, Ricardo. *Cristóbal Colón, genovés? Los testamentos de Colón*. Patronato de Huérfanos de Intendencia é Intervención Militares, 1925.
Bergreen, Laurence. *Columbus: The Four Voyages 1492–1504*. Penguin, 2012.
Bernáldez, Andrés. *Memorias del reinado de los Reyes Católicos*. Manuel Gómez-Moreno and Juan de Mata Carriazo, eds. Real Academia de la Historia, 1962.
Berry, Steve. *The Columbus Affair*. Ballantine, 2012.
Blakemore, Steven. *Joel Barlow's Columbiad: A Bicentennial Reading*. University of Tennessee Press, 2007.
Bloy, Léon. *Le Révélateur du Globe: Christophe Colomb et sa Béatification Future*. Paris: A. Sauton, 1884 [LoC].
Boorstin, Daniel J. *The Discoverers: A History of Man's Search to Know His World and Himself*. Random House, 1983.
Borrman, Kristina. "Naming, Blaming, and Claiming: The Columbus Monument and the Struggle for Diversity Rights in Syracuse, New York." *Panorama* 8, no. 2 (Fall 2022), 22 pages, doi.org/10.24926.24716839.15172.
Boyd, William. *The Romantic: The Real Life of Cashel Greville Ross. A Novel*. Penguin, 2023.
Boyle, David. *Toward the Setting Sun: Columbus, Cabot, Vespucci, and the Race for America*. New York: Walker, 2008.
Bradley, Charlie. "Archaeology breakthrough as Christopher Columbus' first tomb found: 'We got it!' " *Express* (April 9, 2022), express.co.uk/news/world/1593704.
Brocken, Baron van. *Des Vicissitudes Posthumes de Christophe Colomb, et de sa Beatification Possible*. Leipzig and Paris, 1865 [LoC].
Burton, Richard Francis. *Wanderings in West Africa from Liverpool to Fernando Po*. 2 vols. London: Tinsley Brothers, 1863 [TU/HT].

Bushman, Claudia L. *America Discovers Columbus: How an Italian Explorer Became an American Hero.* University Press of New England, 1992.
Bustos y Bustos, Marqués de Corvera, Alfonso de. *La patria de Cristóbal Colón.* Talleres Voluntad, S.A., 1927.
Butel, Paul. *The Atlantic.* Routledge, 1999.
Butler, James D. *Portraits of Columbus: A Monograph.* Madison: n.p., 1883 [JCB].
Calzada, Rafel. *La patria de Colón.* Buenos Aires: Librería "La Facultad," 1920.
Campbell, Phillip. "The nineteenth century Movement to Canonize Columbus." In *Catholic Exchange* (October 10, 2022), catholicexchange.com/the-19th-century-movement-to-canonize-columbus/.
Campbell, Colin, and Emily Opilo. "Christopher Columbus statue near Little Italy brought down, tossed into Baltimore's Inner Harbor." *Baltimore Sun* (July 4, 2020), baltimoresun.com/maryland/baltimore-city/bs-md-ci-columbus-statue-20200705-xc4bhthfhjaflifz72org2lrhy-story.html.
Campe, Joachim Heinrich. *Die Entdeckung von Amerika: Ein Unterhaltungsbuch für Kinder und junge Leute.* Reutlingen: Mäden'schen Buchhandlung, 1820 [TULAL]; 1825 [HCAR].
Campe, Joachim Heinrich. *The Discovery of America; for the Use of Children and Young Persons.* London: J. Johnson, 1799 [PSUL].
Canel, Eva. *La cuna de Colon: Conferencia dada en la catedra de historia crítica del Perú de la Facultad de letras, el 8 de julio de 1913.* Lima: Revista universitaria, 1913 [PSUL].
Canoutas, Seraphim G. *Christopher Columbus: A Greek Nobleman.* St. Mark's Printing Co., 1943.
Caraci, Ilaria Luzzana. *Colombo vero e falso: La costruzione delle historie fernandine.* Sagep, 1989.
Caraci, Ilaria Luzzana, ed. *The History of the Life and Deeds of the Admiral Don Christopher Columbus: Attributed to his Son Fernando Colón.* Geoffrey Symcox and Blair Sullivan, trans. *Repertorium Columbianum,* Geoffrey Symcox, general ed., Volume XIII. Brepols, 2004.
Carpentier, Alejo. *The Harp and the Shadow* [1979]. Mercury, 1990.
Carbia, Rómulo D. *La patria de Cristóbal Colón: examen crítico de las fuentes históricas en que descansan las aseveraciones itálicas e hispánicas, acerca del origen y lugar de nacimiento del descubridor de América.* Buenos Aires: Talleres Jacobo Preuser, 1923.
Carrillo, Jesús, ed. *Oviedo on Columbus.* Diane Avalle-Arce, trans. Anthony Pagden, pref. *Repertorium Columbianum,* Geoffrey Symcox, general ed., Volume IX. Brepols, 2000.
Casa di Colombo. Tourist brochure for "Columbus' House" in Genoa. Comune di Genova and Musei di Genova, n.d.
Casanova de Pioggiola, Martin. *La vériteé sur l'origine et la patrie de Christophe Colomb.* Bastia, Corsica: Veuve Ollagnier, 1880 [LoC]. *Édition considérablement augmenté*: Ajaccio, Corsica: B. Robaglia, 1889 [TULAL].
Cateras, Spyros. *Christopher Columbus Was a Greek Prince: And His Real Name Was Nikolaos Ypshilantis from the Greek Island Chios.* Self-published (Ergatis), 1937.
Catz, Rebecca. *Christopher Columbus and the Portuguese, 1476–1498.* Greenwood, 1993.

Cesarini de Paoli de Silvareccio, Antionette Eduina. *Cristóbal Colón Nació en Corcega en la Ciudad de Calvi el año 1441.* Montevideo: Morales y Risero, 1924 [TU/HT].
Chambers, Ann Baldessarini, ed. *Columbus in the Capitol: Commemorative Quincentenary Edition.* Washington, DC: 102nd Congress, House Doc. 102–319, 1992.
Chiari, Joseph. *Christopher Columbus: A Play.* Gordian Press, 1979.
Cinotto, Simone, ed. *Making Italian America: Consumer Culture and the Production of Ethnic Identities.* Fordham University Press, 2014.
City of Genoa, Commissione Colombiana. *Christopher Columbus: Documents and Proofs of His Genoese Origin.* English-German edition. Officine Dell' Istituto Italiano D'Arti Grafiche, 1932 [Italian edition 1931; cited as CGCC].
Clarke, Richard H. *Old and New Lights on Columbus.* New York: Richard H. Clarke (printed by author), 1893 [JCB].
Cocchia, fray Rocco. "Acta del descubrimiento." Spanish translation of pastoral letter of 18 September 1877, in (see) La Real Academia de la Historia, *Los restos de Colón,* 181–91.
Cock Hincapié, Olga. *Historia del nombre de Colombia.* Instituto Caro y Cuervo, 1998.
Cohen, Michele. "The Preservation Dilemma: Confronting Two Controversial Monuments in the United States Capitol." In Sierra Rooney et al., eds., *Teachable Monuments: Using Public Art to Spark Dialogue and Confront Controversies,* 131–42. Bloomsbury, 2021.
Coleman, David. *Creating Christian Granada: Society & Religious Culture in an Old-World Frontier City, 1492–1600.* Cornell University Press, 2003.
Colmeiro, Manuel. *Los restos de Colón.* See Real Academia de la Historia.
Colombo, Fernando. *Historie del Signor D. Fernando Colombo, Nelle quali si ha particolare, e vera relazione della vita, e de' fatti dell' Ammiraglio D. Christoforo Colombo suo Padre.* Venice: per il Lovisa, 1728 [JCB].
Colón, Cristóbal. *Libro Copiador.* Facsimile. Ministerio de Cultura, 1989.
Colón, Diego. *Memorial de Don Diego Colon, Virrey y Almirante de las yndias a S. C. C. Magd el Rey don Carlos sobre la conversion e consvació de las gentes de las yndias.* Henry Stevens, ed. London: Carlos Whittingham Chiswick Press, 1854 [HCAR; also see HCAR, CC #127].
Colón, Hernando. *Vida del Almirante Don Cristobal Colon, escrita por su hijo.* Ramón Iglesia, ed. Fondo de Cultura Económica, 1947.
Conn, Steven. "Italy on Display: Representing Italy in the 1876 Centennial and the 1926 Sesquicentennial." In Andrea Canepari and Judith Goode, eds., *The Italian Legacy in Philadelphia: History, Culture, People, and Ideas,* 167–74. Temple University Press, 2021.
Connell, William J., and Fred Gardaphé, eds. *Anti-Italianism: Essays on a Prejudice.* Palgrave Macmillan, 2010.
Cormaney, Ava. "Columbus Monument, Columbus Circle, NYC." *Clio: Your Guide to History* (May 11, 2020), theclio.com/entry/100336.
Corradi, Juan. *Descubrimiento y Conquista de la América, ó Compendio de la Historia General del Nuevo Mundo.* 3 vols. Madrid: Imprenta de doña Catalina Piñuela, 1817 [TULAL].
Costa-Amic, Bartomeu. *Colom: Catalán de Mallorca, sobrino de los Reyes Católicos.* Costa-Amic Editores, 1989.

Cortesão, Armando. 1937. "The Mystery of Columbus." *Contemporary Review* 151 (January 1), 322–30.
Cronau, Rodolfo. *América: Historia de su Descubrimiento desde los Tiempos Primitivos Hasta los Más Modernos*. Barcelona: Montaner y Simon, Editores, 1892 [HCAR].
Crosby, Alfred W. *The Columbian Exchange: Biological and Cultural Consequences of 1492* [1972]. 30th Anniversary edition. Praeger, 2003.
Crosby, Alfred W. *The Measure of Reality: Quantification and Western Society, 1250–1600*. Cambridge University Press, 1997.
Curtis, William Eleroy. *The Relics of Columbus*. Washington, DC: William H. Lowdermilk Co., 1893 [LoC].
Dalton, David S., and David Ramírez Plascencia, eds. *Imagining Latinidad: Digital Diasporas and Public Engagement Among Latin American Migrants*. Brill, 2023.
David, Maurice. *Who was "Columbus"? His Real Name and Real Fatherland: A Sensational Discovery Among the Archives of Spain*. Research Publishing Co., 1933.
Davidson, Miles H. *Columbus Then and Now: A Life Reexamined*. University of Oklahoma Press, 1997.
Deagan, Kathleen. *En Bas Saline: A Taíno Town Before and After Columbus*. University of Florida Press, 2023.
De la Torre y del Cerro, José. *Beatriz Enríquez de Harana y Cristóbal Colón*. Compañia Ibero-Americana de Publicaciones, 1933.
Delaney, Carol Lowery. *Columbus and the Quest for Jerusalem: How Religion Drove the Voyages that led to America*. Free Press, 2001.
Dell, Melissa, et al. "American Stories: A Large-Scale Structured Text Dataset of Historical U.S. Newspapers." Open access at arxiv.org/abs/2308.12477 (August 2023) and dell-research-harvard.github.io/resources/americanstories (August 2024).
Dennis, Matthew. "The Eighteenth-Century Discovery of America." In William Pencak, Matthew Dennis, and Simon P. Newman, eds., *Riot and Revelry in Early America*, 205–28. Pennsylvania State University Press, 2002.
Desimoni, Cornelio. *Di Alcuni Recenti Giudizi Intorno Alla Patria di Cristoforo Colombo*. Genoa: R. Istituto Sordo-Muti, 1890. [JCB].
Dickey, J. M. *Christopher Columbus and His Monument Columbia*. Chicago and New York: Rand, McNally & Co., 1892 [PSUL].
Dillehay, Thomas D. *The Settlement of the Americas: A New Prehistory*. Basic Books, 2000.
Dotson, John, and Aldo Agosto, eds. *Christopher Columbus and His Family: The Genoese and Ligurian Documents. Repertorium Columbianum*, Geoffrey Symcox, general ed., Volume IV. Brepols, 1998.
Dozier, Thomas. "The Controversy on Whereabouts of Columbus' Body." *Smithsonian* 5, no. 7 (October 1974), 92–99.
Drayson, Elizabeth. *The Moor's Last Stand: How Seven Centuries of Muslim Rule in Spain Came to an End*. Profile, 2017.
Duff, Charles. *The Truth about Columbus and the Discovery of America*. Jarrolds, 1936.
Dugard, Martin. *The Last Voyage of Columbus: Being the Epic Tale of the Great Captain's Fourth Expedition, Including Accounts of Swordfight, Mutiny, Shipwreck, Gold, War, Hurricane, and Discovery*. Little, Brown, 2005.
Durnin, Richard G. *Columbus on Postage Stamps*. Booklet. Middlesex County Cultural and Heritage Commission, 1992.

Eatough, Geoffrey, ed. and trans. *Selections from Peter Martyr. Repertorium Columbianum*, Geoffrey Symcox, general ed., Volume V. Brepols, 1998.
Eco, Umberto. *Serendipities: Language and Lunacy*. Harcourt Brace, 1998.
Elliott, Walter. "The Catholic Publication Society." In David Turley, ed., *American Religion: Literary Sources and Documents*, chap. 118 (accessed online). Routledge, 1998.
Espinoza G., Ricardo, and Carlos González E. "La enfermedad del Almirante Mayor del Mar Océano, don Cristóbal Colón." *Revista Médica de Chile* 125: 6 (June 1997), 732–37.
"Exhuman 7 ósos do proxecto 'Colón galego' no cemiterio de San Salvador de Poio" [no author]. *Galicia Confidencial* (September 11, 2022), galiciaconfidencial.com/noticia/213610-exhuman-7-osos-proxecto-colon-galego-cemiterio-san-salvador-poio.
Farina, Luciano F., and Robert W. Tolf, eds. *Columbus Documents: Summaries of Documents in Genoa*. Omnigraphics, 1992.
Fernández, Rafael Diego. *Capitulaciones Colombinas (1492-1506)*. El Colegio de Michoacán, 1987.
Fernández-Armesto, Felipe. *Before Columbus: Exploration and Colonization from the Mediterranean to the Atlantic, 1229–1492*. University of Pennsylvania Press, 1987.
Fernández-Armesto, Felipe. *Columbus*. Oxford University Press, 1992.
Fernández-Armesto, Felipe. *Columbus and the Conquest of the Impossible* [1974]. Phoenix Press, 2000.
Fernández-Armesto, Felipe. *Columbus on Himself*. Folio Society, 1992. Reprinted Hackett, 2010.
Fernández-Armesto, Felipe. "Cristóbal Colón y los libros de caballería." In Carlos Martínez Shaw and Celia Parcero Torre, eds., *Cristóbal Colón*, 115–28. Junta de Castilla y León, 2006.
Fernández-Armesto, Felipe. *Ferdinand and Isabella*. Taplinger, 1975.
Fernández-Armesto, Felipe. "Maritime History and World History." In Daniel Finamore, ed., *Maritime History as World History*, 7–34. University Press of Florida and Peabody Essex Museum, 2004.
Fernández-Armesto, Felipe. *The Canary Islands After the Conquest: The Making of a Colonial Society in the Early Sixteenth Century*. Clarendon Press and Oxford University Press, 1982.
Fernández-Armesto, Felipe, and Manuel Lucena Giraldo. *How the Spanish Empire was Built: A 400-Year History*. Reaktion Books, 2024.
Ferro, Gaetano. *Liguria and Genoa at the Time of Columbus*. Anne Goodrich Heck, trans. *Nuova Raccolta Colombiana*, Volume III. Istituto Poligrafico e Zecca dello Stato, 1992.
Filkins, Dexter. "Borderline Chaos." *New Yorker* (June 19, 2023), 30–41.
Flint, Valerie I. J. *The Imaginative Landscape of Christopher Columbus*. Princeton University Press, 1992.
Franzoni, Domenico. *La Vera Patria di Cristoforo Colombo: Giustificata a Favore de' Genovesi*. Rome: Luigi Perego Salvioni, 1814 [JCB; BM; LoC/HH].
Freedman, Russell. *Who Was First? Discovering the Americas*. Clarion Books, 2007.
Freneau, Philip Morin. *A Poem, on the rising glory of America, being an exercise delivered*

at the public commencement at Nassau-Hall, September 25, 1771. Philadelphia: Joseph Crukshank, 1772 [LoC].

Freneau, Philip Morin. *The Poems of Philip Freneau, Written chiefly during the late war.* Philadelphia: Francis Bailey, 1786 [PSUL].

Fryd, Vivien Green. *Art and Empire: The Politics of Ethnicity in the United States Capitol, 1815–1860.* Ohio University Press and the Unites States Capitol Historical Society, 2001.

Games, Allison. "Atlantic History: Definitions, Challenges, and Opportunities." *American Historical Review* (June 2006), 741–57.

Gaspar da Naia, Alexandre. *Cristóbal Colón, instrumento da política expansionista portuguesa do ultramar.* Self-published (Lisbon), 1950.

Gaspar da Naia, Alexandre. *D. João II e Cristóbal Colón. Factores complementares na consecuçao de um mesmo objectivo.* Self-published (Lisbon), 1951.

Genetin-Pilawa, C. Joseph. "A Curious Removal: Leta Myers Smart, *The Rescue*, and *The Discovery of America*." *Capitol Dome* 52: 1 (2015), 2–9.

Geronimi, Victor. *Christophe Colomb Corse du Niolu, 1445–1506.* Self-published (Corsica), 2017.

Gil, Juan. *Columbiana: Estudios sobre Cristóbal Colón, 1984–2006.* Academia Dominicana de la Historia, 1984.

Gil, Juan. *Mitos y utopías del descubrimiento: I. Colón y su tiempo.* Alianza Editorial, 1989.

González, Serxio. "El sepulcro del Proyecto Colón Galego mantiene las cuñas que lo sellaron en 1496." *La Voz de Galicia* (November 13, 2022), lavozdegalicia.es/noticia/arousa/2022/11/13/sepulcro-proyecto-colon-galego-mantiene-cunas-sellaron-1496/0003_202211A13C1991.htm.

Gopnik, Adam. "The Great Interruption." *New Yorker* (April 24 and May 1, 2023), 62–66.

Greenberg, Amy S. *A Wicked War: Polk, Clay, Lincoln, and the 1846 U.S. Invasion of Mexico.* Knopf, 2012.

Greenberg, Amy S. *Lady First: The World of First Lady Sarah Polk.* Knopf, 2019.

Greenwood, Rosie. *I Wonder Why Columbus Crossed the Ocean and Other Questions About Explorers.* Kingfisher, 2005.

Griffin, Nigel, ed. and trans. *Las Casas on Columbus: Background and the Second and Fourth Voyages.* Anthony Pagden, intro. *Repertorium Columbianum*, Geoffrey Symcox, general ed., Volume VII. Brepols, 1999.

Guglielmo, Jennifer, and Salvatore Salerno, eds. *Are Italians White? How Race Is Made in America.* Routledge, 2003.

Guglielmo, Thomas A. *White on Arrival: Italians, Race, Color, and Power in Chicago, 1890–1945.* Oxford University Press, 2003.

Harrisse, Henry. *Christophe Colomb et la Corse: Observations sue un décret recent du gouvernement français.* Paris: Ernest Leroux, 1883.

Harrisse, Henry. *Christophe Colomb: son origine, sa vie, ses voyages, sa famille & ses descendents.* 2 vols. Paris: Ernest Leroux, 1884 [JCB and LoC].

Harrisse, Henry. "Les restes mortels de Christophe Colomb." *Revue Critique d'Histoire et de Littérature* 12: 1 (January 5, 1877), 14–23.

Hart, Avery. *Who Really Discovered America? Unraveling the Mystery and Solving the Puzzle.* Williamson, 2001.

Heaney, Christopher. *Empires of the Dead: Inca Mummies and the Peruvian Ancestors of American Anthropology.* Oxford University Press, 2023.

Heck, Thomas. "The Operatic Christopher Columbus: Three Hundred Years of Musical Mythology." *Annali d'Italianistica* 10 (1992), 236–78.
Hedgecoe, Guy. "'A cock-and-bull story': Christopher Columbus DNA tests aim to end dispute over birthplace." *Irish Times* (January 22, 2023), irishtimes.com/world/europe/2023/01/22/.
Henderson, Helen W. *The Art Treasures of Washington*. Boston: L. C. Page & Co., 1912 [AOC].
Henige, David P. *In Search of Columbus: The Sources for the First Voyage*. University of Arizona Press, 1991.
Hensel, William Uhler. "An Italian Artist in Old Lancaster (Louigi Persico-1820)." *Journal of the Lancaster County Historical Society* 12: 3 (1912), 67–101 [PSUL].
Hernandez, Zeke. *The Truth About Immigration: Why Successful Societies Welcome Newcomers*. St. Martin's Press, 2024.
Herrera, Antonio de. *Historia General de los Hechos de los Castellanos en las Islas, y Tierra firme del mar Oceano*. 8 vols. [*décadas*]. Madrid: La Emplenta [*sic*] Real, 1601–15 [JCB and LoC].
Herrera, Antonio de. *Historia General de los Hechos de los Castellanos en las Islas, y Tierra firme del mar Oceano*. 8 vols. [*décadas*]. Madrid: La Officina Real de Nicolás Rodríguez Franco, 1726–30 [PSUL].
Horta y Pardo, Constantino de. *La verdadera cuna de Cristóbal Colón*. New York: Imprenta de J. B. Jonathan Co., 1912 [TULAL and PSUL].
Horwitz, Tony. *A Voyage Long and Strange: Rediscovering the New World*. Henry Holt, 2008.
Hsu, Tiffany. "It Supercharges Conspiracy Theories." *New York Times* (May 5, 2024), Sunday Business, 4.
Huyghe, Patrick. *Columbus Was Last: From 200,000 BC to 1492, a Heretical History of Who Was First*. MJF Books, 1992.
In laudem Serenissimi Ferdinandi Hispaniae regis Bethicae & regni Granatae obsidio Victoria & triumphus / Et de Insulis in mari Indico nuper inuentis. Rome: [Stephan Plannck], 1494 [HCAR].
Introvigne, Massimo. "Modern Catholic Millennialism." In Catherine Wessinger, ed., *The Oxford Handbook of Millennialism*, 549–66. Oxford University Press, 2011.
Irizarry, Estelle. *Christopher Columbus's Love Letter to Queen Isabel*. Ediciones Puerto, 2012 [2011 as *La Carta de Amor de Cristóbal Colón a la Reina Isabel Sola*].
Irizarry, Estelle. *Christopher Columbus: The DNA of His Writings*. Ediciones Puerto, 2009.
Irving, Washington. *A History of the Life and Voyages of Christopher Columbus*. New York: G. & C. & H. Carvill, 1828 [PSUL; rev. edition, 1831, JCB; modern edition, Twayne, 1981].
Irving, Washington. *Histoire de la Vie et des Voyages de Christophe Colomb*. 4 vols. Paris: Charles Gosselin, 1828 [TULAL].
Irving, Washington. *The Life of George Washington*. 3 vols. New York: G. P. Putnam & Co., 1855–59 [HCAR; 1871 edition in TU/HT].
Irving, Washington. *Works: Tales of the Alhambra* [1832]; *Conquest of Granada* [1829]; *Conquest of Spain* [1830?]; *Spanish Voyages of Discovery, aka Voyages and Discoveries of the Companions of Columbus* [1831]. Chicago: Donohue, Henneberry & Co., ca. 1895 [my collection].
Isaac, Barry L. *Aztec Cannibalism: A Critical Assessment*. Self-published, 2025.

Jacobson, Matthew Frye. *Whiteness of a Different Color: European Immigrants and the Alchemy of Race.* Harvard University Press, 1998.

Justice, V. B. "The Columbian Stamps." *Philatelic Journal of Great Britain* 3: 26 (February 10, 1893), 18–19 [APRL].

Kadir, Djelal. *Columbus and the Ends of the Earth: Europe's Prophetic Rhetoric as Conquering Ideology.* University of California Press, 1992.

Kamen, Henry. *Imagining Spain: Historical Myth and National Identity.* Yale University Press, 2008.

Kauffman, Christopher J. *Patriotism and Fraternalism in the Knights of Columbus: A History of the Fourth Degree.* 2nd edition. Crossroad, 2001.

Keegan, William F. *Taíno Indian Myth and Practice: The Arrival of the Stranger King.* University Press of Florida, 2007.

Keen, Benjamin, trans. and ed. *The Life of the Admiral Christopher Columbus by his son Ferdinand* [1959]. Rutgers University Press, 1992.

Kelly, Annie. "How QAnon Broke Conspiracy Culture." *New York Times* (August 6, 2023), Sunday Opinion, 6–7.

Killgrove, Kristina. "DNA Testing of Skeleton May Prove Christopher Columbus Was Really Portuguese." *Forbes* (January 19, 2018), forbes.com/sites/kristinakillgrove/2018/01/19.

Kinstler, Linda. "Defining an Erasure." *New York Times Magazine* (August 25, 2024), 20–23, 45.

Klass, Perri. *A Good Time to Be Born: How Science and Public Health Gave Children a Future.* Norton, 2020.

Kritzler, Edward. *Jewish Pirates of the Caribbean: How a Generation of Swashbuckling Jews Carved Out an Empire in the New World in Their Quest for Treasure, Religious Freedom—and Revenge.* Doubleday, 2008.

Kubal, Timothy. *Cultural Movements and Collective Memory: Christopher Columbus and the Rewriting of the National Origin Myth.* Palgrave Macmillan, 2008.

Ladero Quesada, Miguel Ángel. *Granada: Historia de un país islámico (1232–1571).* Editorial Gredos, 1969.

Ladero Quesada, Miguel Ángel. "I Genovesi a Siviglia e nella sua regione: elementi di permanenza e di radicamento (secoli XIII-XVI)." In Mario del Treppo, ed., *Sistema di rapporti ed élites economiche in Europa (secoli XII-XVII)*, 211-230. Liguori Editore, 1994.

LaGumina, Salvatore J. *Wop! A Documentary History of Anti-Italian Discrimination* [1973]. 2nd edition. Guernica, 1999.

Lamartine, A. de. *Christophe Colomb.* Paris: L. Hachette et Cie., 1859 [TU/HT].

Lane, Kris. *Pillaging the Empire: Global Piracy on the High Seas, 1500–1700.* 2nd edition. Routledge, 2016.

Larkin, Tanya. *Christopher Columbus.* PowerKids Press, 2001.

Lardicci, Francesca, ed. *A Synoptic Edition of the Log of Columbus's First Voyage.* Cynthia L. Chamberlain and Blair Sullivan, trans. *Repertorium Columbianum*, Geoffrey Symcox, general ed., Volume VI. Brepols, 1999.

Las Casas, Bartolomé de. *Historia de las Indias.* 3 vols. Fondo de Cultural Económica, 1951.

Lawson, Elisa. "Cecil Roth and the Imagination of the Jewish Past, Present and Future in Britain, 1925–1964." PhD dissertation, University of Southampton, 2005.

Leo XIII, Pope. *Quarto Abeunte Saeculo*. Rome: July 16, 1892, vatican.va/content/leo-xiii/en/encyclicals/documents/hf_l-xiii_enc_16071892_quarto-abeunte-saeculo.html.
Lester, Toby. *The Fourth Part of the World: The Race to the Ends of the Earth, and the Epic Story of the Map That Gave America Its Name*. Free Press, 2009.
Leupp, Francis E. *Walks About Washington*. Boston: Little, Brown, 1915 [AOC].
Levy, Shawn, dir. *Night at the Museum*. Feature film. Twentieth Century Studios, 2006.
Lollis, C. de, ed. *Raccolta di documenti e studi pubblicati dalla Commissione Colombiana*. 6 parts, 14 vols. Rome: Ministero della Pubblica Istruzione, 1891–96 [cited as RDSP].
Lopez, Barry. *The Rediscovery of North America* [1990]. Vintage, 1992.
Lugo, Americo. *Los Restos de Colón (Publicado por la primera vez en "Clio" en 1934)*. Librería Dominicana, 1950.
Macdonald, Fiona. *You Wouldn't Want to Sail with Christopher Columbus!* Franklin Watts, 2004.
Macías Hernández, Antonio M. "Expansión Ultramarina y Economía Vitivinícola: El Ejemplo de Canarias (1500–1550)." *Investigaciones de Historia Económica* 3: 8 (2007), 13–44.
Madariaga, Salvador de. *Christopher Columbus, Being the Life of the Very Magnificent Lord Don Cristobal Colón* [1940]. Hollis and Carter, 1949 [Reprinted, Frederick Ungar Publishing Co., 1967].
Madley, Benjamin. *An American Genocide: The United States and the California Indian Catastrophe*. Yale University Press, 2016.
Maheshwari, Sapna. "In Big Sky Country, a Battle Against Big Tech." *New York Times* (September 3, 2023), Sunday Business, 1, 6.
Maineri, Ronald J., and Charles O. Hamill. *Checklist of Christopher Columbus Related Postage Stamps*. Booklet. Christopher Columbus Philatelic Society, 1987.
Mann, Charles C. *1491: New Revelations of the Americas Before Columbus*. Knopf, 2005.
Mantle, Burns. "Gay Columbus Played Too Dully for Gotham Audience and Critics." *Chicago Daily Tribune* (June 12, 1932), F1, F7.
Manzano Manzano, Juan. *Colón y su secreto: El predescubrimiento* [1976]. Ediciones de Cultura Hispánica, 1989.
Manzano Manzano, Juan, and Ana María Manzano Fernández-Heredia. *Los Pinzones y el descubrimiento de América*. 3 vols. Ediciones de Cultúra Hispanica, 1988.
Marinari, Maddalena. *Unwanted: Italian and Jewish Mobilization against Restrictive Immigration Laws, 1882–1965*. University of North Carolina Press, 2020.
Markey, Lia. "Stradano's Allegorical Invention of the Americas in Late Sixteenth-Century Florence." *Renaissance Quarterly* 65: 2 (Summer 2012), 385–442.
Marlowe, Stephen. *The Memoirs of Christopher Columbus*. Charles Scribner's Sons, 1987.
Mascarenhas Barreto, Augusto. *"Colombo" Português: Provas Documentais*. 2 vols. Nova Arraniada, 1997.
Mascarenhas Barreto, Augusto. *O Português Cristóvão Colombo, agente secreto do Rei Dom Joáo II*. Edições Referendo, 1988.
Mascarenhas Barreto, Augusto. *The Portuguese Columbus: Secret Agent of King John II*. St. Martin's Press, 1992.
Masters, Adrian. *We, the King: Creating Royal Legislation in the Sixteenth-Century Spanish New World*. Cambridge University Press, 2023.
Matsumoto, Seicho. *Tokyo Express* [1958]. Penguin, 2022.

Mattei, Roberto de. "Columbus noster est!" In *Rorate Caeli* (June 17, 2020), rorate-caeli.blogspot.com/2020/06/de-mattei-columbus-noster-est.html.
Maurer, Pattrice. "Christopher Columbus & The AIDS Quilt." *Agenda* (University of Michigan) 7: 6 (October 1992), 7.
McKenzie, Precious. *Manatees*. Rourke, 2010.
Mellgren, Doug. "Was Columbus Really a Norwegian Son?" *The Journal* (Alexandria, VA) (March 7, 1991).
Menéndez, Marta. "Isabel la Católica se enamoró de Cristóbal Colón e inentó matar a su amante." *El Independiente* (June 8, 2022), elindependiente.com/tendencias/2022/06/08.
Mentz, Steve. *Ocean*. Bloomsbury, 2020.
Milhou, Alain. *Colón y su Mentalidad Mesianica en el Ambiente Franciscanista Española*. Casa-Museo de Colón, 1983.
Mira Caballos, Esteban. *El Indio Antillano: repartimiento, encomienda y esclavitud (1492–1542)*. Muñoz Moya Editor, 1997.
Molloy, Charles. *De Jure Maritimo et Navali: or, A Treatise of Affaires Maritime, And of Commerce*. London: John Bellinger et al., 1676 [Bodl].
Morales, Manuel. "Christopher Columbus: A product of his time or guilty of genocide?" English version by Heather Galloway. *El País* (November 19, 2018), english. elpais.com/elpais/2018/11/16/inenglish/1542382462_591917.html.
Moreno Ortíz, Alejandro. "En busca del origen Gallego de Cristóbal Colón." *Sevilla Actualidad* (November 28, 2022), sevillaactualidad.com/espana/503759-en-busca-del-origen-gallego-de-cristobal-colon.
Morison, Samuel Eliot. *Christopher Columbus: Admiral of the Ocean Sea*. Oxford University Press, 1942.
Morison, Samuel Eliot. *Portuguese Voyages to America in the Fifteenth Century*. Harvard University Press, 1940.
Morison, Samuel Eliot. *The European Discovery of America: The Southern Voyages, 1492–1616*. Oxford University Press, 1974.
Morton, Thomas. *Columbus: or, A World Discovered. An Historical Play*. London: W. Miller, 1792 [BL; JCB; TU/HT].
Moscoso, Francisco. *Caciques, aldeas y población taína de Boriquén (Puerto Rico), 1492–1582*. Academica Puertorriqueña de la Historia, 2008.
Mosquiera Manso, José M. *La Cuna Gallega de Cristóbal Colón*. Editorial Citania, 1961 [Reprinted, Ediciós do Castro, 1992].
Moya Pons, Frank. *Después de Colón: Trabajo, sociedad y política en la economía del oro*. Alianza Editorial, 1987.
Mucher, Christian. *Before American History: Nationalist Mythmaking and Indigenous Dispossession*. University of Virginia Press, 2023.
Muldoon, James. *The Americas in the Spanish World Order: The Justification for Conquest in the Seventeenth Century*. University of Pennsylvania Press, 1994.
Muldoon, James, and Felipe Fernández-Armesto, eds. *The Medieval Frontiers of Latin Christendom: Expansion, Contraction, Continuity*. Vol. 1 of *The Expansion of Latin Europe, 1000–1500*. Ashgate, 2008.
Muñoz, Juan Bautista. *Historia del Nuevo-Mundo*. Madrid: La Viuda de Ibarra, 1793 [two copies of Vol. I, the only volume written and published, JCB].

Muratore, Joseph R. *The Remains of Christopher Columbus*. Muratore Agency, 1972.
Murray, John O'Kane. *Lives of the Catholic Heroes and Heroines of America*. New York: J. Sheehy, 1879 [TU/HT].
Murrell, David. "Who Was the Real Christopher Columbus?" *Philadelphia* magazine (July 29, 2020), phillymag.com/news/2020/07/29/christopher-columbus-statue-philly-matthew-restall/.
Nader, Helen, trans. and ed. *The Book of Privileges Issued to Christopher Columbus by King Fernando and Queen Isabel, 1492–1502. Repertorium Columbianum*, Geoffrey Symcox, general ed., Volume II. University of California Press, 1996.
Navarrete, Martín Fernández de, ed. *Colección de los viajes y descubrimientos que hicieron por mar los españoles desde fines del siglo XV*. 5 vols. Editorial Guaranía, 1945–46 [cited as CVD].
Navarrete, Martín Fernández de, ed. *Viages de Colón: Almirantazgo de Castilla*. Vol. I of *Coleccion de los Viages y Descubrimientos, que hicieron por mar los españoles desde fines del siglo XV*. 5 vols. Madrid: La Imprenta Real, 1825 [HCAR].
Odoardi, Giovanni. "Il processo di beatificazione di Cristoforo Colombo." In *Studi Colombiani* vol. III, 261–72. Civico Istituto Colombiano, 1951.
Oliver, José R. *Caciques and Cemí Idols: The Web Spub by Taíno Rulers between Hispaniola and Puerto Rico*. University of Alabama Press, 2009.
Orique, David Thomas, and Rady Roldán-Figueroa, eds. *Bartolomé de las Casas, O.P.: History, Philosophy, and Theology in the Age of European Expansion*. Brill, 2019.
Otero Sánchez, Prudencio. *España: Patria de Colón*. Biblioteca Nueva, 1922.
Oviedo y Valdés, Gonzalo Fernández de. *Historia General y Natural de las Indias, Islas y Tierra Firme del Mar Océano*. 1535 and 1557 [JCB].
Oviedo y Valdés, Gonzalo Fernández de. *Historia General y Natural de las Indias, Islas y Tierra Firme del Mar Océano*. 3 vols. Madrid: La Real Academia de la Historia, 1851 [TULAL].
Paul, Heike. *The Myths that Made America: An Introduction to American Studies*. Transcript Verlag, 2014.
Paulmier, Hilah, and Robert Haven Schauffler, eds. *Columbus Day*. Our American Holidays series. Dodd, Mead & Company, 1938.
Payne, Antony, ed. *The Spanish Letter of Columbus: A Facsimile of the Original Edition Published by Bernard Quaritch in 1891*. Felipe Fernández-Armesto, intro. Quaritch, 2006.
Pennock, Caroline Dodds. *On Savage Shores: How Indigenous Americans Discovered Europe*. Weidenfeld & Nicolson, 2022.
Pereira, Eduardo C. N. "Cristóvão Colombo no Porto Santo e na Madeira." *Das Artes e da História da Madeira* 4: 22 (1956), 20–27.
Peris, Jorge. "Yo soy el Cristóbal Colón número 20." *El Tiempo* (June 25, 2019), eltiempo.com/bocas/ yo-soy-el-cristobal-colon-numero-20-380400.Pestana, Manuel G. (aka Pestana Junior). *D. Cristóbal Colón, ou Symam Palha na História e na Cabala*. Imprensa Lucas & Ca., 1928.
Pettegree, Andrew, and Arthur Der Weduwen. *The Library: A Fragile History*. Basic Books, 2021.
Philiponus, Honorius. *Nova typis transacta navigation*. Linz: n.p., 1621 [JCB].
Phillips, Carla Rahn. "Exploring from Early Modern to Modern Times." In Daniel

Finamore, ed., *Maritime History as World History*, 62–81. University Press of Florida and Peabody Essex Museum, 2004.

Phillips, Seymour. "European Expansion Before Columbus: Causes and Consequences." *Haskins Society Journal: Studies in Medieval History* no. 5 (1993), 45–59. Reprinted in James Muldoon and Felipe Fernández-Armesto, eds., *The Medieval Frontiers of Latin Christendom: Expansion, Contraction, Continuity*. Vol. 1 of *The Expansion of Latin Europe, 1000–1500* (Ashgate, 2008), 327–41.

Phillips, William D., Jr. "Maritime Exploration in the Middle Ages." In Daniel Finamore, ed., *Maritime History as World History*, 47–61. University Press of Florida and Peabody Essex Museum, 2004.

Phillips, William D., Jr., ed. *Testimonies from the Columbian Lawsuits*. Mark D. Johnson and Anne Marie Wolf, trans. *Repertorium Columbianum*, Geoffrey Symcox, general ed., Volume VIII. Brepols, 2000.

Phillips, William D., Jr., and Carla Rahn Phillips. *The Worlds of Christopher Columbus*. Cambridge University Press, 1992.

Pike, Ruth. "Sevillian Society in the Sixteenth Century: Slaves and Freedmen." *Hispanic American Historical Review* 47: 3 (August 1967), 344–59.

Pohl, Frederick Julius. *The New Columbus*. Security Dupont Press, 1986.

Ponce de León, Nestor. *The Columbus Gallery: The Discoverer of the New World, as represented in portraits, monuments, statues, medals, and paintings*. N. Ponce de León, Publisher, 1893 [PSUL].

Posse, Abel. *The Dogs of Paradise*. Margaret Sayers Peden, trans. Atheneum, 1989.

Prescott, William H. *History of the Reign of Ferdinand and Isabella*. C. Harvey Gardiner, ed. Heritage Press, 1962.

Prescott, William H. *The History of the Reign of Ferdinand and Isabella the Catholic*. 3 vols. Cambridge: Folsom, Wells, and Thurston, 1838 [TULAL].

Provost, Foster. *Columbus: An Annotated Guide to the Scholarship on His Life and Writings, 1750–1988*. Omnigraphics, Inc., 1991.

Racine, Karen. *Francisco de Miranda: A Transatlantic Life in the Age of Revolution*. Scholarly Resources, 2003.

Raff, Jennifer. *Origins: A Genetic History of the Americas*. Twelve, 2022.

Ramos, Demetrio. "Los Contactos Trasatlanticos Decisivos, Como Precedentes del Viaje de Colón." Cuadernos Colombinos #2. Casa-Museo de Colón, 1972.

Razaf, Andy. "Christopher Columbus." Song lyrics. Music by Leon "Chu" Berry. 1936. Classic recordings by Fats Waller (1936) and Dinah Washington (1957).

Real Academia de la Historia, La. *Los restos de Colón: Informe de la Real Academia de la Historia al gobierno de S. M. sobre el supuesto hallazgo de los verdaderos restos de Cristóval Colón en la iglesia cathedral de Santo Domingo*. Madrid: M. Tello; el Ministerio de Fomento, 1879 [written by Manuel Colmeiro but no author credit on title page] [JCB].

Regal, Brian. *The Battle Over America's Origin Story: Legends, Amateurs, and Professional Historiographers*. Palgrave Macmillan, 2022.

Reséndez, Andrés. *The Other Slavery: The Uncovered Story of Indian Enslavement in America*. Houghton Mifflin, 2016.

Restall, Matthew. "Montezuma Surrenders in the Capitol." *Capitol Dome* 53: 2 (Fall 2016), 2–10.

Restall, Matthew. *When Montezuma Met Cortés: The True Story of the Meeting That Changed History.* Ecco, 2018.
Restall, Matthew. *Seven Myths of the Spanish Conquest* [2003]. Updated edition. Oxford University Press, 2021.
Restall, Matthew. "The Trouble with 'America.'" *Ethnohistory* 67: 1 (January 2020), 1–28.
Restall, Matthew, Amara Solari, John F. Chuchiak IV, and Traci Ardren. *The Friar and the Maya: Diego de Landa and the Account of the Things of Yucatan.* University Press of Colorado, 2023.
Ribeiro, Patrocínio. *A Nacionalidade Portuguesa de Cristovam Colombo: Solução do Debatidissimo Problema de Sua Verdadeira Naturalidade, Pela Decifração Definitiva da Firma Hieroglífica / The Portuguese Nationality of Christopher Columbus: The Much-Discussed Problem of His Actual Nationality at Last Disposed of through the Conclusive Decifration of His Hieroglyphic Sign* [sic]. Libraria renascença, J. Cardoso, 1927.
Ribot García, Luís Antonio, ed. *El tratado de Tordesillas y su época.* 3 vols. Sociedad V Centenario del Tratado de Tordesillas, 1995.
Robbins, Frank, trans. *The Columbus Letter of 1493.* The Clements Library, 1952.
Robertson, William. *The History of America.* 2 vols. London: W. Strahan, T. Cadell, and J. Balfour, 1777 [JCB].
Rocca, Al M. *A Historical Geography of Christopher Columbus's First Voyage and his Interactions with Indigenous Peoples of the Caribbean.* Routledge, 2024.
Rogers, Francis M. "The Vivaldi Expedition." *Annual Report of the Dante Society,* no. 73 (1955), 31–45.
Roncari, Alexander John James. *The Nine Arguments in Defense of Christopher Columbus: The Untold Truth.* Self-published (Hamilton), 1982.
Roop, Peter and Connie, eds. *I, Columbus: My Journal, 1492–3.* Walker and Co., 1990.
Roselly de Lorgues, Antoine François Félix. *Christophe Colomb, histoire de sa vie et de ses voyages d'après des documents authentiques tires d'Espagne et d'Italie.* 2 vols. Paris: Didier et Compagnie, 1856 [TULAL; 2nd edition, 1859, in JCB].
Roselly de Lorgues, Antoine François Félix. *Histoire posthume de Christophe Colomb.* Paris: Didier et Compagnie, 1885 [LoC].
Roselly de Lorgues, Antoine François Félix. *Monumento a Colón: Historia de la Vida y Viajes de Cristóbal Colón.* 3 vols. Barcelona: Jaime Seix, 1878 [TULAL].
Roselly de Lorgues, Antoine François Félix. *Satan contre Christophe Colomb.* Paris: Didier et Compagnie, 1876 [LoC].
Roth, Cecil. "Who *Was* Columbus? In the Light of New Discoveries." *Menorah Journal* XXVIII: 3 (October–December 1940), 279–295.
Royal, Robert. *1492 and All That: Political Manipulations of History.* Ethics and Public Policy Center, 1992.
Rumeu de Armas, Antonio. *El "Portugues" Cristóbal Colón en Castilla.* Ediciones Cultúra Hispanica, 1987.
Rusconi, Roberto, ed. *The Book of Prophecies Edited by Christopher Columbus.* Blair Sullivan, trans. *Repertorium Columbianum,* Geoffrey Symcox, general ed., Volume III. University of California Press, 1997.
Russell, Jeffrey Burton. *Inventing the Flat Earth: Columbus and the Modern Historian.* Prager, 1991.

Russell, William. *The History of America, from Its Discovery by Columbus to the Conclusion of the Late War*. 2 vols. London: Fielding and Walker, 1778 [JCB].

Rydell, Robert W. *All the World's a Fair: Visions of Empire at American International Expositions, 1876–1916*. University of Chicago Press, 1984.

Sabelko, Rebecca. *Indigenous Peoples' Day*. Bellweather, 2023.

Sabellicus, Marcantonio Coccio. *Secunda pars enneadum ab orbe condito ad inclinatonem Romani Imperii*. Venice: Bernardinum et Matheum Venetos, qui vulgo dicuntur "Li Albanesoti," 1498 [LoC].

Sable, Martin. "Columbus, Marrano Discoverer from Mallorca." Pamphlet. Milwaukee, WI: printed by author, 1992 [JCB].

Sale, Kirkpatrick. *The Conquest of Paradise: Christopher Columbus and the Columbian Legacy*. Knopf, 1990.

Salinas, Brenda. "'Columbusing': The Art of Discovering Something That Is Not New." NPR, "Code Switch" (July 6, 2014), npr.org/sections/codeswitch/2014/07/06/328466757/.

Sanguineti, Angelo. *La canonizzazione di Cristoforo Colombo*. Genoa: R. Istituto Sordo-Muti, 1875 [TULAL].

Sanguineti, Angelo. *Vita di Cristoforo Colombo*. 2nd edition. Genoa: R. Istituto Sordo-Muti, 1891 [JCB].

Sannes, Tor Borch. *Christopher Columbus—en europeer fra Norge?* Oslo: Norsk Maritimt Forlag, 1991 [also at nb.no/items/URN:NBN:no-nb_digibok_2007111900008].

Sarna, Jonathan D. "Columbus & the Jews." *Commentary* 94: 5 (November 1992), 38–41.

Scheirer, Walter J. *A History of Fake Things on the Internet*. Stanford University Press, 2023.

Schlereth, Thomas J. "Columbia, Columbus, and Columbianism." *Journal of American History* 97 (December 1992), 937–68.

Schoenrich, Otto. *The Legacy of Christopher Columbus: The Historic Litigations Involving His Discoveries, His Will, and His Descendants*. 2 vols. Arthur H. Clark, 1949–50.

Schreiber, Dan. *The Theory of Everything Else: A Voyage into the World of the Weird*. William Morrow, 2022.

Schuessler, Jennifer. "Smithsonian's Latino Museum, Unbuilt, Raises Political Furor." *New York Times* (September 24, 2023), 1, 22.

Scott, Ridley, dir. *1492: Conquest of Paradise*. Feature film. Paramount Pictures, 1992.

Scott, William. *The Log of Christopher Columbus's first Voyage to America in the year 1492* [1938]. Linnet, 1989.

Seguin, Charles Franklin. "Making a National Crime: The Transformation of US Lynching Politics, 1883–1930." PhD dissertation, University of North Carolina at Chapel Hill, 2016.

Seguin, Charles, and Sabrina Nardin. "The Lynching of Italians and the Rise of Anti-lynching Politics in the United States." *Social Science History* 46 (2022), 65–91.

Seguin, Charles, and David Rigby. "National Crimes: A New National Data Set of Lynchings in the United States, 1883–1941." *Socius* 5 (May 2019), doi.org/10.1177/2378023119841780.

Sehgal, Parul. "Tell No Tales." *New Yorker* (July 10 and 17, 2023), 68–72.

Shimoni, Youval. *MiBa'ad LaKarka'it HaShkufa* (*Beyond the Transparent Bottom*). Am Oved, 2021.
Shin, Youjin, Nick Kirkpatrick, Catherine D'Ignazio, and Wonyoung So. "Columbus Monuments are Coming Down, but He is Still Honored in 6,000 Places Across the US." *Washington Post* (October 11, 2021), washingtonpost.com/history/interactive/2021/christopher-columbus-monuments-america-map/.
Sobral, Ana. "Christopher Columbus and Jamaican Jews: History into Memory." In Sina Rauschenbach and Jonathan Schorsch, eds., *The Sephardic Atlantic: Colonial Histories and Postcolonial Perspectives*, 247–75. Palgrave Macmillan, 2018.
St. George, Zach. "The Long Tale." *New York Times Magazine* (March 10, 2024), 30–35, 45.
Stage, Jeff. "Many Fairs Commemorated on U.S. Stamps." *American Philatelist* 137: 7 (July 2023), 621.
Stavans, Ilan. *Imagining Columbus: The Literary Voyage*. Palgrave, 1993.
Stevens-Arroyo, Anthony M. "The Inter-Atlantic Paradigm: The Failure of Spanish Medieval Colonization of the Canary and Caribbean Islands." *Comparative Studies in Society and History* 35 (1993), 515–43; reprinted in James Muldoon and Felipe Fernández-Armesto, eds., *The Medieval Frontiers of Latin Christendom: Expansion, Contraction, Continuity*. Vol. 1 of *The Expansion of Latin Europe, 1000–1500*, 343–71. Ashgate, 2008.
Stone, Erin Woodruff. *Captives of Conquest: Slavery in the Early Modern Caribbean*. University of Pennsylvania Press, 2021.
Stories of Great Men. Boston: Educational Publishing, 1895 [LoC].
Sued Badillo, Jalil. *Caribe taíno: ensayos históricos sobre el siglo XVI*. Editorial Luscinia, 2020.
Summerhill, Stephen J., and John Alexander Williams. *Sinking Columbus: Contested History, Cultural Politics, and Mythmaking during the Quincentenary*. University Press of Florida, 2000.
Symcox, Geoffrey, ed. *Italian Reports on America, 1493–1522: Letters, Dispatches, and Papal Bulls*. Giovanna Rabitti, Italian text ed. Peter D. Diehl, trans. *Repertorium Columbianum*, Geoffrey Symcox, general ed., Volume X. Brepols, 2001.
Symcox, Geoffrey, and Jesús Carrillo, eds. *Las Casas on Columbus: The Third Voyage*. Michael Hammer and Blair Sullivan, trans. *Repertorium Columbianum*, Geoffrey Symcox, general ed., Volume XI. Brepols, 2001.
Symcox, Geoffrey, and Luciano Formisano, eds. *Italian Reports on America, 1493–1522: Accounts by Contemporary Observers*. Theodore J. Cachey, Jr., and John C. McLucas, trans. *Repertorium Columbianum*, Geoffrey Symcox, general ed., Volume XII. Brepols, 2002.
Taviani, Paolo Emilio. *Christopher Columbus: The Great Design*. Orbis, 1985 (English edition of *Cristoforo Colombo: la genesi della grande scoperta*, 1974, rev. 1980, 1982).
Taviani, Paolo Emilio. *Columbus: The Great Adventure: His Life, His Times, and His Voyages*. Orion, 1991.
Taviani, Paolo Emilio. *The Voyages of Columbus: The Great Discovery*. Vols. 1–2. Instituto Geografico de Agostini, 1991.
Thacher, John Boyd. *Christopher Columbus: His Life, His Work, His Remains*. 3 vols. New York: G. P. Putnam's Sons and The Knickerbocker Press, 1903 [#26-of-100

7-volume "Collector's Edition" in JCB; citation pagination same as for 3-vol. edition].

"The Canonization of Columbus." *Sacramento Daily Record-Union* 3: 148 (October 4, 1877), 2, accessed through the *California Digital Newspaper Collection* (cdnc .ucr.edu).

"The Voyager in Marble: Unveiling of the Great Columbus Monument." *New York Times*, October 13, 1892.

Tobia, Darren. "As Anti-Racism Protests Sweep the Nation, Italians Stand Their Ground on Columbus Issue" (June 26, 2020), jerseydigs.com/ christopher-columbus-statue-removed-washington-park-newark/.

Toro, Alfonso. *Historia Colonial de la América Española*. Vol. I: *Los Viajes de Colón*. Editorial Patria, 1946.

Ugorji, Basil. "Christopher Columbus: A Controversial Monument in New York." 2021 Paper available at icermediation.org/the-columbus-monument/.

Ulloa, Luis. *Cristòfor Colom Fou Català: La Veritable Gènesi del Descobriment*. Libreria Catalonia, 1927 [TULAL].

United Nations General Assembly. 37th Session, Official Records, 83rd Plenary Meeting, A/37/PV.83 (November 29, 1982).

Valen, Nino. *Being the Heart of the World: The Pacific and the Fashioning of the Self in New Spain, 1513–1641*. Cambridge University Press, 2023.

Van Deusen, Nancy E. *Global Indios: The Indigenous Struggle for Justice in Sixteenth-Century New Spain*. Duke University Press, 2015.

Van Duzer, Chet. "Columbus and the Nature of a New World." In John W. Hessler, Daniel De Simone, and Chet Van Duzer, *Christopher Columbus: Book of Privileges*, 1–26. Levenger Press and Library of Congress, 2014.

Van Dyke, Henry. *Counsel Upon the Reading of Books*. Boston: Houghton, Mifflin and Co., 1902 [LoC].

Varela, Consuelo, ed. *Cristóbal Colón: Textos y documentos completos*. Alianza Universidad, 1982.

Varela, Consuelo, ed. *Diario del primero y tercer viaje de Cristóbal Colón: Fray Bartolomé de Las Casa, Obras Completas*. Alianza Editorial, 1989.

Varela, Consuelo, ed. *La caída de Cristóbal Colon: El juicio de Bobadilla*. Marcial Pons, 2006.

Vignaud, Henry. *Toscanelli and Columbus: The Letter and Chart of Toscanelli*. London: Sands & Co., 1903 [JCB].

Wade, Mary Dodson. *Christopher Columbus: Famous Explorer*. Capstone Press, 2007.

Wanamaker, John. "Columbian Adhesive Postage Stamps: Circular to Postmasters." *Philatelic Journal of America*, no. 97 (January 1893), 5–6 [APRL].

Warren, Louis S. *The Hunter's Game: Poachers and Conservationists in Twentieth-Century America*. Yale University Press, 1997.

Wassermann, Jakob. *Christoph Columbus: Der Don Quichote des Ozeans. Eine Biographie*. S. Fischer, 1929 (English edition, Little, Brown, 1930).

Watson, Bruce. *Sacco and Vanzetti: The Men, the Murders, and the Judgement of Mankind*. Viking, 2007.

Watt, Mary Alexandra. *Dante, Columbus and the Prophetic Tradition: Spiritual Imperialism in the Italian Imagination*. Routledge, 2017.

Watts, Pauline W. "Prophecy and Discovery: On the Spiritual Origins of Christopher Columbus's 'Enterprise of the Indies.'" *American Historical Review* 90: 1–2 (1985), 73–102.
Wey Gómez, Nicolás. *The Tropics of Empire: Why Columbus Sailed South to the Indies.* MIT Press, 2008.
Whitley, Alvin. "Hood and Dickens: Some New Letters." *Huntington Library Quarterly*, 14: 4 (August 1951), 385–413.
Wiesenthal, Simon. *Sails of Hope: The Secret Mission of Christopher Columbus* [1968]. Macmillan, 1973.
Willingham, Elizabeth Moore. *The Mythical Indies and Columbus's Apocalyptic Letter: Imagining the Americas in the Late Middle Ages.* Sussex Academic Press, 2016.
Wilson-Lee, Edward. *The Catalogue of Shipwrecked Books: Christopher Columbus, His Son, and the Quest to Build the World's Greatest Library.* Scribner, 2019.
Wimberly, Lowry Charles. "Spook English." *American Speech* 1: 6 (March 1926), 317–21.
Winchester, Simon. *Atlantic: Great Sea Battles, Heroic Discoveries, Titanic Storms, and a Vast Ocean of a Million Stories.* Harper, 2010.
Winsor, Justin. *Christopher Columbus, and How He Received and Imparted the Spirit of Discovery.* Boston: Houghton, Mifflin, and Co., 1892 [LoC; PSUL].
Wolanin, Barbara. "Columbus in the Art of the Capitol." *Capital Dome* 27: 1 (January 1992), 6–7.
Wyatt, Valerie. *Who Discovered America?* Kids Can Press, 2008.
Yates, Clinton. "Columbusing black Washington." *Washington Post* (October 8, 2012), washingtonpost.com/blogs/therootdc/post/nouveau-columbusing-black-washington/2012/10/08/62b43084-10db-11e2-a16b-2c11003151 4a_blog.html.
Zahniser, David. "L.A. City Council replaces Columbus Day with Indigenous Peoples Day on city calendar." *Los Angeles Times* (August 30, 2017), latimes.com/local/lanow/la-me-ln-indigenous-peoples-day-20170829-story.html.
Zamora, Margarita. *Reading Columbus.* University of California Press, 1993.
Zavala, Silvio. "Destrucción y daños causados a monumentos públicos en México" (1994), letraslibres.com/vuelta/destruccion-y-danos-causados-a-monumentos-publicos-en-mexico/.

Illustration Credits

"The Mediterranean and Atlantic Worlds of Columbus's Lives" (map), page x: Created by the author
"Said to Be," page xvii: Courtesy of the Metropolitan Museum of Art, New York
"Homeland," page xxvi: Photograph by the author
"Commemorative," page 2: Fair use
"Sea Views," page 3: Wikimedia Commons
"Another Sea View," page 4: Photograph by the author
"Ridiculed," page 17: Illustrated London News (October 8, 1892)
"All at Sea," page 28, top: Courtesy of the National Postal Museum, Smithsonian Institution
"All at Sea," page 28, bottom: Fair use / photograph by the author
"Naked and Afraid?," page 34: Courtesy of the John Carter Brown Library
"A Fishy Tale," page 38: Courtesy of the John Carter Brown Library
"Presenting People," page 48: Courtesy of the National Postal Museum, Smithsonian Institution
"The Climb," page 56: Photograph by the author
"Remainders," page 62: Fair use / photograph by the author
"Dust to Ashes," page 70: Photograph by the author
"Dark Lighthouse," page 71, top and bottom: Photograph by the author
"Columbian Viceroy," page 80: Wikimedia Commons
"Saintly," page 88, top: Wikimedia Commons (courtesy of the Museo Lázaro Galdiano)
"Saintly," page 88, bottom: Photograph by the author
"Prophesied," page 105: Used with permission of Capstone Publishing, Mankato, Minnesota (copyright 2007 by Capstone Press)
"Holy Hero," page 107: Courtesy of the John Carter Brown Library
"Imagined," page 111: Fair use
"F**ker," page 118, left: Courtesy of former TikTok user @basicrightsbec
"F**ker," page 118, right: Courtesy of TikTok user @che.jim
"Pledging," page 126: Courtesy of the National Postal Museum, Smithsonian Institution
"Ridden," page 139: Courtesy of the John Carter Brown Library
"Born Again," page 142, top: Fair use
"Born Again," page 142, bottom: Wikimedia Commons (photograph courtesy of Jean-Pol Grandmont)
"The Greek," page 150: Fair use

"We Are Sailing," page 159: Wikimedia Commons
"It Is Believed," page 162, top: Photograph courtesy of Lucía Lago
"It Is Believed," page 162, bottom: Photograph courtesy of Fermín Rodríguez
"Pointed," page 173: Wikimedia Commons
"Capitol Columbuses," page 186: Courtesy of the Library of Congress
"Arrival," page 189: Courtesy of the Architect of the Capitol
"Welcome," page 190: Courtesy of the Architect of the Capitol
"Progenitor," page 196: Courtesy of the Library of Congress (PAGA 7, no. 4114)
"History Men," page 202: Courtesy of the National Portrait Gallery, Washington, DC
"Nine Pins," page 209: Courtesy of the Architect of the Capitol
"A Small World," page 210: Illustrated London News (April 19, 1845)
"Half-Lives," page 212, left: Courtesy of the Architect of the Capitol
"Half-Lives," page 212, right: Courtesy of the Architect of the Capitol
"On a Pedestal," page 216: Courtesy of the Library of Congress (LC-DIG-stereo-1s07505)
"Five Heads," page 223: Courtesy of the Columbus Monument Corporation, Syracuse
"Genoa–New York," page 226: Courtesy of the Library of Congress
"Italy in New York," page 227: Courtesy of the Library of Congress
"Boxed In," page 234, top and bottom: Photograph by the author
"Face Down," page 239: Photograph courtesy of Brad Sigal

Index

Page numbers in *italics* refer to illustrations.
"CC" = Christopher Columbus.

Abreu, João Carlos, 158, 160, 161
Afonso V (king of Portugal), 151
African Americans. *See* Black Americans
African coastal trade, 12, 22, 37, 41
Albornoz, Miguel, 18–19
Alexander VI (Pope), 50–51
Al-Farghani, 21
alternate discovery claims, 149, 150, 151, 235–37, 291n27, 291nn24,27
Alvarado, Pedro de, 133
Álvarez de Soutomaior, Pedro "Madruga," xiv, 163–64
Amazons, 136, 139
American Columbus, 187–215
   alternate discovery claims and, 236
   anti-Spanish views and, 192–93
   Black Legend and, 192–93, 286n9
   CC as agent of God and, 191, 192, 195, 200
   Centennial and, *196*
   children's literature and, 203–4, 287n26
   Columbus Day and, 190, 195, 204, 205
   Discovery of America (Persico) and, *186*, 188, 207–8, 210–15, *210*, *212*, 284n29, 288nn33,38
   early U.S. writings on, 191–95, 286nn9,12
   English influences, 286n13
   flaws and, 193–94
   great men theory of history and, 200, 201–2
   imperialism/colonialism and, 189, 190, 198–99
   Irving and, 189, 199–201
   Italian American Columbus and, 233, 235
   localism and, 215
   Manifest Destiny and, 208–11, 215
   monuments and images and, 187–91, *189*, *190*, 207–9, 210–12, 285nn1,5–6
   naming and, 195–98, 286–87n14–15
   presidential inaugurations and, 209–10, *210*
   Quadricentennial and, 201, 204–5
   Quincentennial and, 206, 213–14, 215
   1792 celebration and, 194–95, 286n12
   U.S. Bicentennial and, 213–14, 215
American Indian Movement (AIM), 237, *239*
Americanus, Sylvanus (Samuel Nevill), 191, 285n7
Amler, Jane Francis, 181
Andersen, Hans Christian, 138
Andre, Marius, 283n24
anti-Columbus protests, xii, xv
   homosexuality possibility and, 137–38
   Indigenous people and, 195, 224, 238–39, *239*
   Italian American Columbus and, 220–22, 223–25, 237–41, 291–92n34
   Mexico, 184–85
   monuments to CC and, 184–85, 211–12, 220–21, 223–24, 231, *239*, 288nn38,40, 289n9
Antilles, 53
anti-Semitism, 177–78, 179, 284n32

Aragon, 23–24, 25, 261n50
Arana, Diego de, 276n7
Arana, Marie, 182
Aristizábal, Gabriel de, 68
Arribas, Enrique, 171
Asensio, José María, 68–69, 112, 113, 268n13
Asociación Colón Gallego, 164, 170
Assassin's Creed II: Discovery, 292n4
Ataíde, António de, 165, 282n3
Atlantic voyaging
  Carrera de Indias and, 43
  chivalric romance and, 37–38
  cod fleets and, 264n19
  collective nature of, 75–76, 264n27
  Columbiana marginalization and, 32
  dangers of, 60
  Danish expedition, 151, 174
  fleet formation, 43
  imperialism/colonialism and, 41, 47, 50
  island lore and, 34, *34*, 35, 37–38, 39–41, 53, 263n16
  motivations for, 47, 50
  navigation and, 38–39, 42–43, 264n24
  ship technology and, 42
  siren mythology and, 138–39
  sugar plantation economy and, 41
  transatlantic slave trade and, 60–61
  uniqueness myth and, xxi, 34, 41, 42, 46–47, 75–76
Avatar, 247
Azores, 22, 39–43
Aztecs, 55, 60, 73, 75, 185, 235

Báez, Buenaventura, 63
Baldwin, James, 231
Ballesteros Beretta, Antonio, 148, 178, 277n15, 281n25
Baltimore, 205, 231, 240–41, 290n22, 291–92n34
Barabino, Nicolò, *17*
Barbeito, Mário, 157–58
Barcelona, 182
  CC in, 29, 32, 35, 99, 106, 127, 174, 188, 201, 262n5
  statue of CC in, 173, *173*, 185, 285n38
Barlow, Joel, 193, 286n15
Barreto, Augusto Mascarenhas, 172
Barron, Charles, 221
Barsotti, Annie, 218–19
Barsotti, Charles (Carlo), 218
Baum, Dwight James, 222
Baxter, Stephen, 247
Belknap, Jeremy, 194–95
Benzoni, Girolamo, 261n43
Berardi family, 260n28
Bergreen, Laurence, 276n6, 278n22
Bernáldez, Andrés, 26
Berry, Chu, 19
Berry, Steve, xx, 181, 247, 284n32
Bianco, Andrea, 40
Biden, Joe, 237
Billini, Francisco Javier, 63, 64, 67
biographies of CC, xviii–xix, 256nn5–7
  See also Columbiana; *specific biographers*
Bissipat, Georgi, 149
Black Americans
  anti-Columbus protests and, 224, 231, 238, 241
  anti-Italian/anti-immigrant prejudice and, 229–30
Black Death, 263n13
Black Legend, 59, 192–93, 200, 286n9
Black Lives Matter (BLM), 224, 231, 238
Bloy, Léon, 91, 113
Boabdil (Sultan Muhammad XI of Granada), 23, 24, 99, 201, 261n50
Bobadilla, Beatriz de, 128–30, 188, 278n22
Bobadilla, Francisco de, 32–33, 52, 278n22
Bolívar, Simón, 286–87n15
Bologna, xvi–xvii
Book of Marco Polo, The (Polo), 20, 21
Book of Privileges, 104
Book of Prophecies, 104–5, *105*, 273–74n27
Boorstin, Daniel, 18, 19
Boyd, William, xviii, 200, 287n19
Boyle, David, 259n28, 264n27

Brendan, Saint, *38*, 39
Broccardi, Eugenio, 114–15, 280n12
Brumidi, Constantino, 190, 191
Brunelleschi, Filippo, 261n43
Buchanan, James, 207, 210–11, 288n33
Buet, Charles, 113
Buigas i Monravà, Gaietà, *173*
Burdick, Clark, 211
Burton, Richard, 155

Cabot, John (Giovanni Caboto), 40, 107, 190, 264n24, 264n27
Calhoun, John C., 207–8
California, 208, 230, 233, 239, 240, 288n40
Calvi, *142*, 143–44, 145, 160, 279nn1,3, 280n5
Campbell, Lawton, 127, 277n17
Campe, Joachim, 108–9, 203, 287n26
Canary Islands, 22, 24, 37–38, 41, 42, 53
Canel, Eva, 168–69
cannibalism myth, 53–54, 56, 57, 130–31, 139
canonization campaign. *See* sainthood campaign
Canoutas, Seraphim G., 280n16
Cánovas del Castillo, Antonio, 175–76
Cão, Diogo, 14, 39
Caonabó, 49, 60–61
Capellano, Antonio, 285n5
Cape Verde Islands, 22
Capitol building, U.S., xiv, *186*, 187–91, 201, 203, 207–15, *210*, 220, 222, 249
Capitulaciones de Santa Fe, 33, 47–48, 52, 73–74, 78–79
  See also CC's titles
Carbajal, Luisa de (Luís Colón's wife), 134–35, 278n35
Carbia, Rómulo D., 283n24
Cardona y Colón, Cristóbal (Fifth Admiral), 77, 270nn31–32, 275n38
Cardona y Colón, María de, 78
Carpentier, Alejo, 90–91, 113, 127, 274n35
Casa di Colombo (Genoa), 255n3
Casa di San Giorgio (Genoa), 5

Casa Museo de Colón (Poio), 164
Casanova de Pioggiola, Martin, 113, 145, 160
Casenove, Guillaume de, 149
Caserio, Barnaba, 59, 267n51
Castelo de Soutomaior, xiii–xiv
Castile, 22, 23–24, 25, 261n50
Castilian War of Succession (1474–79), 22
Castro, Ana de (Diego Colón's wife), 134, 278n34
Castro, Marcial, 84–85, 86
Catalá i Roca, P., 283n18
Catalan Nationality (Prat de la Riba), 173
Catalonia, 172–74, *173*, 258n9, 283n18
Catarina de Bragança (Queen Catherine of England), 154, 281n26
Cateras, Spyros, 150
Catholic Columbus
  CC as agent of God and, 90
  Italian American Columbus and, 231, 232, 235
  sainthood campaign and, 110, 112, 113–14
Catz, Rebecca, 276n4
Causici, Enrico, 285n5
Caxton, William, 19
Columbus, Christopher (CC), as agent of God
  American Columbus and, 191, 192, 195, 200
  Black Legend and, 192–93, 200
  Book of Prophecies and, 104, 273–74n27
  CC's disgrace and, 102, 104, *105*
  CC's grandiosity and, 102–3, 104
  CC's shift to religious focus and, 26, 99–102, 273n23
  Columbiana and, 98, 100, 108–9, 273n18
  crusade obsession and, 64, 103–4, 267n1
  Franciscan influence and, 102, 103
  Historie della Vita on, 91, 93–94

Columbus, Christopher (CC), as agent of
    God (*continued*)
  images of CC and, *88, 107,* 271–72n2,
    274n31
  Indigenous people as subjects for con-
    version and, 45, 92, 96, 101
  Italian culture and, 107–8, *107*
  Las Casas on, 92, 96, 97–98, 100, 101,
    106–7, 192, 273n18
  millenarianism and, 103–4
  Oviedo and, 106
  Quincentennial and, 115
  social ambition and, 99–100, 105–6, 146
  Spanish monarchs on, 98–99, 102
  See also sainthood campaign
CC's campaign for Spanish royal support,
  15–25
  CC's lack of geographical knowledge
    and, 21–22, 25
  Columbus–Isabel affair myth and,
    127–28
  compensation demands and, 25
  Flat Earth myth and, 16–18
  Isabel delay theory, 261n52
  island lore and, 38
  name change and, 16
  political-economic context and, 22–24,
    25
  social ambition and, 15–16, 29
  Talavera Commission/Council of Sala-
    manca and, 20–22, 169
  visionary myth and, 16–17
CC's Genoese life, 1–27
  CC's concealment of, 25–27
  childhood, 3–4
  departure, 6–7
  family background, 2–3
  Genoa's natural harbor and, 2
  Genoese politics and, 6
  historical evidence on, xi, 1, 147,
    256–57n2, 280n12
  Mediterranean Atlantic system and, 5–6
  mystery myth and, 27, 249
  non-Genoese origin claims and, 144,
    145, 147, 148, 152, 153

CC's geographical misunderstandings
  American Columbus and, 194
  CC's campaign for Spanish royal sup-
    port and, 21–22, 25
  island lore and, 40
  João II Atlantic voyage pitch and, 14
  size of earth and, 20–21
  Talavera Commission/Council of Sala-
    manca and, 21–22
  Toscanelli letters and, 12–13
CC's personal characteristics
  chivalric ambition, 11, 38, 259n23
  grandiosity, 26, 64, 102–3, 105–6
  nepotism, 43–44
  obsessiveness, 31, 132
  violent discipline of subordinates, 55,
    266n45
CC's Portugal years
  Bartolomeo Colón and, 9, 43, 259n19
  departure, 14
  frustration, 14–15
  historical evidence on, 9–10
  João II Atlantic voyage pitch, 13–15
  Mediterranean Atlantic system and, 11–12
  pirate myth and, 8–9, 258–59nn17–18
CC's remains, *62,* 63–82
  CC as agent of God and, 108
  chains myth, 6, 267n2
  Columbiana and, 85, 271n42
  Columbus Lighthouse and, 70, *71,* 72
  cultural survival of, 72, 87
  debate on location of, 67–69, 72,
    268nn9,13, 269n17
  Genoa "ashes" vial, 69, *70,* 72,
    268–69n15
  Haitian Revolution and, 66–68
  Havana entombment, 67, 68, 82, 108
  initial placements of, 64–65
  mystery myth and, 86
  Santo Domingo tomb, 63–64, 65–66,
    67–68, 268n4, 269n17
  Seville tomb, 69, 72, 84–85, 165,
    269n17
CC's seafaring
  childhood myths, 3–4, 257n5

Iceland claim, 12
Madeira–Genoa–Lisbon voyage (1478),
　11–12
motivations for, 6
transport of enslaved people and, 12
See also Four Voyages
CC's sexuality, 119–41
　Columbus–Bobadilla myth and,
　　128–30, 278n22
　Columbus–Isabel affair myth, 125–28,
　　277nn15–16,18–19
　conquistador rape culture and, 130–31
　Enríquez relationship and, 122–25,
　　276nn7–10
　homosexuality possibility, 135–38,
　　279n40
　marriage and, 120–22
　sons/grandsons and, 131–35
　TikTok bestiality rumors, *118*, 119–20,
　　138–41
CC's social ambition
　campaign for Spanish royal support
　　and, 15–16, 29
　CC as agent of God and, 99–100, 105–6,
　　146
　concealment of origins and, 25–27
　Enríquez relationship and, 122–23, 125
　Historie della Vita and, 7–8, 146–47
　marriage and, 121–22, 125
　titles and, 29–30
CC's sons
　affection for, 125
　Beatriz Enríquez de Arana and, 124,
　　131–32
　CC's disgrace and, 33
　CC's Genoese life and, 26
　court placement of, 26, 121, 132, 167
　Iberian origin claims and, 167
　social ambition and, 16, 99
　See also Colón, Diego; Colón, Fernando
CC's titles
　acronymic signature and, *28*, 149–50,
　　171, 180, 262n2
　bestowal of, 29–30, 32, 152
　Colón family lawsuits and, 78
　Enríquez relationship and, 124–25
　Indigenous people and, 35
CC's voyages. See Four Voyages
Centurione, Lodovico, 7
Ceuta, 37
Champneys, Benjamin, 288n33
Charles II (king of England), 154
Charles V (Holy Roman Emperor), 45,
　75–76, 95, 272n2
children's literature, 108–9, 203–4, 247,
　287n26, 292n6
Chinese Exclusion Act (United States;
　1882), 290n18
chivalric romance, 11, 37–38, 259n23
Christian III (king of Denmark), 151
Christophe Colomb (Buet), 113
Christophe Colomb (Harrisse), xix
*Christopher Columbus, Being the Life of the
　Very Magnificent Lord Don Cristobal
　Colón* (Madariaga), 175, 176–78,
　283–84nn26–27
"Christopher Columbus" (Berry and
　Razaf), 19
*Christopher Columbus—en europeer fra
　Norge?* (Sannes), 150–51
Christopher Columbus Follies, The, 137–38
*Christopher Columbus and Queen Isabella of
　Spain Consummate Their Relationship,
　Santa Fé, January 1492* (Rushdie), 127
*Christopher Columbus's Love Letter to
　Queen Isabel* (Irizarry), 127
Christopher Comes Across, 137
Cocchia, Rocco, 63–64, 67
Colau, Ada, 285n38
Colomb, Jean, 147
Colombo, Andrea (CC's cousin), 44
Colombo, Antonio (CC's great-
　grandfather), 2
Colombo, Baldassare, 146, 270–71n32
Colombo, Bartolomeo (don Bartolomé
　Colón) (CC's brother), 5, 13, 20, 147,
　177, 188
　arrest of, 52
　CC's Portugal years and, 9, 43, 259n19
　First Voyage and, 43

Colombo, Bartolomeo (*continued*)
    Second and Fourth Voyages and, 46, 52
    monuments to CC and, 188
    remains of, 65–66, 85
Colombo, Bernardo, 271n32
Colombo, Bianchinetta (CC's sister), 258n7
Colombo, Cristoforo. *See* Columbus, Christopher
Colombo, Domenico (CC's father), 1, 3, 6, 257n5
Colombo, Giacomo (don Diego Colón) (CC's brother), 5, 43, 52
Colombo, Giovanni (CC's grandfather), 2–3, 4
Colombo, Giovanni Antonio (CC's cousin), 44
Colombo, Susanna Fontanarossa (CC's mother), 1, 4
"Colombo" Português (Barreto), 172
Colón, Cristóbal (Twentieth Admiral), 84, 128, 246
Colón, Diego (CC's son; Second Admiral)
    CC's marriage and, 9–10
    CC's social ambition and, 16
    Colón family lawsuits and, 72, 74, 75, 95, 269n21
    Colón palace and, 66
    death of, 75, 95, 269n24
    descendants of, 76
    Enríquez relationship and, 123–24
    relationship to Spanish monarchs, 121
    remains of, 64, 65, 67, 267n2
    sexuality of, 132–35
    See also CC's sons
Colón, Felipa, 77, 270n29
Colón, Fernando (CC's son)
    birth of, 123, 276n8
    CC as agent of God and, 104, 274–75n27
    CC's remains and, 65, 268n3
    childhood of, 276n10
    Colón family lawsuits and, 76, 95, 271n34
    Fourth Voyage and, 44
    library of, 91, 94, 131, 132, 182, 278n38
    lost writings of, 95–96, 272n12
    mother and, 111, 132, 275n38, 276n10
    obsessiveness of, 132
    remains of, 85, 165
    women and, 131–32
    See also CC's sons; *Historie della Vita*
Colón, Luís (CC's grandson; Third Admiral)
    Colón family lawsuits and, 75, 76
    Fernando Colón's library and, 94
    Historie della Vita and, 96–97
    marriages of, 133–35, 278nn34–35
    remains of, 64, 65, 67, 68
Colón, Mariano (Fourteenth Admiral), 67, 68, 82
Colón de Carvajal y Maroto, Cristóbal (Nineteenth Admiral), 84
Colón de la Cerda, Cristóbal (Sixteenth Admiral), 83, 160–61, 220
Colón de Portugal, Pedro Nuño (Eighth Admiral), 80, *80*
Colón de Pravia, Diego (Fourth Admiral), 76–77, 135, 270n29
Colón de Pravia, Francisca, 77, 78, 79, 81–82, 270n31
Colón family lawsuits, 72–84
    Capitulaciones de Santa Fe and, 73–74
    Columbus–Enríquez marriage myth and, 274–75n38
    consanguine marriages and, 77, 270n29
    descendants and, 76–80, 270–71nn31–32
    ETA murders and, 84
    first generation, 72–76
    immortality of, 79, 271n34
    imperialism/colonialism and, 73, 75, 269n23
    Jamaica and, 80–81
    Luís Colón's marriages and, 135, 278n35
    nineteenth century, 82–83
    non-Genoese origin claims and, 147, 148

social ambition and, 78–79
Spanish civil war and, 83–84
uniqueness myth and, 74–75
colonialism. *See* European expansion; imperialism/colonialism
Colón palace (Santo Domingo), 66, 268n6
Colón y Mosquera, Felipa, 35
Colón y Mosquera, María, 270n31
Columbiad, The (Barlow), 193
Columbiana
  biographies and, xviii
  CC as agent of God and, 98, 100, 108–9, 273n18
  on CC's disgrace, 33
  on CC's marriage, 10, 276nn4,6
  CC's modern irrelevance and, 249
  CC's remains and, 85, 271n42
  on CC's sexuality, 138
  cognitive dissonance and, 251–52
  coincidence and, 152–53
  Columbus–Isabel affair myth, 125–28, 277nn15–16,18–19
  confirmation bias and, 180
  conspiracy theories and, 86, 251
  defined, xi–xii, 255n1
  details and, 12, 13, 26, 58, 60, 147, 159, 172
  Enríquez relationship and, 124, 274n37, 275n38
  explanations of, 250–52
  faithistory and, xiv, xvi, xxi, 86, 144, 161, 164, 165
  fiction/history blurring and, xviii, xx, 180–81, 200
  First Landing and, 60
  Four Voyages and, 30
  fraud and, 153
  human remains and, xv, 166
  images of CC and, 271n2
  Italian American Columbus and, 108
  jewel-pawning myth, 126, *126*, 277n15
  Jewish origin myth, 172, 176–82, 277n19, 283–84nn24,26–27,31–32
  marginalization of Indigenous people and, 32, *34*, 35

multiple book versions and, 283n19
nationalism and, 167
overview, 249–50
patriotism and, 145–46, 213, 215, 269n17
pattern amalgamation and, 36–37
Pinzón brothers and, 264n26
Quadricentennial and, 58, 114
racist perceptions of Indigenous people and, *34*, 56–57, 263n10
as rough corner, xxii–xxiii, 251–52
sainthood campaign and, 110, 114, 116–17
science and, 164–65
social media and, *118*, 119–20, 140–41, 170
Spanish settlements and, 49–50
See also specific myths
Columbian Exchange, 140, 262n10
Columbian Exposition (Chicago, 1892), 83, 190, 204–6, 219–20, 235
Columbian Series stamps, *28*, *48*, 126, *126*, 189, 201, 206, 277n15
Columbia University, 197, 286n14
Columbus, Christopher (CC)
  acronymic signature of, *28*, 149–50, 171, 180, 262n2
  appearance of, xvii
  birth of, xi, 1
  crusade obsession, 64, 103–4, 267n1
  cultural survival of, xi, xii, xv, 72, 86–87
  death in neglect myth, 64
  death of, 64
  fame of, xi
  health issues, 31, 64, 267n1
  historical evidence on, xi, xix–xx, 9–10, 147, 199–200, 256–57nn9,2, 280n12
  marriage of, 9–10, 11, 120–22
  modern irrelevance of, 245–47, 248–49, 292n6
  name versions, xi, 16, 152
  wealth of, 33
Columbus Affair, The (Berry), xx, 247, 284n32

Columbus Day (United States)
  American Columbus and, 190, 195, 204, 205
  anti-Columbus protests and, 239
  Italian American Columbus and, 220–21, 222, 231, 232, 233, 290–91n22
Columbus Doors (Rogers) (U.S. Capitol), *186*, 188, 222
Columbus–Enríquez marriage myth, 111–12, *111*, 114, 115, 123, 274–75nn37–38
Columbus Fountain (Taft), 285n1
Columbus House (Porto Santo), 158
Columbus and the Indian Maiden (Brumidi), 191
"Columbusing" term, 248, 293n8
Columbus Lighthouse (Faro a Colón), 70, *71*, 72
Columbus's Egg myth, 261n43
Columbus's Jewish Roots (Amler), 181
confirmation bias, 180
Constantinople, Ottoman seizure of (1453), 5, 37
Corsica, *142*, 143–44, 145, 160, 279nn1,3, 280n5
Cortés, Hernando, 55, 107, 131
Cosa, Juan de la, 53, 54–55
Cristòfor Colom Fou Català (Ulloa), 174
Cronau, Rodolfo, 268n13
Crosby, Alfred, 140, 262n10
Cuban wars of independence, 69
Cuneo, Michele da, 61
Cuneo, Michele de, 46, 129–31, 264n28
Cuomo, Andrew, 221

D'Aguiar, Fred, 61
d'Ailly, Pierre, 20, 21, 103
Dante Alighieri, 107
Dark Ages myth, xxi, 16–21, 110, 201
David, Maurice, 284n29
Davidson, Miles H., 256n6, 259n28, 276n6, 281n25
de Blasio, Bill, 221
De Jure Maritimo (Molloy), 154, 277n15, 281n25

Delaney, Carol Lowery, 273nn18,23
dell'Altissimo, Cristofano, 255n3
d'Esmenaut, Jean (João Esmeraldo), 156, 158
determinism. *See* responsibility myth
Deza, Diego, 136–37, 279n40
Día de la Hispanidad (Spain), 86, 176
Diario (Columbus), 31, 45, 50, 59, 96, 101, 262n5
Dias, Bartolomeu, 13, 14–15, 39
Dias, Vicente, 39
Díaz de Isla, Ruy, 140, 279n46
Díaz de Solis, Juan, 46
Dickens, Charles, 248
Dinegro, Paolo, 7
Discoverers, The (Boorstin), 18
Discovery of America (Persico), *186*, 188, 207–8, 210–15, *210*, *212*, 284n29, 288nn33,38
DNA evidence
  CC's remains and, 85, 86, *162*
  non-Genoese origin claims and, 164–65, 167, 170, 174, 181
Doctor Who, 247
Dogs of Paradise (Posse), 180
Dole, Bob, 213
Dominican Republic, 63, 69–70, *71*, 72, 184, 283n21
Drake, Francis, 65
Duarte (king of Portugal), 171
Dueñas Judgment (1534), 75
Dugard, Martin, 264n26

Elmina, 12
England, 176, 191, 194, 286n15
  Atlantic voyaging and, 35, 40–41, 59, 264n24
  Madeira and, 153–55, 159–60
  monarchs of, 40, 59, 81, 281n26
Enneads (Sabellicus), 8
Enríquez de Arana, Beatriz
  CC's sexuality and, 122–25, 276nn7–10
  CC's sons and, 124, 131–32
  Jewish origin myth and, 177
  marriage myth, 111–12, *111*, 114, 115, 274–75nn37–38

Enríquez de Arana, Rodrigo, 123, 276n7
enslavement
　cannibalism myth and, 53–54
　CC's worldview and, 53
　First Voyage and, 51, 52
　Granada conquest and, 24, 53
　imperialism/colonialism and, 47, 51, 55
　Indigenous systems of, 52–53, 265–66n39
　mortality and, 60–61
　Pinzón brothers and, 46
　Santángel letter on, 54
　Second Voyage and, 53
　Spanish monarchs and, 52
　Spanish system of, 54–55, 266n43
　sugar plantation economy and, 41
　transatlantic slave trade, 60–61, 194
　transport of enslaved people, 12
Entdeckung von Amerika, Die (Campe), 108–9
Erikson, Leif, 150, 236, 237, 291n27
Escobar, Mario, 128
Esmeraldo House (Madeira), 156–57
Esteban, Eduardo, 170
ETA, 84
eugenics, 229
European expansion
　CC's worldview and, 20
　nepotism and, 44
　pattern amalgamation and, 37
　Portuguese–Castilian competition and, 41
　timespan of, 263n13
　See also Atlantic voyaging; imperialism/colonialism; responsibility myth

Fairchild, Lucius, 255n3
faithistory, xiv, xvi, xxi, 86, 116, 144, 161, 164, 165, 249
Fava, Francesco Saverio, 219
Fernandes, Gonçalo, 39
Fernández-Armesto, Felipe, xiv
　on CC as agent of God, 273n24
　on CC's campaign for Spanish royal support, 127–28
　on CC's grandiosity, 105–6
　on CC's marriage, 122, 276nn4,6
　on CC's sexuality, 135
　on images of CC, 271–72n2
　on Jewish origin myth, 176–77
　on lionization of, CC, 32
　nonpartisan approach of, xviii–xix
　on Toscanelli letters, 259n28
　on uniqueness myth, 41
Fernando (king of Aragon). *See* Spanish monarchs
fictography, 200, 287n19
First Crusade, 20
First Landing, xviii, 31, 60, 195, 233
　See also First Voyage
First Voyage
　CC as agent of God and, 99, 101
　celebratory responsibility myth and, 58–59
　Columbiana and, 30
　commemorations of, *28*
　enslavement and, 51, 52
　expulsion of Jews and, 178
　funding for, 15
　Hispaniola settlement and, 49, 52
　Indigenous people and, 49
　length of, 31
　mermaid/siren mythology and, 139
　navigation and, 42
　nepotism and, 43
　Pinzón brothers and, 36, 44–45
　Unknown Pilot myth, 35–37, 263n12
　See also Diario
Flanagan, Peggy, *239*
Flanders, Michael, 154, 281n23
Flat Earth myth
　CC's campaign for Spanish royal support and, 16–18
　CC's lack of knowledge and, 19–20
　early United States and, 18, 260n38
　popular culture and, 19, 260n40
　Talavera Commission/Council of Salamanca and, 21–22, 201
　untruth of, 19
　visionary myth and, xxi
Fleres, Ugo, 218
Flint, Valerie I. J., 278n38

Florence, 258n9
Florescano, Enrique, 184
Flower, Roswell, 218
Forcia, Mike, *239*
1492: Conquest of Paradise (Scott), 18, 19
Fourth Voyage, 32, 44, 46, 64, 105
Four Voyages
  Capitulaciones de Santa Fe and, 47–48
  CC's disgrace and, 32–33, 52, 74, 102, 104, *105*, 128
  CC's title and, 29–30, 32
  collaboration and, 46, 264n28
  Columbiana and, 30
  commemorations of, *28*
  First Landing, 31, 60
  Las Casas on, 31, 262n5
  lionization of CC and, 32
  nepotism and, 43–44
  ocean as antagonist and, 29, 30
  Pinzón brothers and, 36, 44–46, 75, 264n26
  settlements and, 49–50, 52
  timespan of, 31
  Unknown Pilot myth, 35–37, 263n12
  See also First Voyage; responsibility myth
France
  Genoese factional politics and, 6
  non-Genoese origin claims and, 143–49, 153, 167
  sainthood campaign and, 91, 109–15
Franco, Francisco, 84, 176, 183
Franzoni, Domenico, 148
Freneau, Philip Morin, 193–94
Frieze of American History (Brumidi), 190, *190*
Fuentes, Carlos, 184

Gagliano, Anthony, 213
Gallego, Pedro, 266n45
Gallego origin claims, xiii–xiv, xv, *162*, 163–65, 168–70, 179, 283n24
Gallo, Antonio, 259n19
Gamboa, Isabel de, 132, 133
García de la Riega, Celso, 168–69, 179, 282n6, 283n24, 284n29

Garibaldi, Giuseppe, 108
gender, 131, 191
Genoa, *3*
  Castilian War of Succession and, 22
  cheese mongers in, 16, 245–46, 258n7
  CC's birth in, 1
  CC's invisibility in, 245–46
  CC's remains and, 69, *70*, 72, 268–69n15
  Corsica and, 144, 146
  European expansion and, 37
  factional politics, 6
  Granada conquest, 24
  Italian immigration to United States and, 231
  Mediterranean Atlantic system and, 5–6
  mercantile empire of, 5
  monument to CC, *xxvi*, 1, *2*, *4*
  natural harbor of, 2
  nepotism and, 44
  sainthood campaign and, 114–15
  See also CC's Genoese life
genocide
  defined, 266n48
  Spanish colonialism as, xxii, 48, 266–67n48
  See also responsibility myth
Geography (Ptolemy), 20, 21
George Washington (Greenough), *186*, 187–88, 285n1
George Washington University, 286n14
Geronimi, Victor, 280n5
Gershwin, Ira, 260n40
Ghirlandaio, 255n3
Giustiniani, Agostino, 272n11
Gleave, Joseph Lea, 70
González, Felipe, 183
Gorricio, Gaspar, 104, 274–75n27
Granada conquest, 22, 23–24, 25, 53, 98–99, 102, 201, 261n50
Gran Colombia, 197, 286–87n15
great men theory of history, 200, 201–2
Greek origin claims, 8–9, 148–50, *150*, 258nn17–18, 280n16

Greenough, Horatio, *186*, 187–88, 208, 285n11
Grodys, Svein-Magnus, 151
Guacanagarí, 49
Guerra, Cristóbal, 54–55
Gulf Stream, 43
gyres and currents. *See* navigation

Haitian Revolution (1795), 66–68
Hakluyt, Richard, 191
Harp and the Shadow, The (Carpentier), 90–91, 113, 127, 274n35
Harrison, Benjamin, 204, 218, 226–27
Harrisse, Henry
   American Columbus and, 233
   on CC's marriage, 121, 122
   on CC's remains, 271n42
   on *Historie della Vita*, 93
   on Irving, 256n7
   on non-Genoese origin claims, 144, 145, 146, 279n3
   nonpartisan approach of, xix, 203
Hatuey, 286n9
Havana, entombment in, 67, 68, 82, 108
Heaton, A. G., 285n6
Henige, David P., 256n6
Hernandez, Zeke, 290n18
Herrera, Antonio de, 16, 147, 275n38, 277n15
Hill, Sidney, 224
Historia de las Indias (Las Casas). *See* Las Casas, Bartolomé de
Historia de las Indias (López de Gómara), 263n12
Historia del Mondo Nuovo (Benzoni), 261n43
Historia del Nuevo-Mundo (Muñoz), 200, 287n20
Historia general y natural de las Indias (Oviedo), 35–36, 106
Historia Rerum ubique Gestarum (Pius II), 20
Historie della Vita (attr. Colón), 256n5
   American Columbus and, 194
   authorship of, 94–95, 96, 272nn7,11, 276n4

   on CC as agent of God, 91, 93–94
   CC's concealment of origins and, 26
   on CC's Iceland voyage, 12
   on CC's marriage, 11, 121, 259n24, 276n4
   CC's sexuality and, 120, 274n4
   CC's social ambition and, 7–8, 146–47
   on Enríquez relationship, 276n10
   fiction/history blurring and, xviii
   inaccuracies in, 93, 94–95, 97
   influence of, 92–93
   Irving and, 199
   jewel-pawning myth and, 277n15
   Las Casas and, 96
   Luís Colón and, 96–97
   non-Genoese origin claims and, 146–47, 148, 149, 150, 280n8
   pirate myth and, 8–9, 149
   Plutarch and, 278n38
   portraits of CC and, 255n2
   sainthood campaign and, 91, 93–94
   on Toscanelli letters, 259–60n28
   uniqueness myth and, 97
History of the Life and Voyages of Christopher Columbus (Irving), 189
   See also Irving, Washington
Hogarth, William, 261n43
Holocaust, 179, 237, 266n48
Horwitz, Tony, xix
Huevo de Colón (Seville), 261n43
Humboldt, Alexander, 267n51

Iardella, Francesco, 285n5
Iberian origin claims, 163–85
   Catalan, 172–74, *173*, 283n18
   DNA evidence and, 86, *162*
   Gallego, xiii–xiv, xv, *162*, 163–65, 168–70, 179, 283n24
   human remains and, 165–66
   Jewish origin myth and, 178, 182, 283n24
   Majorca, 174, 283n19
   nationalism and, 167–68, 170–72, 175–76

Iberian origin claims (*continued*)
    Quincentennial and, 174–75, 182–85
    sainthood campaign and, 110
Ibiza, 283n19
Iceland, 12, 40
Iceland myth, 12
Image du monde, 19
images of CC
    American Columbus and, *186*, 187–88, 189–90, *189*, 285n1
    CC as agent of God and, *88*, *107*, 271–72n2, 274n31
    Columbiana and, 271n2
    imagined nature of, xvi–xvii, *xvii*, 255nn2–3
    Irving and, 201
    Jewish origin myth and, 180, 284n31
Imago Mundi (d'Ailly), 20, 103
Immigration Acts (United States), 225, 290–91n10
Immoral Isabella? (Campbell), 127
imperialism/colonialism
    American Columbus and, 189, 190, 198–99
    Atlantic voyaging and, 41, 47, 50
    Black Legend on, 59, 192–93, 200, 286n9
    bottom-up structure of Spanish, 73
    Capitulaciones de Santa Fe and, 47–48, 52
    Castilian War of Succession and, 22, 24
    CC's worldview and, 24, 47, 53
    Colón family lawsuits and, 73, 75, 269n23
    enslavement and, 47, 51, 55
    as genocide, xxii, 48, 266–67n48
    Granada conquest and, 24, 53, 98–99, 102
    human remains and, 165–66
    Italian American Columbus and, 222
    Las Casas on, 92, 96, 192
    manatee stories and, 139–40
    Manifest Destiny, 208–11, 215
    Mexico and, 261n51
    millenarian view and, 98–99, 102–3
    non-Genoese origin claims and, 166–67
    protests against, 183–85, 206, 285n38

    Quincentennial and, 183–85
    responsibility myth and, 51–52, 57–58, 60, 206
    Santángel letter and, 51
    settler colonialism, 57–58
    Spanish monarchs and, 52, 265n38
    tribute system and, 50, 52, 57–58, 269n23
    U.S. racism and, 230
    violence as intrinsic to, 58
    See also racist perceptions and portrayals of Indigenous people
Indian Removal Act (1830), 210
Indian Reorganization Act (1934), 211
indigenismo movement, 237
Indigenous people
    alternate discovery claims and, 237
    American Columbus and, 187–88, 189, 198–99
    anti-Columbus protests and, 195, 224, 238–39, *239*
    Black Legend and, 193, 286n9
    cannibalism myth and, 53–54, 56
    CC's "presentation" of, *48*, 55
    CC's violence against, 49, 55, 57, 58, 59
    Columbiana marginalization of, 32, *34*, 35
    Columbiana portrayals of, *34*, 56–57, 263n10
    Columbus Day and, 195
    enslavement of, 51, 52, 54–55
    enslavement systems of, 52–53, 265–66n39
    importance to Four Voyages of, 50
    indios term for, 49, 265n32
    manatee stories and, 139, *139*
    monuments to CC and, *56*, 187–88, 208, 211, 222, 288n38
    monuments to, 185
    rape of, 130–31
    as source of wealth, 50–51, 53
    Spanish colonialism as genocide and, 48, 266–67n48
    as subjects for conversion, 49, 51, 92, 96, 101, 266n43

tribute system and, 50, 52, 57–58,
    269n23
  See also racist perceptions and portrayals
    of Indigenous people
Inspiration of Christopher Columbus, The
    (Obregón), 18
internet, xx, 86, 140, 250–51, 291n24
Irish immigrants/Irish Americans, 195,
    227, 231–32, 236
Irizarry, Estelle, 127, 277nn18–19
Irving, Washington, *202*
  alternate discovery claims and, 236
  American Columbus and, 189,
    199–201
  anti-Spanish views and, 193
  on CC's disgrace, 33
  Flat Earth myth and, 260n38
  Harrisse on, 256n7
  historical evidence and, 199–200
  on *Historie della Vita*, 92–93
  influences on, 191–92, 199, 287n20
  monuments to CC and, 222
  mystery myth and, 199, 287n17
  non-Genoese origin claims and, 148,
    175, 283n21
  responsibility myth and, 59
Isabel (queen of Castile)
  Columbus affair myth, 125–28,
    277nn15–16,18–19
  jewel-pawning myth, 126, *126*,
    277n15
  royal support delay and, 261n52
  sainthood campaign for, 128, 176
  See also Spanish monarchs
island lore, 34, *34*, 35, 37–38, *38*, 39–41, 53,
    263n16
Italian American Columbus, 217–43
  American Columbus and, 233, 235
  anti-Columbus protests and, 220–22,
    223–25, 237–41, 291–92n34
  anti-Italian/anti-immigrant prejudice
    and, 219, 220, 224, 226–30, 232,
    240, 289–90nn9–10,14,18
  assimilation and, 230–31, 240–42
  Columbiana and, 108

Columbus Day and, 220–21, 222, 231,
    232, 233, 290–91n22
  imperialism/colonialism and, 222
  Knights of Columbus and, 231–32, 233,
    291n22
  Latino Americans and, 241–43,
    292nn37–38
  migration experience and, 218, 225–26,
    226, 227
  monuments to CC and, *216*, 217–19,
    220–24, *223*, 233, *234*, 285n1
  Quadricentennial and, 217–18
  Spanish–Italian competition and,
    219–20
Italy
  CC as agent of God and, 107–8,
    107
  non-Genoese origin claims and, *142*,
    143–44, 145, 146–47, 148
  unification of, 63–64, 108
  See also CC's Genoese life; Genoa

Jamaica, 66, 80–81, 181
Jean de Béthencourt, 37
Jefferson, Thomas, xvi
jewel-pawning myth, 126, *126*, 277n15
Jewish origin myth, 86, 172, 176–82,
    277n19, 283–84nn24,26–27,31–32
Jewish Pirates of the Caribbean (Kritzler),
    181
João II (king of Portugal), 13–15,
    267n51
John of Holywood, 260n41
John of Mandeville, 19, 20
John Paul II (Pope), 184
Joseph, Jacob, 179
Jovellanos, Gaspar Melchor de, 82
Joven de Amajac, La, 185
Juan Carlos I (king of Spain), 183
Julius II (Pope), 23
Junior, Pestana, 171

Kensington Stone, 236
King, William Rufus, 288n33
Kinstler, Linda, 266n48

Knights of Columbus, 231–32, 233, 285n1, 291n22
Kritzler, Edward, 181
Ku Klux Klan, 224, 229–30, 289n9

La Coruña Judgment (1520), 75
Landing of Columbus (Vanderlyn), 189–90, *189*, 218, 249, 263n10
Landrieu, Moon, 213
Larreátegui, Pedro Isidoro de, 82
Las Casas, Bartolomé de
  background of, 91–92
  Black Legend and, 192, 200
  on CC as agent of God, 92, 96, 97–98, 100, 101, 106–7, 192, 273n18
  on CC's João II Atlantic voyage pitch, 13
  on CC's marriage, 11, 121–22, 158
  on CC's violence against Indigenous people, 55, 59
  Colón family lawsuits and, 269n23
  on enslavement, 54
  on Four Voyages, 31, 262n5
  Historie della Vita and, 96
  on imperialism/colonialism, 92, 96, 192
  on Indigenous people as subjects for conversion, 92, 96
  Irving and, 199
  on island lore, 263n16
  Jewish origin myth and, 177
  on mermaid/siren mythology, 139
  non-Genoese origin claims and, 155–56, 158
  Oviedo and, 106–7
  on Pinzón brothers, 45
  on pirate myth, 9
  on settlements, 49
  on Toscanelli letters, 259–60n28
  on tribute system, 57
  on Unknown Pilot myth, 35
Latin America, 197, 198, 286–87n15
Latini, Brunetto, 19
Latino Americans, 241–43, 292nn37–38
Leif Erikson Day, 237, 291n27

Lemkin, Raphael, 266n48
Leo XIII (Pope), 89, 114, 115
León-Portilla, Miguel, 184
Lewis, C. Douglas, 214
Li Livres dou Trésor (*Book of Treasure*) (Latini), 19
Lincoln, Abraham, 256n4
Little Mermaid, The (Andersen), 138
Lives of the Catholic Heroes and Heroines of America, 112
Lives of the Noble Greeks and Romans (Plutarch), 135–36, 278n38
Lopez, Barry, 59
López, Oriol, 285n38
López de Gómara, Francisco, 139, *139*, 246, 263n12
Lorente Acosta, José Antonio, 85, 86, *162*, 164, 174
lynching, 219, 229–30

Machado Roma, Carlos, 144
Madariaga, Salvador de, 26, 175, 176–78, 180–81, 185, 283–84nn26–27
Madeira/Porto Santo, 153–60, *159*, 281n35
Magalhães, Fernão de (Magellan), 46, 147, 267n51
Magnalia Christi Americana (Mather), 191
Majorca, 174, 283n19
Malaga, 23–24
Maldonado de Talavera, Rodrigo, 20–21
Malocello, Lancelotto, 37
manatees, *118*, 119–20, 138–41, *139*
Manifest Destiny, 208–11, 215
Marconi, Guglielmo, *234*
Mariño de Soutomaior, Xohán, 163
Marlowe, Stephen, 180–81
Martínez, Leonor, 132
Martorell, Joanot, 37–38, 259n23
Marvin, Rolland, 222
Massachusetts, 58, 194, 209, 213, 236
Mather, Cotton, 191
Mathews, George, 206–7
Means, Russell, 237–38

Medici portrait (dell'Altissimo), xvi, 255n3
Medieval Warm Period, 263n13
Memoirs of Christopher Columbus (Marlowe), 180–81
Mentz, Steve, 262n10
mermaid/siren mythology, 138–39
Mexican-American War (1846–48), 193, 207, 208
Mexico, 75, 80, 81, 83, 107, 133, 184–85, 235, 237, 261n51, 283n21
micropatriotism, xxii, 197–98
   non-Genoese claims and, 142–75, 185, 251
Milan, 6
Miranda, Francisco de, 286n15
Mirrour of the World, The (Caxton), 19
Molloy, Charles, 154, 277n15, 281n25
Moniz, Filipa (CC's wife), 10, 65, 120–21, 125, 259n20
Monument al Descobriment d'Amèrica (Ibiza), 261n43
monuments to CC
   American Columbus and, *186*, 211–12, 285n1
   anti-Columbus protests and, 184–85, 211–12, 220–21, 223–24, 231, *239*, 288nn38,40, 289n9
   Columbus's Egg myth and, 261n43
   Discovery of America (Persico), *186*, 188, 207–8, 210–15, *210*, *212*, 284n29, 288nn33,38
   Genoa, *xxvi*, 1, 2, 4
   George Washington (Greenough), *186*, 187–88, 285n1
   Indigenous people and, *56*, 187–88, 208, 211, 222, *223*, 288n38
   Italian American Columbus and, *216*, 217–19, 220–24, *223*, 233, *234*, 285n1
   Quadricentennial and, 2
   Santo Domingo, *88*
   U.S. number of, 289n8
Morison, Samuel Eliot
   on CC's marriage, 11, 276n4
   on CC's sexuality, 274n4
   on Columbiana, 18
   on *Historie della Vita*, 93
   jewel-pawning myth, 277n16
   on Jewish origin myth, 284n27
   on Unknown Pilot myth, 36
Morton, Levi, 190
Morton, Thomas, 286n13
Mosquera, María de (Diego Colón's wife), 133, 134
Mujeres que Lucha, Las, 185
Muliart, Miguel, 266n45
Muñoz, Juan Bautista, 200, 287n20
Museu Biblioteca Mário Barbeito de Vasconcelos (Madeira), 157
Mussolini, Benito, 222, 241
mystery myth, xi–xii, xx–xxi
   American Columbus and, 199
   CC's concealment of origins and, 25–27
   CC's Genoese life and, 27, 249
   CC's remains and, 86
   fiction/history blurring and, xx
   Irving's perpetuation of, 199, 287n17
   Jewish origin myth and, 180, 181
   non-Genoese origin claims and, 146, 148, 152–53, 163, 164–65, 167, 170, 172
   untruth of, xix, 256–57nn9,2
   See also Columbiana
mythistory, xx
   See also Columbiana; *specific myths*

Nader, Helen, 273n24
National Hispanic Quincentennial Commission, 175
nationalism
   non-Genoese origin claims and, 166–68, 172, 175–76
   Quincentennial and, 182–84, 246
   Spanish monarchs and, 176
National Museum of the American Latino, 242, 292n37
National Origins Act (1924; United States), 229, 290n18
Navarrete, Martín Fernández de, 199, 200

navigation, 38–39, 42–43, 60, 264n24
Newark, New Jersey, 233, 241
Newman, George, 228
Night at the Museum, 246
Nine Arguments in Defense of Christopher Columbus, The: The Untold Truth (Roncari), 115
Nixon, Richard M., 290n22
non-Genoese origin claims, 143–61
    Calvi, *142*, 143–44, 145, 160, 279nn1,3, 280n5
    Catalan, 172–74, *173*, 283n18
    Colón family lawsuits and, 147, 148
    Cuccaro, 146, 147
    Denmark, 151
    France, 145, 147, 148
    Gallego, xiii–xiv, xv, *162*, 163–65, 168–70, 179, 283n24
    Greece, 8–9, 148–50, *150*, 258nn17–18, 280n16
    Greek pirate myth, 8–9, 148–49, 258nn17–18
    Historie della Vita and, 146–47, 148, 149, 150, 280n8
    imperialism/colonialism and, 166–67
    Madeira/Porto Santo, 153–60, *159*, 281n35
    motivations for, 153
    mystery myth and, 146, 148, 152–53, 163, 164–65, 167, 170, 172
    nationalism and, 166–68, 172, 175–76
    Norway, 150–51
    number of, 144, 279n3
    patriotism and, 145–46, 153, 167, 172
    Portugal, 170–72, 282n15
    tourism and, 159–60, *159*, 169, 170, 281n35
    United States and, 175, 283n21
    See also Iberian origin claims
Norway, 235, 264n24
Nova typis transacta navigation (Philiponus), *38*

Obregón, José María, 18
Ocean Sea. *See* Atlantic voyaging
O'Connor, James, 211

Odoardi, Giovanni, 115
O'Farrell, Mitch, 238, 239, 240
Ojeda, Alonso de, 61
Ojibwe, 239
O Mistério Colombo Revelado (Rosa), 172
Ornelas, Agostinho de, 157
Orozco, María de, 133, 278n34
Ottoman Empire, 5, 37
Oviedo, Gonzalo Fernández de, 35–36, 55, 57, 75, 106, 266n43

Padilla Guevara, Luís, 277n18
Pagden, Anthony, 59
Parry, J. H., 264, 264, 264n19
Passing of the Great Race, The, 229
Penn, William, 65–66, 81
Peraza, Hernando, 129
Perestrelo, Bartolomeu, 10–11, 120–21, 122, 259nn22,24
Perry, Bliss, 248
Persico, Luigi, *186*, 188, 207–8, 210–15, *210*, *212*, 284n29, 288nn33,38
Peter Martyr (Pietro Martire), 32, 139, *139*, 191
Philadelphia, 204, 205, 226, 233, *234*, 240, 248
Philiponus, Honorius, *38*
Phillips, Carla Rahn, 256n5, 259n28, 265nn29,38, 276n6
Phillips, William D., Jr., 256n5, 259n28, 265nn29,38, 276n6
Piniés, Jaime de, 126
Pinzón, Arias, 46
Pinzón, Francisco Martín, 44, 46
Pinzón, Juan, 46
Pinzón, Martín Alonso, 36, 44, 45, 75, 101, 264n26
Pinzón, Vicente Yáñez, 44, 45–46
Pinzón brothers, 36, 44–45, 75, 101, 264n26
Piombo, Sebastiano del, xvi, *xvii*, 255n2
pirate myth, 8–9, 148–49, 258nn17–18
Pius II (Pope), 20
Pius IX (Pope), 109, 112, 115, 274n35
Pledge of Allegiance, 204

Plutarch, 135–36, 278n38
Poio, *162*, 164
Polk, James, 209–10
Polo, Marco, 20, 21
Ponce de León, Néstor, 189–90
Pontevedra origin myth, 168–69, 171
portraits, 190
   of Beatriz Enríquez, *111*
   of CC, xvi–xviii, *xvii*, *17*, *48*, *88*, 90, *105*, *107*, *150*, 180, *189*, 190, *190*, *196*, 256nn2–3
   of Eighth Admiral, *80*
Portugal
   African coastal trade, 12, 22, 37
   Atlantic voyaging and, 38–40, 42
   Castilian War of Succession and, 22
   non-Genoese origin claims, 170–72, 282n15
   See also CC's Portugal years
Portugal y Ayala, Pedro Nuño II de (Tenth Admiral), 81
Portugal y Colón, Álvaro de, 78, 270n31
Portugal y Colón, Catalina de (Eleventh Admiral), 81
Portugal y Córdova, Jorge Alberto de, 270n32
Portuguese Columbus myths. *See* Iberian origin claims
Posse, Abel, 180
postage stamps, 1, *2*, *28*, 48, *48*, 126, *126*, 189, 201, 206, 277n15
postmodernism, xx
Praça de Colombo (Madeira), 158
Prat de la Riba, Enric, 173
Prescott, William, 175, 193, 201, 202, 283n21
Preston, William C., 207–8
Provost, Foster, 255n1
pseudoarchaeology, 291n24
Ptolemy, 20, 21
Purchase, Samuel, 191

Quadricentennial (1892)
   American Columbus and, 201, 204–5
   CC's remains and, 69
   children's socialization and, 204
   Flat Earth myth and, 201
   Iberian origin claims and, 168
   images of CC and, xvi
   Italian American Columbus and, 217–18
   jewel-pawning myth and, 126, *126*, 277n15
   Jewish origin myth and, 179
   monuments to CC and, *2*
   non-Genoese origin claims and, 155, 157, 172–73
   responsibility myth and, 58
   sainthood campaign and, 89–90, 113, 114
Quincentennial (1992), 257n4
   alternate "discoveries" and, 291n24
   American Columbus and, 206, 213–14, 215
   anti-Columbus protests and, 237–38
   CC homosexuality possibility and, 137–38
   CC's remains and, 70
   commemoration conflicts, 174–75
   First Voyage commemorations and, *28*
   Flat Earth myth and, 18–19
   imperialism/colonialism and, 183–85
   jewel-pawning myth and, 126, *126*
   non-Genoese origin claims and, 158, 172, 174–75, 182–85
   sainthood campaign and, 115
   Spanish nationalism and, 182–84, 246
   visionary myth and, 18–19
Quintero, Cristóbal, 44

racist perceptions and portrayals of Indigenous people, 48
   alternate "discoveries" and, 235–36, 291n24
   American Columbus and, 187–88
   Columbiana and, *34*, 56–57, 263n10
   gender and, 131, 187
   Indigenous people as subjects for conversion and, 51
   island lore and, 35, 50
   monuments to CC and, *56*, 187–88, 208, 222, *223*

rape, xv, 55, 59, 130–31, 278n27
Raynal, Abbé, 192
Razaf, Andy, 19
Real Academia de la Historia (Spain), 199
Recall of Columbus, The (Heaton), 285n6
Regal, Brian, 291n24
Rescue, The (Greenough), 208, 211–12, 288n38
responsibility myth, xi, xxi–xxii
    Black Legend and, 59
    CC's "presentation" of Indigenous people and, 55
    CC's sexuality and, 137–38, 141
    CC's violence against Indigenous people and, 55–56, 59
    CC's violent discipline of subordinates and, 55, 266n45
    celebratory versions of, 58–59, 267n51
    imperialism/colonialism and, 51–52, 57–58, 60, 206
    Quincentennial and, 206
    syphilis and, 140, 279n46
    tribute system and, 57–58
    untruth of, xxii, 47, 51–52, 60, 61
    See also anti-Columbus protests
Révélateur du Globe, Le: Christophe Colomb et sa Béatification Future (Bloy), 113
Ribeiro, Patrocínio, 171
Robertson, William, 192, 193, 200
Rocca, Al M., 264n26
Rogers, Randolph, 188, 222
Romanticism, 126, 148
Roosevelt, Franklin Delano, 229
Rosa, Manuel, 172
Roselly de Lorgues, Comte de (Antoine François Félix Valalette), 109–13, 115, 274n35, 275n38
Roth, Cecil, 178
Rushdie, Salman, 127
Russell, Jeffrey Burton, 19, 200, 260n40, 286n9
Russell, William, 192, 193
Russo, Gaetano, *216*, 218

Sabellicus, 8
Sacco, Nicola, 228
sainthood campaign, 109–17
    Catholic Columbus and, 110, 112, 113–14
    Columbiana and, 110, 114, 116–17
    Columbus–Enríquez marriage myth and, 111–12, *111*, 114, 115
    fiction on, 90–91
    Genoa and, 114–15
    great men theory of history and, 202
    Historie della Vita and, 91, 93–94
    Italian American Columbus and, 232
    Quadricentennial and, 89–90, 113, 114
    Quincentennial and, 115
    Roselly de Lorgues and, 109–13, 115, 274n35, 275n38
    See also CC as agent of God
Saint-Méry, Moreau de, 66
Salamanca Council/Talavera Commission, 16–17, *17*, 20–22, 169, 201, 260n38
Sale, Kirkpatrick, 259n28
Sanguineti, Angelo, 114
Sannes, Tor Borch, 150–51
Santa María de las Cuevas (Seville), 64–65, 269n17
Santana e Vasconcelos, João de, 158
Santángel, Luís de, 277n15
    See also Santángel letter
Santángel letter (1493), *34*
    on enslavement, 54
    enslavement and, 51, 54
    on *indios* term, 265n32
    misinformation and, 265n36
    on motivations for Atlantic voyaging, 50
    non-Genoese origin claims and, 174
    Unknown Pilot myth and, 35
    wide dissemination of, 262–63n10, 265n34
Santo Domingo
    monument to CC in, *88*
    tomb in, 63–64, 65–66, 67–68, 268n4, 269n17

Sawyer, William, 209
Schussele, Christian, *202*
Scolvus, Joannes, 151, 172, 174
Scott, Ridley, 18, 19
Second Voyage, 31, 42, 43, 49, 53, 55
settler colonialism, 57–58
Seville
    Castilian War of Succession and, 22
    tomb in, 69, 72, 84–85, 165, 269n17
Seville Judgment (1511), 74, 95
Shawnee people, 198–99
Shea, John Gilmary, 113–14
Sheinbaum, Claudia, 185
Shimoni, Youval, 284n32
ship technology, 6, 42
siren/mermaid mythology, 138–39
Six Years' War (1868–74), 63
social media, xv, *118*, 119–20, 138–41, 170
Society of Tammany, 195
Solari, Juan, 168
Spain
    Atlantic voyaging and, 43
    Black Legend on, 59, 192–93, 200, 286n9
    CC's nepotism and, 44
    European expansion and, 44
    Genoese factional politics and, 6
    Granada conquest, 22, 23–24, 25, 53, 98–99, 102, 201, 261n50
    Mexico conquest, 261n51
    patriotism, 36
    Quincentennial and, 182–84, 246
    See also CC's campaign for Spanish royal support; Iberian origin claims; imperialism/colonialism; Spanish monarchs
Spanish-American War (1898), 69, 83, 193
Spanish Civil War, 83–84, 176
Spanish monarchs
    bestowal of CC's titles and, 32
    Black Legend and, 192
    CC as agent of God and, 98–99, 102
    CC's disgrace and, 33, 102, 104, *105*, 128
    CC's marriage and, 121, 122
    CC's "presentation" of Indigenous people to, *48*, 55
    Columbus–Isabel affair myth, 125–28, 277nn15–16,18–19
    crusade ambitions and, 103
    enslavement and, 52
    Granada conquest and, 22, 23–24, 25, 53, 98, 102, 201, 261n50
    great men theory of history and, 201, 202
    Hispaniola settlement and, 49
    imperialism/colonialism and, 52, 265n38
    Irving on, 201
    nationalism and, 176
    non-Genoese origin claims and, 173, 174
    Pinzón brothers and, 45
    See also CC's campaign for Spanish royal support
Spanish origin claims. *See* Iberian origin claims
Sphere, The (John of Holywood), 260n41
"Star-Spangled Banner, The," 197
Stewart, George, 212–13
Stigliani, Tommaso, 107–8
Stradano, Giovanni, 107–8, *107*, 274n31
Stuart y Silva, Carlos Fernando (Thirteenth Admiral), 81–82
sugar plantation economy, 41
Sumner, Charles, *209*, 213
Suñol, Jerónimo, 220
Swann, Donald, 281n23
syphilis, 140, 279n46
Syracuse, New York, 222–24, *223*

Taft, Lorado, 285n1
Taíno people, 49, 52–53, 55, 60–61, *139*, 265–66n39, 286n9
    See also Indigenous people
Talavera, Hernando de, 20, 260–61n43
    See also Talavera Commission/Council of Salamanca
Talavera Commission/Council of Salamanca, 16–17, *17*, 20–22, 169, 201, 260n38

Tasso, Torquato, 107–8
Taviani, Paolo
  background of, 257n4
  on Beatriz de Bobadilla, 129, 130
  on CC's Genoese life, 3
  on CC's marriage, 11, 120
  on Molloy, 281n25
  on non-Genoese origin claims, 144, 145, 175
  on sainthood campaign, 115
  on Toscanelli letters, 12
Teive, Diogo de, 39
Tetreault, Paul, 256n4
Thacher, John Boyd, 135
Third Voyage, 31–33, 44, 46, 100, 101–2
Thurmond, Strom, 207
Tirant lo blanch (Martorell), 37–38, 259n23
Toledo, María de (Diego Colón's wife), 75, 76, 133
Toro, Alfonso, 274n35
Torre, Juana de la, 103, 273n24, 277n19
Torres, Antonio de, 265n39, 273n24
Toscanelli, Paolo dal Pozzo, 12
  See also Toscanelli letters
Toscanelli letters, 12–13, 259–60n28
tourism
  non-Genoese origin claims and, 153, 155, 158–60, *159*, 164, 169–70, 281n35
  Quincentennial and, 183
transatlantic slave trade, 60–61, 194
Travels (John of Mandeville), 19, 20
Trump, Donald, 237

Ulloa, Alfonso, 97
Ulloa, Luís, 174
uniqueness myth
  Atlantic voyaging and, xxi, 34, 41, 42, 46–47, 75–76
  collaboration and, 46, 264n27
  Colón family lawsuits and, 74–75
  Fernando Colón on, 97
  navigation and, 42
  untruth of, 46–47
United States
  Catholic Columbus and, 90, 110, 113–14
  CC homosexuality possibility and, 137
  Colón family lawsuits and, 83
  Flat Earth myth and, 18, 260n38
  Italian immigration to, 108, 290–91n10
  Jewish origin myth in, 178–79, 284n29
  non-Genoese origin claims and, 175, 283n21
  number of monuments to CC in, 289n8
  Quincentennial and, 174–75
  racism in, 227–30, 232, 235–36
  Spanish–Italian competition in, 219–20, 289n3
  See also American Columbus; Irving, Washington
Unknown Pilot myth, 35–37, 263n12

Valladolid, CC's death in, xi, 11, 64, 68, 128, 143, 267n1
Valladolid Judgment (1527), 75
Vanderlyn, John, 189–90, *189*, 218, 263n10
Vanzetti, Bartolomeo, 228
Vasari, Giorgio, 261n43
Velasco, Pedro de, 39
Verd, Gabriel, 174, 283n19
Verrazzano, Giovanni da, 107
Vespucci, Amerigo, 45, 61, 107, 147, 148, 196, 264n27
Vignaud, Henry, 13, 93
Viking, 235
Viking discovery claims, 236–37, 291n27
Vikings, 263n13
Villareal, Alonso de, 134, 278n34
Vinland, 236, 291n27
Virgen de Cristóbal Colón, La, 88, 90, 271–72n2
visionary myth, xxi, 18, 201, 292n6
  See also Flat Earth myth
Vita (Colón). *See Historie della Vita*
Vita di Cristoforo Colombo (Sanguineti), 114
Vivaldi, Ugolino, 6
Vivaldi, Vadino, 6

Wade, Mary Dodson, *105*
Waller, Fats, 19
Walz, Tim, *239*
Washington, D.C., 175, 187–91, 197, 213–14, 248
Washington, George, 286n14
Washington Irving and His Literary Friends at Sunnyside (Schussele), *202*
Wassermann, Jakob, 283–84n26
Watts, Pauline W., 273n23
weavers/wool merchants, 4–5, 6–7, 16, 26, 32, 125, 256n2
Welsh, 235–36
Wey Gómez, Nicolás, 259n28
When Montezuma Met Cortés (Restall), xv
White, George, 206–7, 213, 214

White League, 229
Whittaker, Bob, 214
Wiesenthal, Simon, 179, 181, 284n32
Wilson-Lee, Edward, 259n28
winds. *See* navigation
Winsor, Justin, 113, 203, 233, 235
World Display'd, The, 200, 203
World's Fair/Exposition events, xvi, 83, *173*, 190, 204–6, 219–20, 235, 277n15
World Wars I and II, 69, 211, 225, 230
Wright, Frank Lloyd, 70

Zarco family origin myth, 171, 172, 282n15
Zea, Leopoldo, 184